NATURE'S EXPERIMENTS AND THE SEARCH FOR SYMBOLIST FORM

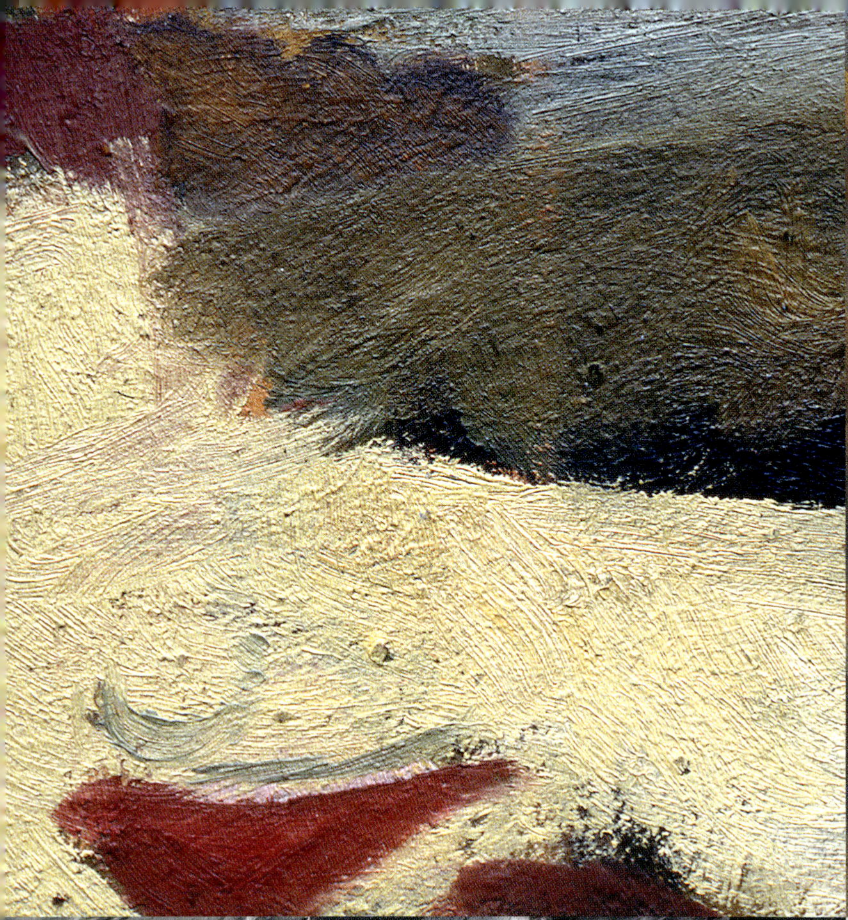
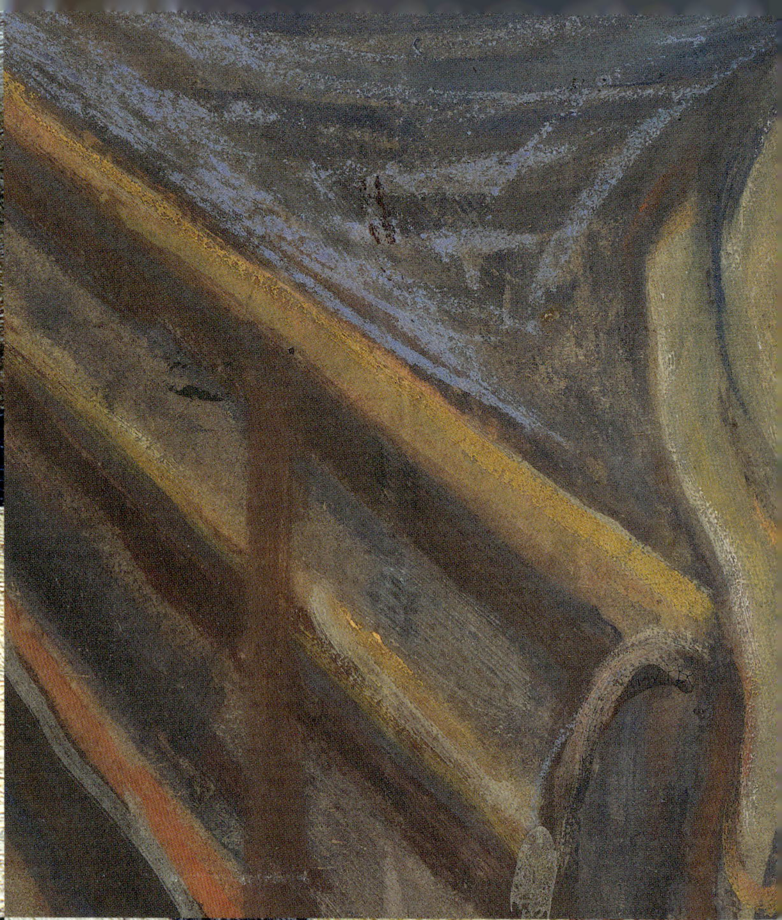
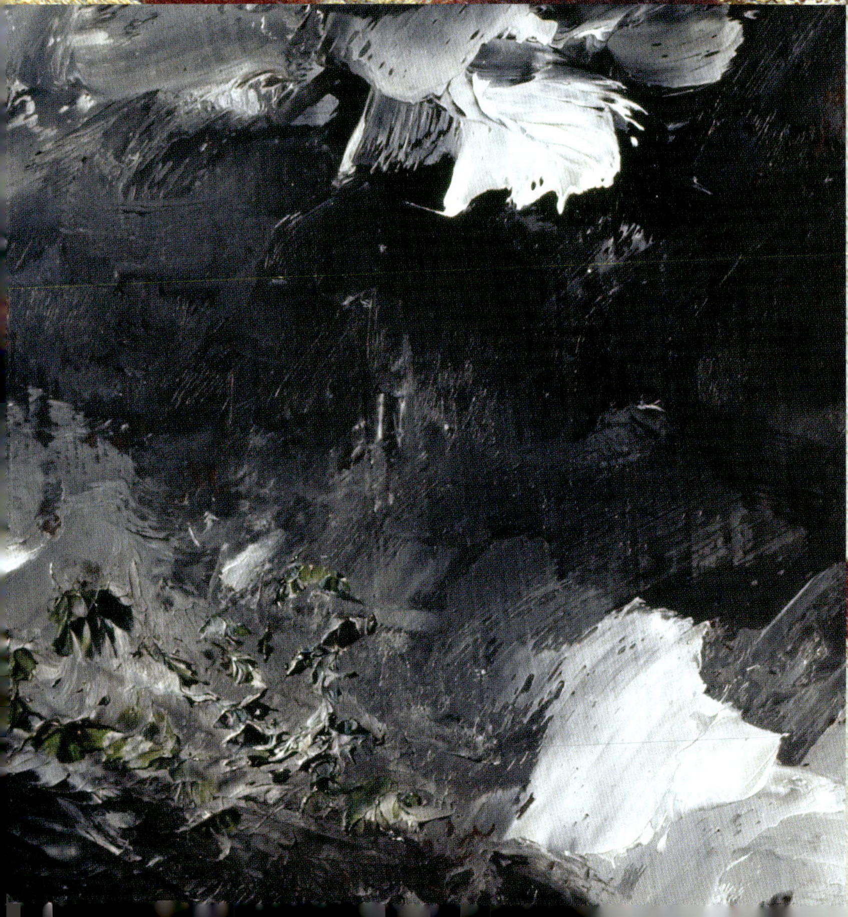
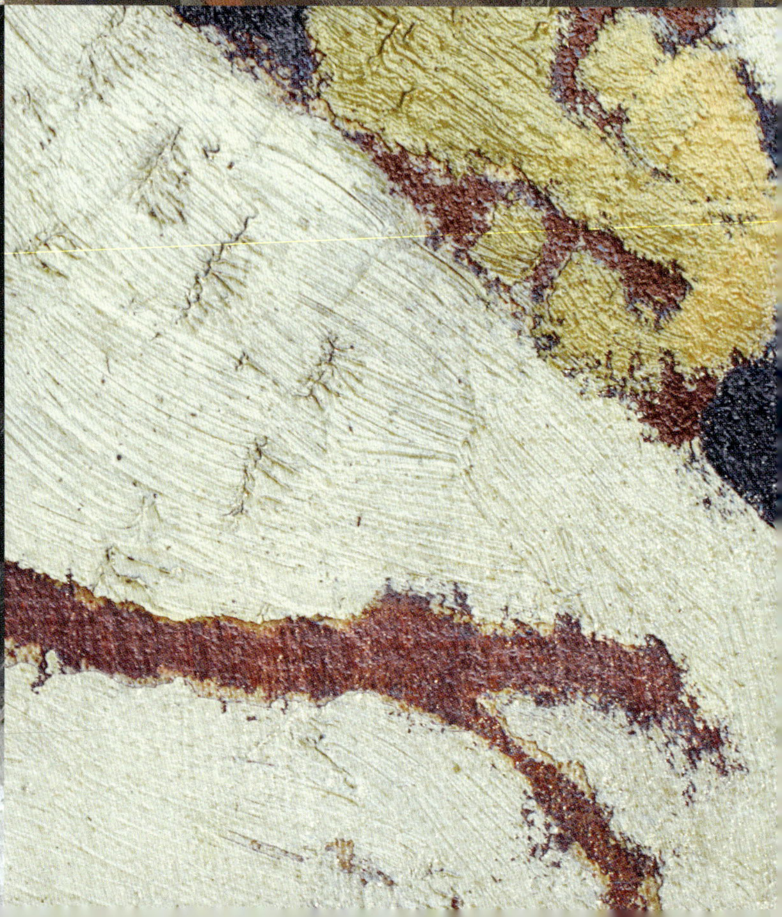

NATURE'S EXPERIMENTS AND THE SEARCH FOR SYMBOLIST FORM

Allison Morehead

The Pennsylvania State University Press
University Park, Pennsylvania

Additional credits: Pages ii–iii, clockwise from upper left: details of Maurice Denis, study for *Décor*, ca. 1890–91 (fig. 3); Edvard Munch, *The Scream*, 1893 (fig. 6); Édouard Vuillard, *Woman Sleeping*, 1892 (fig. 43), and August Strindberg, *Night of Jealousy*, 1893 (fig. 5). Pages vi–vii, detail of Edvard Munch, *The Scream*, 1893 (fig. 6). Page 24: detail of advertisement for Émile Zola, *Nana* and "Le roman expérimental," *Le Voltaire*, no. 467 (15 October 1879): 4 (fig. 13). Page 46: detail of Maurice Denis, study for *Décor*, ca. 1890–91 (fig. 21). Page 78: detail of Édouard Vuillard, *Woman Sleeping*, 1892 (fig. 43). Page 110: detail of August Strindberg, *Wonderland*, 1894 (fig. 72). Page 136: detail of Edvard Munch, *Women in Hospital*, 1897 (fig. 96).

Illustrations in this book were funded in part or in whole by a grant from the Meiss/Mellon Author's Book Award of the College Art Association.

An earlier version of chapter 2 appeared as "Defending Deformation: Maurice Denis's Positivist Modernism," in *Art History* 38, no. 5 (2015): 890–915, and I am grateful to the Association of Art Historians for granting me permission to reproduce this material here.

Chapter 5 includes revised excerpts from "'Are There Bacteria in the Rooms of Monte Carlo?' The Roulette Paintings, 1891–93," in *Munch Becoming "Munch": Artistic Strategies, 1880–1892*, exh. cat., edited by Ingebjørg Ydstie and Mai Britt Guleng (Oslo: Munch Museum, 2008), 120–36; "The *Musée de la folie:* Collecting and Exhibiting *chez les fous*," *Journal of the History of Collections* 23, no. 1 (2011): 101–26; and, with Elizabeth Otto, "Representation in the Age of Mediumistic Reproduction, from Symbolism to the Bauhaus," in *The Symbolist Roots of Modernist Art*, edited by Michelle Facos and Thor J. Mednick (Burlington, Vt.: Ashgate, 2015), 155–68.

Library of Congress Cataloging-in-Publication Data
Names: Morehead, Allison, author.
Title: Nature's experiments and the search for symbolist form / Allison Morehead.
Other titles: Refiguring modernism.
Description: University Park, Pennsylvania : The Pennsylvania State University Press, [2017] | Series: Refiguring modernism | Includes bibliographical references and index.
Summary: "Examines the influence of experimental science, concerned with the workings of the body, the mind, and their various pathologies, on the works of late nineteenth-century artists Maurice Denis, Édouard Vuillard, August Strindberg, and Edvard Munch"—Provided by publisher.
Identifiers: LCCN 2016025159 | ISBN 9780271076744 (cloth : alk. paper)
Subjects: LCSH: Symbolism (Art movement) | Art and science—History—19th century. | Denis, Maurice, 1870–1943—Criticism and interpretation. | Vuillard, Édouard, 1868–1940—Criticism and interpretation. | Strindberg, August, 1849–1912—Criticism and interpretation. | Munch, Edvard, 1863–1944—Criticism and interpretation.
Classification: LCC N6465.S9 M67 2017 | DDC 704.9/46—dc23
LC record available at https://lccn.loc.gov/2016025159

Printed in China
Published by The Pennsylvania State University Press,
University Park, PA 16802–1003

The Pennsylvania State University Press is a member of the Association of American University Presses.

It is the policy of The Pennsylvania State University Press to use acid-free paper. Publications on uncoated stock satisfy the minimum requirements of American National Standard for Information Sciences—Permanence of Paper for Printed Library Material, ANSI Z39.48–1992.

For Derek, my friend

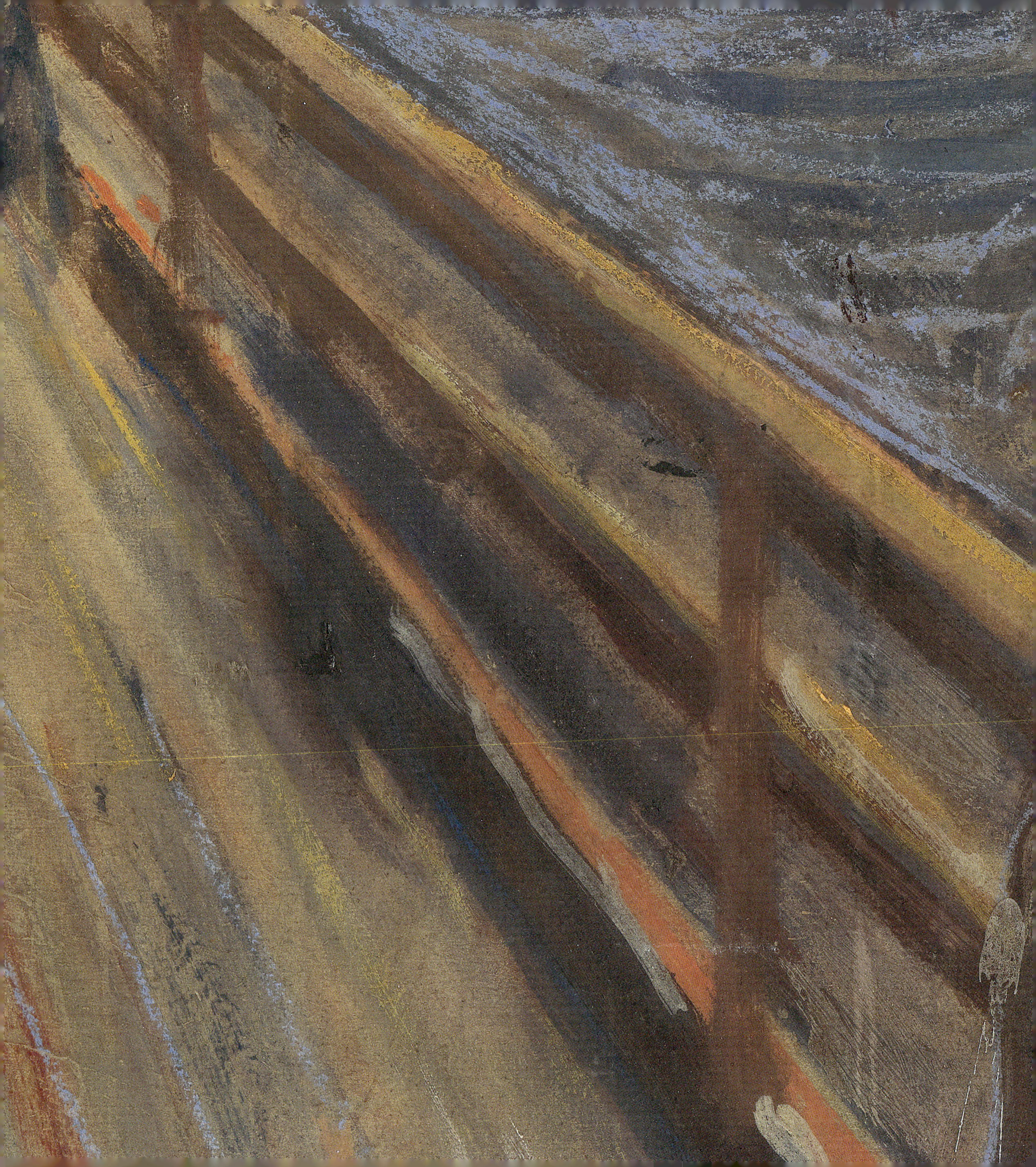

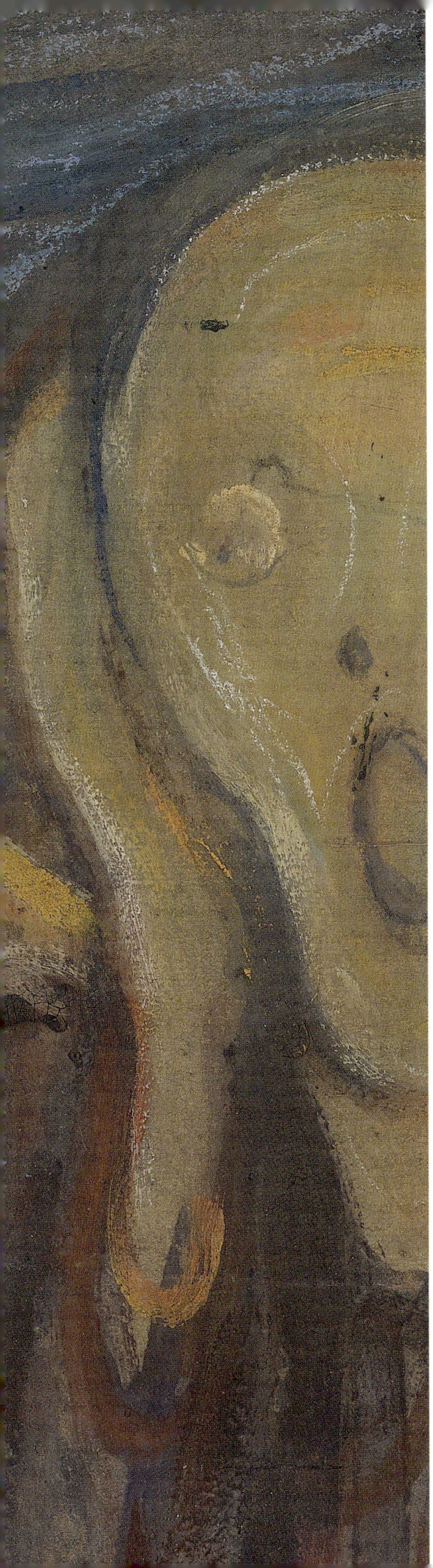

Contents

List of Illustrations | ix

Acknowledgments | xiv

Introduction: Symbolism and Nature's Experiments | 1

1 Toward an Experimental Symbolism: Ideas and Ideals | 25

2 Defending Deformation: Maurice Denis's Positivist Modernism | 47

3 Édouard Vuillard's Experimental Arabesques | 79

4 August Strindberg's Naturalistic Symbolism | 111

5 Madness as Method: The Pathological Experiments of Edvard Munch | 137

Conclusion | 173

Notes | 177

Bibliography | 207

Index | 229

Illustrations

1 H. Lebourgeois, *The Experimental Novel,* from *L'oeuvre de Zola,* 1899–1902. The Thomas Fisher Rare Book Library, University of Toronto. Photo: The Thomas Fisher Rare Book Library. 2

2 Léon Lhermitte, *The Lesson of Claude Bernard, or Session at the Vivisection Laboratory,* 1889. Académie de médecine, Paris. Photo: Bridgeman Images. 4

3 Maurice Denis, study for *Décor,* ca. 1890–91. Private collection. Photo: Catalogue raisonné Maurice Denis, Olivier Goulet. 10

4 Édouard Vuillard, *Aurélien-Marie Lugné-Poë* (1869–1940), 1891. Memorial Art Gallery of the University of Rochester, Rochester, New York, Gift of Fletcher Steele, 72.18. Photo: Memorial Art Gallery. 11

5 August Strindberg, *Night of Jealousy,* 1893. Strindberg Museum, Stockholm. Photo: Strindberg Museum. 12

6 Edvard Munch, *The Scream,* 1893. National Museum of Art, Architecture, and Design, Oslo, NG.M.00939. Photo: National Museum of Art, Architecture, and Design. Photo: National Museum of Art, Architecture, and Design (Børre Høstland). 13

7 Pierre Bonnard, *La revue blanche,* 1894. Yale University Art Gallery, New Haven, Connecticut, University Purchase, 1951.9.19. Photo: Yale University Art Gallery. 16

8 Maurice Denis, *Four Women* or *The Crying Women,* frontispiece for *La revue blanche* 5 (October 1893). Bibliothèque nationale de France. Photo: BnF. 16

9 Édouard Vuillard, *The Seamstress* (*La Couturière*), frontispiece for *La revue blanche* 6 (January 1894). University of Michigan Museum of Art, Museum Purchase, 1959/1.106. Photo: University of Michigan Museum of Art. 17

10 Edvard Munch, *The Scream,* 1895. Lithograph, reproduced in *La revue blanche* 9 (1 December 1895): 528. Bibliothèque nationale de France. Photo: BnF. 17

11 August Strindberg, "L'Exposition d'Edward Munch," *La revue blanche* 10 (1 June 1896): 525. Bibliothèque nationale de France. Photo: BnF. 17

12 "Quelques opinions sur l'oeuvre de H. Taine," with vignette by Félix Vallotton, *La revue blanche* 13 (15 August 1897): 263. Bibliothèque nationale de France. Photo: BnF. 18

13 Advertisement for Émile Zola, *Nana* and "Le roman expérimental," *Le Voltaire,* no. 467 (15 October 1879): 4. Bibliothèque nationale de France. Photo: BnF. 19

14 H. Lebourgeois, *Doctor Pascal,* from *L'oeuvre de Zola,* 1899–1902. The Thomas Fisher Rare Book Library, University of Toronto. Photo: The Thomas Fisher Rare Book Library. 31

15 Claude Monet, *Boulevard des Capucines,* 1873. Pushkin Museum, Moscow. Photo: Bridgeman Images. 32

16 Félix Vallotton, *Jules Laforgue,* in Remy de Gourmont, *Le livre des masques* (Paris: Mercure de France, 1896), 204. Bibliothèque nationale de France. Photo: BnF. 36

17 Félix Vallotton, *Gustave Kahn,* in Remy de Gourmont, *Le livre des masques* (Paris: Mercure de France, 1896), 242. Bibliothèque nationale de France. Photo: BnF. 41

18 Page from the catalogue for the Salon des Indépendants, 1891, with notations in the

hand of Maurice Denis. Musée départemental Maurice Denis, Saint-Germain-en-Laye. Photo by the author. 52
19 Maurice Denis, study after a cast of a faun, 1890. Musée départemental Maurice Denis, Saint-Germain-en-Laye, Gift of the Family of Maurice Denis, 1976, PMD 976.1.17. Photo © Raffaële Scapecchi / Gabrielle Montarnal, Musée départemental Maurice Denis. 53
20 Maurice Denis, *Symbolist Nude* (study for *Décor*), ca. 1890–91. Musée départemental Maurice Denis, Saint-Germain-en-Laye, Gift of the Family of Maurice Denis, 1976, PMD 976.1.17. Photo © rights reserved. 54
21 Maurice Denis, study for *Décor,* ca. 1890–91. Private collection. Photo: Catalogue raisonné Maurice Denis, Olivier Goulet. 55
22 Maurice Denis, study for *Décor,* ca. 1890–91. Private collection. Photo: Catalogue raisonné Maurice Denis, Olivier Goulet. 55
23 Maurice Denis, study for *Décor,* ca. 1890–91. Private collection. Photo: Catalogue raisonné Maurice Denis, Olivier Goulet. 56
24 Maurice Denis, study for *Décor,* ca. 1890–91. Private collection. Photo: Catalogue raisonné Maurice Denis, Olivier Goulet. 56
25 Maurice Denis, study for *Décor,* ca. 1890–91. Private collection. Photo: Catalogue raisonné Maurice Denis, Olivier Goulet. 57
26 Pierre Bonnard, *The Mantelpiece,* 1916. Private collection. Photo: akg-images. 58
27 Pierre Bonnard, *Young Woman at Her Toilette* (study for *The Mantelpiece*), 1916. Margaret and Eugene McDermott Collection, Dallas, Texas. Image courtesy of the Dallas Museum of Art (Brad Flowers). 58
28 Paul Gauguin, *The Loss of Virginity,* 1890–91. Chrysler Museum of Art, Norfolk, Virginia, Gift of Walter P. Chrysler Jr., 71.510. Photo: Chrysler Museum of Art. 59
29 Paul Cézanne, *Portrait of the Painter Achille Emperaire,* ca. 1868. Musée d'Orsay, Paris. Photo: Bridgeman Images. 60
30 Caricature of Paul Cézanne with two paintings rejected by the Salon jury of 1870, from *Album Stock* (Paris: Dufour et Cie, 1870). Photo reprinted from John Rewald, *The Paintings of Paul Cézanne: A Catalogue Raisonné,* 2 vols. (New York: Harry N. Abrams, 1996), 1:118. 60
31 Émile Bernard, *Bathers with a Red Cow,* 1887. Musée d'Orsay, Paris, RF1984–21. Photo © RMN-Grand Palais / Art Resource, New York (Hervé Lewandowski). 61
32 Maurice Denis, *Catholic Mystery,* 1890. Private collection. Photo: Catalogue raisonné Maurice Denis. 63
33 Maurice Denis, *On the Theme of Trinitarian Evening by Ad. Retté: Ornamental Painting,* 1891. Private collection. Photo: Catalogue raisonné Maurice Denis. 64
34 Maurice Denis, study for *Trinitarian Evening*. Private collection. Photo reprinted from Peter Weiermair, ed., *The Nude: Ideal and Reality from Neoclassicism to Today* (Bologna: Galleria d'Arte Moderna, 2004), 92. 65
35 Maurice Denis, *Nude Woman, Seen from Behind, Turned to the Right* (study for *Trinitarian Evening*), 1891. Musée d'Orsay, Paris, RF14915. Photo © RMN-Grand Palais / Art Resource, New York (Hervé Lewandowski). 65
36 Maurice Denis, *Portrait of a Young Girl in an Evening Décor* (also called *Symbolist Marthe*), 1892. Niigata Prefectural Museum of Modern Art, Niigata. Photo: Catalogue raisonné Maurice Denis. 68
37 Paul Sérusier, *The Talisman, the Aven River at the Bois d'Amour,* 1888. Musée d'Orsay, Paris. Photo: Bridgeman Images. 70
38 Vincent van Gogh, *The Night Café,* 1888. Yale University Art Gallery, New Haven, Connecticut, Bequest of Stephen Carlton Clark, B.A. 1903, 1961.18.34. Photo: Yale University Art Gallery. 72
39 Paul Gauguin, *Café at Arles,* 1888. Pushkin Museum, Moscow. Photo: Bridgeman Images. 72

40 Édouard Vuillard, *Reclining Nude,* ca. 1891. Private collection. Photo: Cindy Momchilov / Camera Work, Inc. 75

41 Édouard Vuillard, *Reclining Nude,* ca. 1891. Current whereabouts unknown. Photo reprinted from Antoine Salomon and Guy Cogeval, *Vuillard: The Inexhaustible Glance; Critical Catalogue of Paintings and Pastels,* 3 vols. (Milan: Skira Editore and Wildenstein Institute, 2003), 1:181. 76

42 Édouard Vuillard, *Reclining Nude,* ca. 1891. Private collection. Photo reprinted from Guy Cogeval, *Édouard Vuillard* (Washington, D.C.: National Gallery of Art, 2003), 101. 76

43 Édouard Vuillard, *Woman Sleeping* (otherwise known as *Sleep*), 1892. Musée d'Orsay, Paris. Photo: Bridgeman Images. 80

44 Caricature from Mirliton, "Troisième exposition des peintres impressionistes et symbolistes," *Le journal,* no. 9, literary supplement (26 November 1892), 1. Photo reprinted from Antoine Salomon and Guy Cogeval, *Vuillard: The Inexhaustible Glance; Critical Catalogue of Paintings and Pastels,* 3 vols. (Milan: Skira Editore and Wildenstein Institute, 2003), 1:145. 80

45 Édouard Vuillard, *Woman in Bed,* 1891. National Gallery of Art, Washington, D.C., Ailsa Mellon Bruce Collection, 1970.17.179. Photo: National Gallery of Art. 81

46 Édouard Vuillard, journal. Institut de France, MS 5396(1): *Carnet* I.1, fol. 10r [21 November 1888]. Photo © RMN-Grand Palais / Art Resource, New York (Mathieu Rabeau). 84

47 Édouard Vuillard, journal. Institut de France, MS 5396(1): *Carnet* I.1, fol. 32v [December 1888]. Photo © RMN-Grand Palais / Art Resource, New York (Mathieu Rabeau). 88

48 Édouard Vuillard, *Portrait of Marie Vuillard,* 1888. Photo reprinted from Antoine Salomon and Guy Cogeval, *Vuillard: The Inexhaustible Glance; Critical Catalogue of Paintings and Pastels,* 3 vols. (Milan: Skira Editore and Wildenstein Institute, 2003), 1:45. 89

49 Édouard Vuillard, *Schematized Portrait of Marie Vuillard,* 1890. Private collection. Photo: Bridgeman Images. 90

50 Édouard Vuillard, journal. Institut de France, MS 5396(1): *Carnet* I.1, fol. 57r [September 1890]. Photo © RMN-Grand Palais / Art Resource, New York (Mathieu Rabeau). 94

51 Édouard Vuillard, *Self-Portrait with Walking-Stick and Boater,* ca. 1891–92. Private collection, United States. Photo reprinted from Antoine Salomon and Guy Cogeval, *Vuillard: The Inexhaustible Glance; Critical Catalogue of Paintings and Pastels,* 3 vols. (Milan: Skira Editore and Wildenstein Institute, 2003), 1:93. 95

52 Édouard Vuillard, journal. Institut de France, MS 5396(2): *Carnet* I.2, fol. 3v, ca. 1890. Photo © RMN-Grand Palais / Art Resource, New York (Mathieu Rabeau). 97

53 Édouard Vuillard, journal. Institut de France, MS 5396(2): *Carnet* I.2, fol. 5v, ca. 1890. Photo © RMN-Grand Palais / Art Resource, New York (Mathieu Rabeau). 98

54 Édouard Vuillard, journal. Institut de France, MS 5396(2): *Carnet* I.2, fol. 12v, ca. 1890. Photo © RMN-Grand Palais / Art Resource, New York (Mathieu Rabeau). 98

55 Édouard Vuillard, *Grisélidis,* 1891. Private collection. Photo reprinted from Guy Cogeval, *Édouard Vuillard* (Washington, D.C.: National Gallery of Art, 2003), 103. 100

56 Édouard Vuillard, *Lugné-Poë,* ca. 1891. Private collection. Photo reprinted from Isabelle Cahn et al., *Le Théâtre de l'Oeuvre, 1893–1900: Naissance du théâtre moderne* (Paris: Musée d'Orsay, 2005), 86. 100

57 Édouard Vuillard, *In Bed,* 1891. Musée d'Orsay, Paris, RF1977–374. Photo © RMN-Grand Palais / Art Resource, New York (Hervé Lewandowski). 102

58 Édouard Vuillard, *Sleeping Woman,* 1892. Private collection, Washington, D.C. Photo reprinted from Antoine Salomon and Guy Cogeval, *Vuillard: The Inexhaustible Glance; Critical Catalogue of Paintings and Pastels,* 3

vols. (Milan: Skira Editore and Wildenstein Institute, 2003), 1:146. 103
59 Édouard Vuillard, *Kettle and Pot*, ca. 1892. Formerly Museum of Modern Art, New York. Photo reprinted from Antoine Salomon and Guy Cogeval, *Vuillard: The Inexhaustible Glance; Critical Catalogue of Paintings and Pastels*, 3 vols. (Milan: Skira Editore and Wildenstein Institute, 2003), 1:265. 105
60 Édouard Vuillard, journal. Institut de France, MS 5396(1): *Carnet* I.1, fol. 73r [31 August 1890]. Photo: RMN-Grand Palais / Art Resource, New York. 106
61 Achille Lemot, *Gustave Flaubert Dissecting Madame Bovary*, in *La parodie* (December 1869): 240. Photo: Klassik Stiftung Weimar, Herzogin Anna Amalia Bibliothek/F gr 5297, 2016. 119
62 Front cover of the first issue of *Sphinx: Monatsschrift für die geschichtliche und experimentale Begründung der übersinnlichen Weltanschauung auf monistischer Grundlage* 1, no. 1 (January 1886). Photo: Institut für Grenzgebiete der Psychologie und Psychohygiene e.V. Freiburg i. Br. 122
63 Front cover of *Sphinx: Monatsschrift für Seelen- und Geistesleben* 7, no. 13 (1892). Photo: Institut für Grenzgebiete der Psychologie und Psychohygiene e.V. Freiburg i. Br.
64 "Aus Kerners Kleksographien," *Sphinx* 11, no. 62 (February 1891): 113. Photo: Institut für Grenzgebiete der Psychologie und Psychohygiene e.V. Freiburg i. Br. 122
65 Justinus Kerner, *Klecksograph* (Inkblot), 1850s. Städtische Galerie im Lenbachhaus und Kunstbau, Munich, G18402. Photo: Lenbachhaus und Kunstbau. 123
66 August Strindberg, *Klecksograph*, from *The Occult Diaries*, 23 February 1898. National Library of Sweden, Stockholm, MS SgNM 72, 1898, 54. Photo: Andrea Davis Kronlund / National Library of Sweden. 124
67 August Strindberg, charcoal drawing representing partially burnt coal, 1896. National Library of Sweden, Stockholm, MS SgNM 15:6,5. Photo: Andrea Davis Kronlund / National Library of Sweden. 124
68 August Strindberg, *Flower on the Shore, Dalarö*, 1892. Malmö Art Museum, Malmö, Sweden. Photo © August Strindberg / reproduced by Merja Diaz / Malmö Art Museum. 125
69 August Strindberg, *Palette with Solitary Flower on the Shore*, 1893. Private collection. Photo: Strindberg Museum. 126
70 August Strindberg, *High Seas*, 1894. Sven-Harrys Konstmuseum, Stockholm. Photo: Per Myrehed © Sven-Harrys Konstmuseum. 127
71 August Strindberg, *The Weeping Boy*, 1891. Private collection. Photo: Strindberg Museum. 132
72 August Strindberg, *Wonderland*, 1894. Nationalmuseum, Stockholm. Photo: Nationalmuseum, Stockholm. 133
73 Edvard Munch, *Self-Portrait*, 1893. Munch Museum, Oslo. Photo © Munch Museum. 138
74 Edvard Munch, *Billiard Players*, ca. 1890. Munch Museum, Oslo, T129, p. 43. Photo © Munch Museum. 140
75 Jean Béraud, *The Billiard Players*, before 1919. Whereabouts unknown. Sold by Christie's New York, 27 October 2004. Photo © RMN-Grand Palais / Art Resource, New York (François Vizzavona). 141
76 Edvard Munch, *Billiard Hall*, ca. 1890. Munch Museum, Oslo, T127, p. 13. Photo © Munch Museum. 141
77 Edvard Munch, *Man in a Billiard Hall*, ca. 1889. Munch Museum, Oslo, T126, p. 3. Photo © Munch Museum. 141
78 Jean Béraud, *The Gambling Rooms at Monte Carlo—"Rien ne va plus."* Engraving after *Monte Carlo (Rien ne va plus!)*, 1890. Private collection. Photo © Look and Learn / Illustrated Papers Collection / Bridgeman Images. 143
79 Edvard Munch, *At the Roulette Table*, ca. 1891. Munch Museum, Oslo, T128, p. 32. Photo © Munch Museum. 144

80 Edvard Munch, *Along a Road (The Lonely Ones)* and *At the Roulette Table,* ca. 1891. Munch Museum, Oslo, T129, p. 36. Photo © Munch Museum. 144
81 Edvard Munch, *Roulette I,* 1891–92. Munch Museum, Oslo. Photo © Munch Museum. 145
82 Edvard Munch, *Gamblers in Monte Carlo,* 1892. Private collection. Photo © Munch Museum. 146
83 Edvard Munch, *At the Roulette Table in Monte Carlo,* 1893. Munch Museum, Oslo. Photo © Munch Museum. 148
84 Edvard Munch, *The Roulette Table,* 1903. Munch Museum, Oslo. Photo © Munch Museum. 149
85 *Idiot, 25 Years Old,* plate II in J. G. Spurzheim, *Phrenology, or The Doctrine of the Mind; and of the Relations Between Its Manifestations and the Body*, 3rd ed. (London: Treuttel, Wurtz, and Richter, 1825). The Thomas Fisher Rare Book Library, University of Toronto. 150
86 Edvard Munch, *Death in the Sickroom,* 1893. Munch Museum, Oslo. Photo © Munch Museum. 151
87 Théodore Géricault, *The Woman with Gambling Mania,* 1819–24. Musée du Louvre, Paris. Photo: Bridgeman Images. 152
88 Jean Béraud, *The Insane,* 1885. Whereabouts unknown. Engraving in Henry Havard, *Salon de 1885* (Paris: Librairie d'Art, 1885). Photo © The British Library Board, P.P.1932.fa (between pages 34 and 35). 153
89 *Symbolic Representation of the Soul's Uplifting,* from Gustav Gessmann, "Somnambules Zeichnen," *Sphinx* 6, no. 32 (August 1888): 105. Photo: Institut für Grenzgebiete der Psychologie und Psychohygiene e.V. Freiburg i. Br. 155
90 *A Spirit Head,* from Gustav Gessmann, "Somnambules Zeichnen," *Sphinx* 6, no. 32 (August 1888): 104. Photo: Institut für Grenzgebiete der Psychologie und Psychohygiene e.V. Freiburg i. Br. 155
91 "Gedankenübertragung," *Sphinx* 2, no. 4 (October 1886): 246. Photo: Institut für Grenzgebiete der Psychologie und Psychohygiene e.V. Freiburg i. Br. 156
92 Edvard Munch, *Art,* ca. 1894–96. Munch Museum, Oslo, T408A. Photo © Munch Museum. 158
93 Edvard Munch, *Man's Head,* ca. 1894–96. Munch Museum, Oslo, T408B. Photo © Munch Museum. 158
94 "Gedankenübertragung," *Sphinx* 1, no. 1 (January 1886): 37. Photo: Institut für Grenzgebiete der Psychologie und Psychohygiene e.V. Freiburg i. Br. 161
95 Yvanhoé Rambosson, "Le Salon des Indépendants," *La plume,* 15 May 1897, 311, with reproduction of Edvard Munch, *Death in the Sickroom,* 1896. Bibliothèque nationale de France. Photo: BnF. 161
96 Edvard Munch, *Women in Hospital,* 1897. Munch Museum, Oslo. Photo © Munch Museum. 162
97 Edvard Munch, *Marcel Réja,* 1896–97. Munch Museum, Oslo. Photo © Munch Museum. 163
98 Edvard Munch, *Inheritance,* 1897–99. Munch Museum, Oslo. Photo © Munch Museum. 164
99 Edvard Munch, *Women in the Hospital,* 1896. Munch Museum, Oslo. Photo © Munch Museum. 165
100 After Tony Robert-Fleury, *Philippe Pinel Releasing Lunatics from Their Chains at the Salpêtrière Asylum in Paris in 1795,* 1876–78, British Museum, London. Photo © The Trustees of the British Museum. 166
101 Albert Londe, *Mlle Baranes (hystérie): Tympanite,* ca. 1883. Private collection, France. Courtesy of the Galerie Baudoin Lebon. Photo: Bertrand Huet. 167
102 Edvard Munch, *In the Men's Clinic,* 1896. Munch Museum, Oslo. Photo © Munch Museum. 168
103 Edvard Munch, *Weeping Woman,* 1907. Munch Museum, Oslo. Photo © Munch Museum. 169

Acknowledgments

Writing these acknowledgments overwhelms me with gratitude. In the course of researching and writing this book, I have been the recipient of incredible generosity, kindness, humor, and love. My thanks go first to Martha Ward, Reinhold Heller, and Jan Goldstein, who have supported the project from its inception. If anyone deems me to have come close to the example they set as scholars, teachers, and mentors, I will have achieved more than I ever imagined. My cohort at the University of Chicago, Erin Hazard, Matthew Hunter, Katie Kirtland, and Sarah Miller in particular, offered all kinds of support during the tunnel years and beyond. Sabine Wieber has been an ideal reader, a keen strategist, and a beloved friend. This book and its writer owe more to her than can be condensed into a sentence. I feel similarly about Elizabeth Otto, my vault friend and favorite person to laugh with until the cheeks hurt.

I received institutional support at various stages of the project. I would like to acknowledge here the generosity of the Social Sciences and Humanities Research Council of Canada, the Samuel H. Kress Foundation, the Social Science Research Council's International Dissertation Research Fellowship program, the International Federation of University Women, the Mrs. Giles Whiting Foundation, the University of Chicago Department of Art History and Division of the Humanities, the France Chicago Center, and the American Friends of the Chagall Biblical Message Museum. The project involved research in a number of institutions and private collections in France and Scandinavia, which was made all the more fruitful and enjoyable by the people I met along the way. For assistance beyond their job descriptions, I wish to thank Marie El-Caïdi, now retired from the Musée départemental Maurice Denis, Saint-Germain-en-Laye; Claire Denis and Fabienne Stahl of the Catalogue raisonné Maurice Denis; Mireille Pastoureau and Fabienne Queyroux of the Bibliothèque de l'Institut de France, Paris; Anna Bodin and Arvid Jakobssen, Kungliga Biblioteket, Stockholm; Magne Bruteig, Mai Britt Guleng, Lasse Jacobsen, Ute Kuhlemann-Falck, Gerd Woll, and Ingebjørg Ydstie of the Munch Museum and the National Museum of Art, Architecture, and Design, Oslo; and my research assistants Anna-Maria Moubayed, Emma Richan, Mathieu Sly, Claire Wenngren, and Jennifer Young.

Support comes in all forms, from attentive reading, to the rapid answer to a question, to a timely remark, to ongoing prodding, to unswerving belief, to kindly conveyed disbelief. Some people might be surprised to find their names in this list, but I would be remiss in not thanking Scott Allan, Nell Andrew, Vincent Barras, Patricia G. Berman, Gemma Blackshaw, Jean-Paul Bouillon, Richard Brettell, the late Filiz Burhan, Isabelle Cahn, Jacqueline Carroy, Alison Chang, Rochelle Ching, Mathias Chivot, Jay A. Clarke, Hollis Clayson, Rachael Z. DeLue, Michelle Facos, Francis Frascina, Julia Frey, Gloria Groom, Mary Hunter, Serena Keshavjee, Elizabeth Legge, Patricia Leighten, Alison Matthews-David, Thor Mednick, Sylvie Patry, Rodolphe Rapetti, the late and much missed Claude Ritschard, Richard Shiff, Alison Syme, Belinda and Richard Thomson, and Leslie Topp.

The book was written while at King's College, Cambridge, and at Queen's University, Kingston, Canada. Colleagues and students at those institutions helped more than I suspect they know. I am especially grateful to Jan Allen, Veronica Carter, Una D'Elia, Janice Helland, Cathleen Hoeniger, Deborah Howard, Lynda Jessup, Alyce Mahon, Jean-Michel Massing, Clive Robertson, Katherine Romba, Joan Schwartz, and Allison Sherman. In this age of contingent labor practices, I have been fortunate to be a member of a strong union. The policies of the Queen's University Faculty Association, in particular surrounding parental leave, enabled me to complete this project to the standards that I set for myself, standards that took time to live up to. Eleanor H. Goodman, executive editor at the Pennsylvania State University Press, expertly guided the book through its final stages. It has been a pleasure and a privilege to work with her and her colleagues, including Hannah Hebert and Suzanne Wolk.

I have always had the unquestioning support of my parents, Karon and Clayton (Mac) Morehead. I am thrilled to be able to finally, formally, and publicly thank them for support that can be neither quantified nor qualified. Derek Walker has loved me both for and despite this project. His belief in my value as a thinker, teacher, and human being, his intellectual, emotional, and material support, and his abilities as a father enable me to exist. The book is dedicated to him with all my heart because it simply would not be without him. Our daughters, Eloise and Liliane, knowingly or not, have lived with *Nature's Experiments* their entire lives, during which time they have joyfully manifested their own experimental natures. I hope that they eventually come to see some value in the book, if only as a necessary product of their mother's life.

INTRODUCTION

Symbolism and Nature's Experiments

Experimental art was not always avant-garde. And avant-garde art was not always experimental. In the eighteenth century, long before the emergence of what have been called the historical avant-gardes, *l'art expérimental* was, according to Denis Diderot, quite simply the art of seeking the laws of nature through experience, a method of inquiry increasingly associated with scientific ways of knowing.[1] But by the mid-twentieth century, experimental art had become so synonymous with avant-garde art that Clement Greenberg felt the need to downplay "experiment"—he placed the word in quotation marks—as the avant-garde's primary function.[2] Carrying with it an almost incantatory power, the term *experimental*, in art making, is understood to valorize the new, the radical, the forward-thinking, and the progressive of virtually any time and place, and sometimes in such self-consciously utopic terms as to inspire parody.[3]

But in the early 1890s, at the precise moment that a critical mass of avant-garde strategies was coalescing around groups such as the neoimpressionists and the symbolists, experimental art had yet to stabilize as a discursive formation. In 1892, the poet and critic G.-Albert Aurier, keen to define and to advance symbolist painting, derided experimental art, lambasting what he understood to be the objective and scientific aspirations of a naturalist mode of painting.[4] In Aurier's view, experimental art was crude and materialist, tired and retrograde, the exact opposite of what he hoped symbolist art would be. Had he lived long enough, he would no doubt have sympathized with a turn-of-the-century caricature of Émile Zola flinging fecal matter from a chamber pot onto a canvas, a vulgar representation of Zola's programmatic essay "The Experimental Novel" that suggested that experimentation was, quite literally, a shitty way to go about painting (fig. 1).

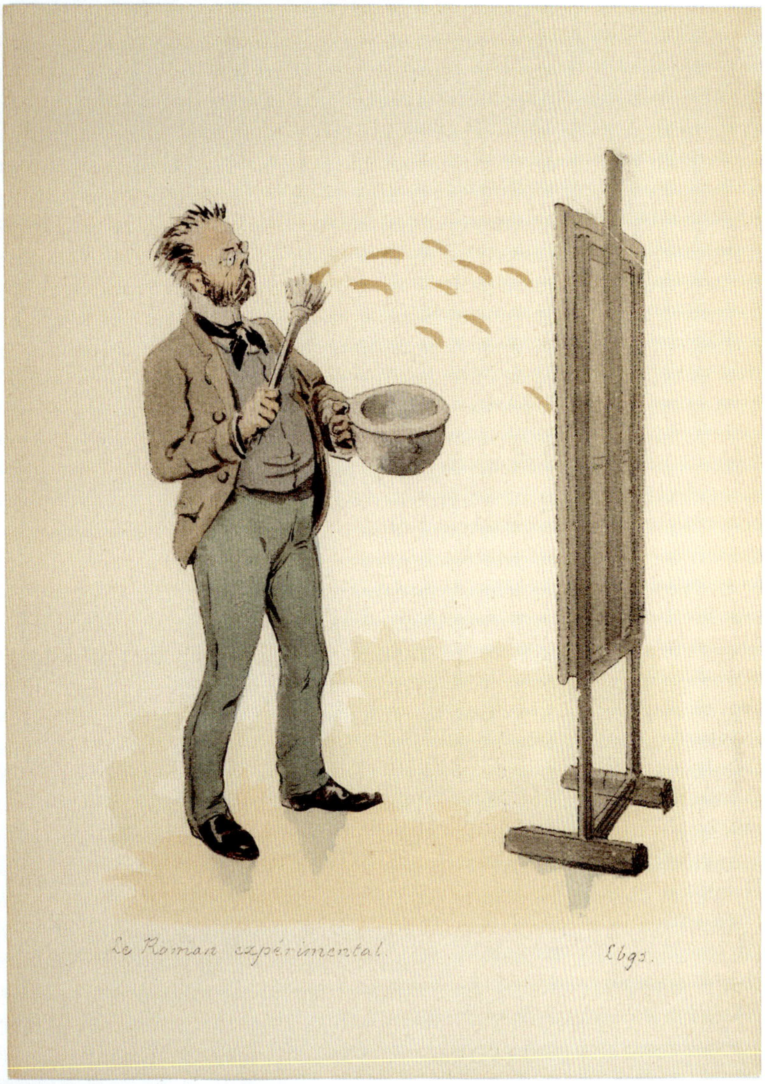

1 H. Lebourgeois, *The Experimental Novel*, from *L'oeuvre de Zola*, 1899–1902. The Thomas Fisher Rare Book Library, University of Toronto.

Over and against what he called experimental art, Aurier sought to promote a vanguard symbolist painting that was idealist (or "ideaist," as he insisted on calling it), radical, and forward-thinking—in short, avant-garde—but definitely not experimental.

Nonetheless, within the context of late nineteenth-century symbolism, when it came to productive alignments between art and experiment, Aurier was in fact an outlier among the many seeking to identify, to define, and to practice symbolist painting at the fin de siècle. Theorists and promoters of symbolism had initially concentrated on reorienting literature, and above all poetry, away from a world of exterior appearances toward a world of interior experiences.[5] Symbolist theory situated symbolist practice as a new form of knowledge production, and called for representational modes that would make understandable or present not the external, visible world but a mysterious, difficult-to-access world within. As such, symbolists adhered to and promoted specious notions of objective truth—"immutable principles," in the words of the painter Paul Sérusier—and psychological universalism, imagining all members of their public to be equally and comparatively susceptible to symbolism's forms, rhythms, and neologisms.[6] Theories of symbolism remained contested terrain, but the critic Gustave Kahn's reduction of symbolism's tenets to a pithy catchphrase both marks out that terrain and, in its very reductiveness, suggests how anxious symbolists were to find clear directives for achieving putatively universal forms of communication. "The essential goal of our art," Kahn wrote in 1886, setting symbolism in opposition to Zola's naturalism, "is to objectify the subjective (the exteriorization of the Idea) instead of subjectifying the objective (nature seen through a temperament)."[7] Experimental science, as Kahn would readily admit, would have a potentially important role to play in the objectifying processes of symbolism.

Experimentalism

I am convinced that when physiology has advanced sufficiently, the poet, the philosopher, and the physiologist will all understand each other.

—CLAUDE BERNARD, 1865

The physiologist Claude Bernard, celebrated in an 1889 painting by Léon Lhermitte, incarnated experimentalism in nineteenth-century France (fig. 2).[8] Henri Bergson would insist that Bernard's *Introduction to the Study of Experimental Medicine*, published in 1865, had been as influential for nineteenth-century ways of thinking as René Descartes's *Discourse on Method* had been for seventeenth- and eighteenth-century thought.[9] Bernard's lofty vision of a future in which the poet, the philosopher, and the physiologist would all speak the same language, a vision shared by many symbolists, was predicated on Bernard's belief that man had just reached the third and final stage of intellectual development, namely, that of experimentalism. While experiment had been installed as the methodological basis for the physical sciences in the early seventeenth century, vitalists had continued to argue that organic matter differed so fundamentally from nonorganic matter that experimental methods could not be applied to the life sciences with any rigor.[10] Bernard's book marked, for many at least, the end of the debate, establishing an experimental physiology as the foundation of medicine, and experimental methods, rather than empirical ones, as the basis for all knowledge about the natural world.[11]

Unlike Auguste Comte, who had argued for a model of human progression through various stages toward the age of positivism, Bernard believed that man cumulated successive stages, with the a priori idea remaining crucial to all, including the age of experiment: "the goal of the experimental method," Bernard wrote, "is to transform [the] a priori concept, founded on an intuition or a vague sense of things, into an a posteriori interpretation established through the experimental study of phenomena."[12] Experiment, he argued, was thus fundamentally "induced observation" (*observation provoquée*); the experimenter observes a phenomenon, formulates an idea based on his observations—a hypothesis—and then subjects that idea to rigorous testing under controlled conditions that he himself creates. Throughout the process, the experimenter must maintain philosophical doubt—not skepticism toward science itself, but a belief that existing theories are at best only partial glimpses of the truth of the natural world. Succinct, engaging, and easy to comprehend, the immensely popular *Introduction* constructed a doctrine for experimental method. By the time the writer and promoter of esotericism Édouard Schuré published his book *The Great Initiates* in 1889, a work of enormous influence for symbolist artists, it was not at all surprising that Bernard's belief in the coming together of the poet, philosopher, and physiologist under the aegis of experiment should appear as the epigraph to the book's preface.

Bernard's *Introduction* not only brought words such as *expérimentalisme* and *expérimentaliste* into use; it contributed to the already shifting meanings of the word *expérience*. Whereas the distinction between the English terms *experience* and *experiment* can be traced back at least to the eighteenth century, both meanings continue to be covered by the single French word *expérience*. From the Latin verb *experiri*, to try, *expérience* was hardly a straightforward term prior to the nineteenth century, but the pace of denotative and connotative change accelerated in the wake of Bernard's *Introduction*.[13] Until 1878, the notoriously outdated dictionary of the French Academy had expressed the dual meaning of *expérience* as, on the one hand, a "trial or test occurring by design or by chance," and, on the other, as expertise, "knowledge of things, acquired through use over a long period." The Academy's previous edition, the sixth, published in the 1830s, had been the first to introduce scientific usage not in a separate entry but with physics and chemistry experiments offered as examples after the first definition. In his 1863 dictionary, the positivist disciple Émile Littré had repeated almost verbatim the Academy's two definitions; but presumably because the idea of a "trial or test occurring . . . *by chance*" (emphasis added) posed problems for positivist science, Littré added a third and charmingly inoffensive definition to cover the scientific usage: "attempt to recognize how something happens," which was followed by the now obligatory examples of physics and chemistry experiments. Littré also recognized two cognate neologisms, *expérimentateur* and *expérimentation*, and two more would appear—*expérimentalisme* and *expérimentaliste*—with credit to Bernard, in the 1870 Larousse dictionary.

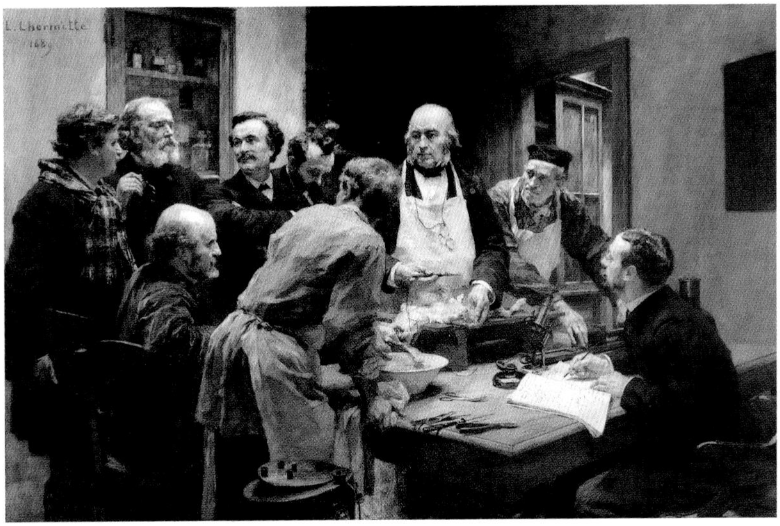

2 Léon Lhermitte, *The Lesson of Claude Bernard, or Session at the Vivisection Laboratory*, 1889. Oil on canvas, 180 × 280 cm. Académie de médecine, Paris.

For his second edition of 1872–77, published after the appearance of the *Introduction*, Littré reworked the entries for *expérience*, adding a fourth, more rigorous definition derived directly from Bernard: "experiment, in the strict sense of the experimental method, denotes a posteriori knowledge gained through the observation of phenomena," and Littré now added physiology experiments to the examples of physics and chemistry. The French Academy finally caught up with Littré and Larousse in the 1930s with its eighth edition, giving the first definition entirely over to science, which necessitated dropping that problematic phrase "by chance": "Trial or test instituted to study the way in which natural phenomena occur, and to determine the laws that govern them by reproducing them artificially." The second definition took on the connotation of "by chance," transforming it into "involuntarily," while relinquishing the notion of expertise by dropping "use over a long period" in favor of "knowledge of things, involuntarily acquired through usage in the world and in life."[14]

This overhaul of *expérience* between the Academy's seventh (1878) and eighth (1932–35) editions, alongside Littré's changes between his first and second editions, signals not only the rapidly increasing authority of science but also the extent to which all meanings of *expérience* were in flux in the second half of the nineteenth century. In other words, it was not simply a case of adding denotations and connotations but of recalibrating existing ones in relation to one another. When "by chance" migrated to the second definition as "involuntarily," this cast one sense of *expérience* as more evidently subjective and passive, over and against a more objective, more active, and more thoroughly "scientific" definition. These recalibrations, which would inflect symbolists' changing understanding not only of experiment but also of the relationship between subjectivity and objectivity, were undoubtedly linked to what Lorraine Daston and Peter Galison have argued was a newly stark but in some ways paradoxical opposition between passive observation and active experimentation that came into play around the middle of the century.[15]

None of these dictionaries included cultural or vanguardist usages for the adjective *expérimental*. Only the Academy's ninth edition, which started appearing in 1986, records experimental art as a discursive formation, while other dictionaries, both French and English, vaguely identify the artistic vanguard usage as a specialized form of the figurative emerging in the early twentieth century.[16] Historians of art and literature have been similarly vague, implying and in a few instances arguing that the conjunction of art and the experimental under the banner of avant-gardism was inevitable in a modern scientific age but largely allegorical.[17]

This book, on the contrary, counters assumptions of a vaguely allegorical relationship between experimentalism and artistic avant-gardism, and argues instead that experimentation and avant-gardism coalesced around symbolism and around a pervasive way of thinking, doing, and knowing in nineteenth-century France. This way of knowing was immensely influential, but it has largely been forgotten. I shall refer to it as the epistemological paradigm of nature's experiments.

Nature's Experiments

In 1888, Friedrich Nietzsche noted, as if it were a widely accepted fact, "It is the value of all morbid states that they show us under a magnifying glass certain states that are normal—but not easily visible when normal." He elaborated on this point by citing Bernard's *Lessons on Animal Heat* (1876): "Health and Sickness are not essentially different. . . . In fact, there are only differences in degree between these two kinds of existence: the exaggeration, the disproportion, the nonharmony of the normal phenomena constitute the pathological state."[18] The belief that the normal and the pathological were quantitatively linked such that the pathological could act like a "magnifying glass" on the normal, and the use of this paradigm as the basis for a scientific methodology—the pathological method—probably has its roots in the late eighteenth century.[19] But it was Comte who credited himself with transferring a nascent version of the pathological method from the biological sciences to a systematic study of "intellectual and moral phenomena" under the banner of experiment.[20] Rejecting vivisection as both a cruel and a useless form of experimentation, Comte recommended pathological cases as the "true scientific equivalent of pure experimentation" for the biological and, by extension, the new sociological sciences. Whereas experimentation, he argued, proved extremely difficult on higher-order organisms, simpler organisms allowed the researcher to establish controlled conditions more easily. As if nature herself had offered up a gift to science, "precisely in the case in which artificial experimentation is the most difficult," Comte argued, "nature fulfills the conditions for us" by providing the opportunity to study quantitative variants of normality in the form of abnormality. Pathological conditions—and for Comte's sociology this meant the state of revolution—thus provided nature's own experimental setup for the investigation of complex organic phenomena and for the extrapolation of laws for normal organisms, including the laws for a stable society.[21]

Despite protests to the contrary, Bernard was deeply indebted to Comte, especially when it came to the experimental value of a continuous relationship between the normal and the pathological.[22] If medicine, Bernard reasoned, by definition the study of disease, was aimed at developing therapies, the scientist must understand the organism's normal state in order to return it to health from the pathological state. Bernard's insistence on this point relied more on what Georges Canguilhem called "monotonous repetition" than on solid evidence, but for the purposes of this study, the fundamental error that Canguilhem identifies—the erroneous assumption about the continuity between the normal and the pathological—is beside the point.[23] The cultish authority surrounding both Comte and Bernard suggests that nature's experiments offered an enticing model for those seeking to establish facts about complex organic phenomena and especially, as was the case with the symbolists, the inner complexities of the subject presumed to be at the apex of normal earthly existence, the white bourgeois man. As men of science increasingly conceived of their own psychological functioning as an object of scientific investigation, the pathological method was ever more frequently summoned to provide psychological studies with the authority of experiment.

Nature's experiments supplied French psychology with both its scientific credentials and its modus operandi in the second half of the nineteenth century. As the philosopher and political theorist Ernest Renan summed it up in 1890:

> Sleep, madness, delirium, somnambulism, and hallucination offer a far more favorable field of experiment than the regular state for studying the psychology of the individual. Phenomena, which in the regular state are almost effaced because of their tenuousness, appear more palpable in extraordinary crises because they are exaggerated. The physicist does not study galvanism in the weak quantities that nature presents, but increases it through experimentation, in order to study it more easily, although the laws in that exaggerated state are identical to those of the natural state. Similarly, human psychology will have to be built around studying the madness of mankind, the dreams, hallucinations, and all those curious absurdities to be found on every page of the history of the human mind.[24]

For much of the nineteenth century, psychology had been the purview of philosophy, and a centerpiece of the eclecticism of Victor Cousin. For Comte, Cousin's introspective method of psychological observation of the internal self was a logical impossibility, since the individual, Comte insisted, could not be at once subject and object of observation.[25] But as positivism gained institutional ground, and especially with the apparent successes of physiology, a scientific psychology seemed to many second-generation positivists to be imaginable. In 1888, the same year that Nietzsche highlighted the value of "morbid states," and after nearly two decades of debate over the name for the new psychology, Théodule Ribot accepted the first chair at the Collège de France in "experimental and comparative psychology."[26] The following year, in Paris, at an international congress timed to coincide with the Universal Exposition, French delegates continued to discuss the name of the new field, rejecting "physiological psychology" as too narrow and "scientific psychology" as "too absolute," and reiterating their approval of "experimental psychology" because it plainly rejected philosophy and explicitly signaled that experimental psychologists would henceforth be "abstaining from any metaphysical questions."[27] Ribot, as Jacqueline Carroy and Régine Plas have argued, recommended that French psychologists eschew what he characterized as a narrow and largely quantitative German experimental psychology for a more descriptive psychology. In doing so, Ribot, along with Hippolyte Taine, chose to derive institutional authority for French experimental psychology from what he understood to be specifically French contributions, above all the new experimentalism and the pathological methods of Comte and Bernard. These proved especially appealing in the first decades of the Third Republic, when anxiety about France's scientific and technological achievements was at its peak.[28]

"The more a phenomenon is bizarre," Taine wrote in his expanded preface to *On Intelligence*, first published in the second edition of 1876, "the more it is instructive." "The most minor, well-selected phenomena" (*tout petits faits bien choisis*), he argued, provoke a heightened state of consciousness, a state that functions, as Nietzsche would later assume, like an optical prosthesis to enhance the capabilities of the human eye and to provide a way to "see" the inner workings of the human mind.[29] As his contribution to a fledgling scientific psychology, Taine offered an exhaustive list of recommended objects of study, which reads as a partial compendium of avant-garde inspiration. Nearly all were what he considered abnormal psychological situations: the language of children, dreams, the experiences of drug addicts, hypnagogic hallucinations, hypnotism, somnambulism, automatism, spirit manifestations, and madness. To this veritable catalogue of the abnormal, he added artistic genius, recommending that "every exceptionally clear-headed painter, poet, and novelist" submit himself to the questions of a "friendly psychologist."[30] Taine lamented the dearth of documents written by the mentally ill themselves, and went so far as to repudiate Comte by recommending the self as an object of study, endorsing a form of self-experimentation on one's own abnormal states.

While Ribot maintained a place within experimental psychology for what he called a "subjective method" akin to Cousin's introspective method, only an "objective method," Ribot argued (what he also called a "morbid psychology"), could establish the discipline as a science.[31] Like Taine, Ribot believed that his predecessors had virtually ignored the abnormal, but that pathological cases offered ideal opportunities to gain knowledge of the most complex, difficult-to-observe phenomena, and could thus constitute the new psychology as objective: "morbid disturbances of the organism that bring about mental disorders; anomalies, monsters in the psychological order, for us these are like experiments prepared by nature and all the more precious for being rare."[32] Ribot subsequently devoted himself to producing a series of monographs that would take "experiments prepared by nature" as their central object of study, publishing *Diseases of Memory* (1881), *Diseases of the Will* (1883), *The Diseases of Personality* (1885), and monographs on attention (1889), the emotions (1896), and creativity (1900), all of which went through numerous editions and were rapidly translated into English, German, and Italian.[33] In each, Ribot took the opportunity to reiterate the critical importance of the pathological method, promoting over and over the triad of "madmen, primitives, and children" as psychology's preeminent objects of study. Ribot thus introduced a dimension of the pathological method only implied in Taine's thinking, namely, an evolutionism derived primarily from Herbert Spencer and John Hughlings Jackson.[34] No longer, Ribot promised, would psychology concern itself only with the higher rungs of the evolutionary ladder, namely, the "white, adult, civilized male." Rather, it would now engage in comparative studies of animals, children, the mentally ill, and "inferior races," all seen as nature's experiments, to be mobilized as objects of investigation. In these putatively pathological cases, "the actions are less numerous, less complicated," Ribot wrote, "but . . . the functioning does not disappear."[35]

Owing in no small part to the writings of Taine and Ribot, French experimental psychology in the late nineteenth century came to be identified with psychopathology, but in the early twentieth century, the discipline began to shift toward a German model.

By 1911, when leading French psychologists dropped the reference to pathological psychology in the subtitles of their influential textbooks, French experimental psychology had ceased to depend upon nature's experiments as its methodological paradigm.[36] A thirty-year reliance on the pathological method, however, meant that readers of French psychology, including many within the emerging avant-gardes, could consider themselves well schooled in altered psychological states and in the mental functioning of "madmen, primitives, and children," which they could understand as continuous with their own mental functioning. Psychopathology had also led to intense debates over the status, function, and nature of the unconscious, raising questions, including legal ones, about how unconscious, automatic activity related to conscious, willed activity.[37] The pathological method had ushered the unconscious, the part of the mind that Taine had argued was usually waiting in the wings, onto scientific psychology's center stage. But the increasing incorporation of the unconscious into an understanding of normal mental life would eventually bump nature's experiments from their starring role as the methodological centerpieces of a scientific psychology. By then, however, which is to say, by the early twentieth century, the avant-garde had taken up what psychology had constructed as pathological alterity, all those altered states and all those "madmen, primitives, and children," as wellsprings for masculinist, colonialist inspiration.[38] While the avant-garde was certainly recapitulating the romantic view of children and "primitives" as original natures,[39] its revaluation of alterities, when we consider its more direct roots in symbolist theory and practice, reveals a scientific cast provided by the epistemological paradigm of nature's experiments.

Symbolism: From Theory to Visual Practice

Consider the following four works, all, at the time of their making in the early 1890s, conceived of, and to a greater or lesser extent received as, instances of symbolist form, and all easily described as experimental: an oil study for Maurice Denis's lost painting *Décor* (ca. 1890–91, fig. 3), Édouard Vuillard's portrait of Aurélien-Marie Lugné-Poë (1891, fig. 4), August Strindberg's *Night of Jealousy* (1893, fig. 5), and Edvard Munch's *The Scream* (1893, fig. 6). They are all extraordinary for their time and extraordinarily hard to characterize. But consider some of their most striking formal aspects: the flattened and distorted forms of the female nude in Denis's sketch, the flattened areas of paint in Vuillard's portrait and its curving arabesques, which are in fact narrow gaps where the painting's support shows through. Consider Strindberg's resolutely painted *Night of Jealousy*, signed on the back "the Symbolist August Strindberg" and only barely registering as a seascape. And consider Munch's iconic *Scream*, with its in-your-face antinaturalism that proclaims its anti-aesthetic avant-gardism as loudly as a Dada sound poem. Are these

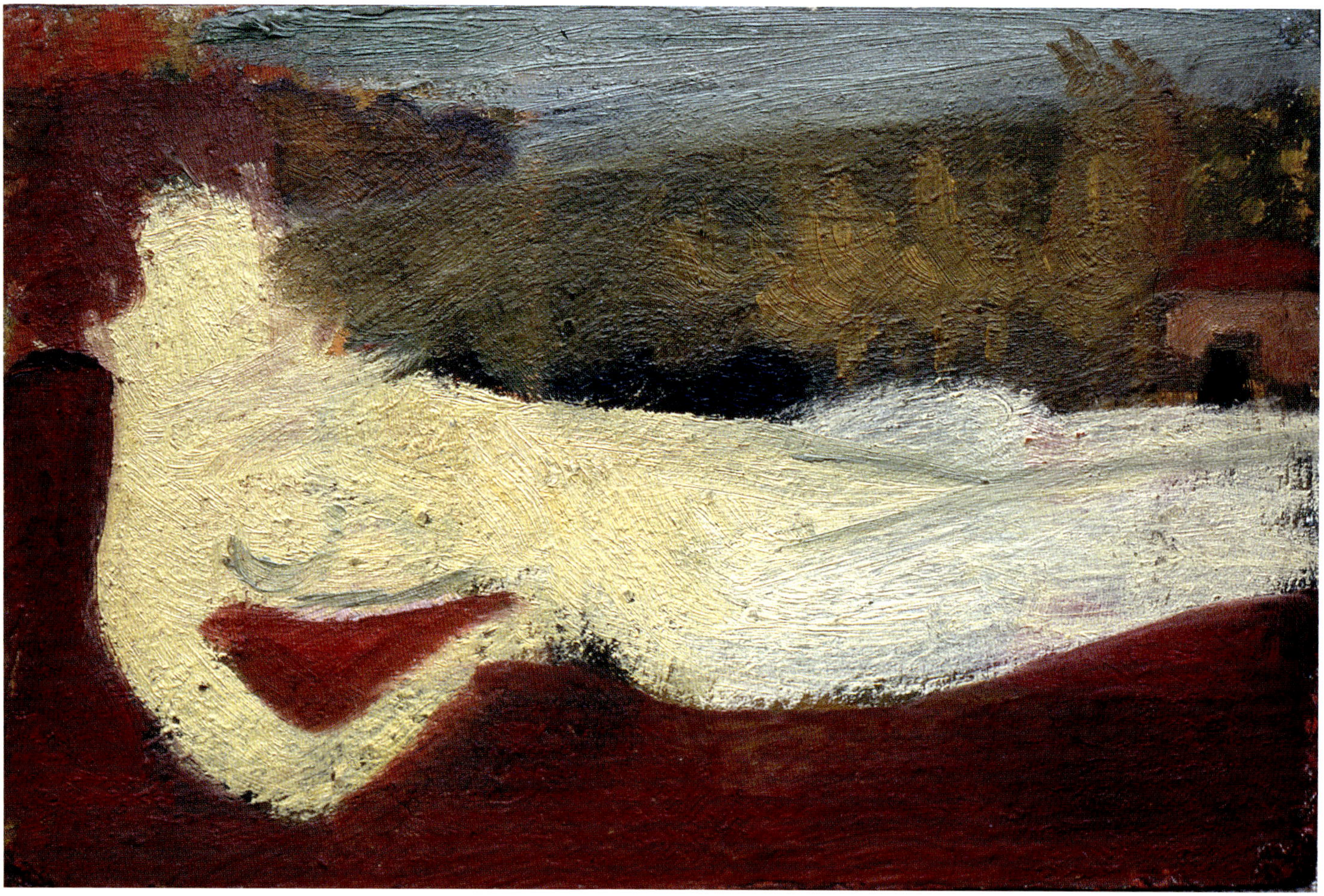

3 Maurice Denis, study for *Décor*, ca. 1890–91. Oil on cardboard, 13.5 × 20.7 cm. Private collection.

works experimental in the sense that they were new or radical in their own time? Are they experimental because they appear to be attempts at something, trials that perhaps are not entirely successful? Are they somehow like scientific investigations? Or are these symbolist works of art connected to experimentalism as it was understood in the late nineteenth century in a more fundamental way?

Following the publication of Jean Moréas's and Kahn's symbolist manifestos in 1886, a number of artists, including Gustave Moreau and Odilon Redon, were proclaimed symbolists *avant la lettre*, extolled for having explored symbolist themes using symbolist modes as early as the 1860s.[40] At the same time, a number of writers, critics, and artists had begun to adapt symbolist literary theory to theories of symbolist visual practice. Artists wishing to make symbolist art—and by the early 1890s there were many—not only faced competing theories designed to describe and disseminate symbolist literary forms,

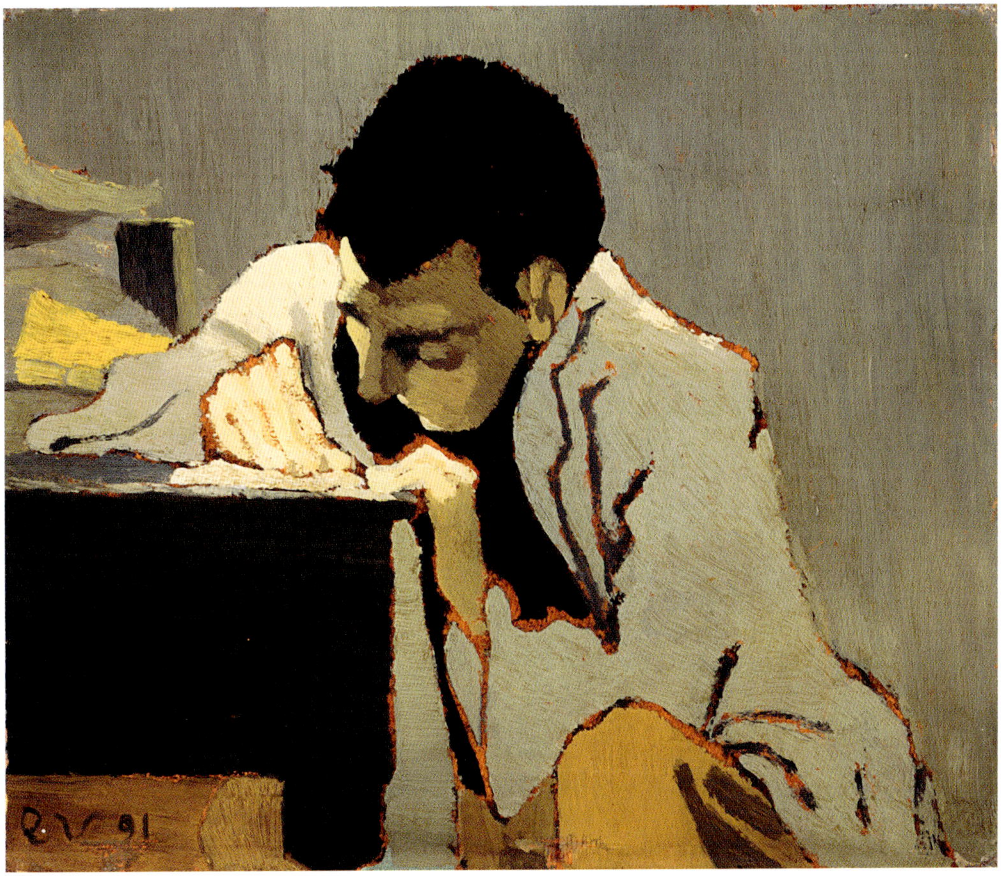

4 Édouard Vuillard, *Aurélien-Marie Lugné-Poë* (1869–1940), 1891. Oil on paper mounted on panel, 22.23 × 26.67 cm. Memorial Art Gallery of the University of Rochester, Rochester, New York, Gift of Fletcher Steele, 72.18.

but increasingly encountered contradictory, even antagonistic statements regarding how that theory might be applied to or transformed for symbolist visual practice.

Symbolism has thus proved an especially slippery thing for art history.[41] Symbolist art lacks the evident stylistic coherence or group dynamics that might enable the art historian to insert it smoothly into canonical narratives, modernist or otherwise. Consequently, the difficulties of defining and categorizing symbolism, and of describing its relationships to synthetism, neoimpressionism, cloisonnism, Stimmungsmalerei, and a host of related movements or terms, has long beset art historians. From Robert Goldwater's *Symbolism* (1979) to more recent publications, such as Rodolphe Rapetti's *Le symbolisme* (2005), Michelle Facos's *Symbolist Art in Context* (2009), and Richard Thomson's collaborative project, *Redefining European Symbolism, 1880–1910* (2010–13), scholars have often taken on the movement or "moment" as a kind of whole, a problematic whole to be sure, but a monograph-worthy one nonetheless.[42] In doing so, definition, redefinition, and even "dedefinition" continue to remain imperative and often productive art-historical strategies.[43] To these have been added contextual and thematic approaches: not only Facos's aptly named text but also the landmark essays in Pierre Théberge and Jean Clair's exhibition catalogue *Lost Paradise: Symbolist Europe*, Patricia Mathews's feminist intervention, *Passionate Discontent: Creativity, Gender, and French Symbolist Art*, and Sharon L. Hirsh's work of social art history, *Symbolism and Modern Urban Society*.

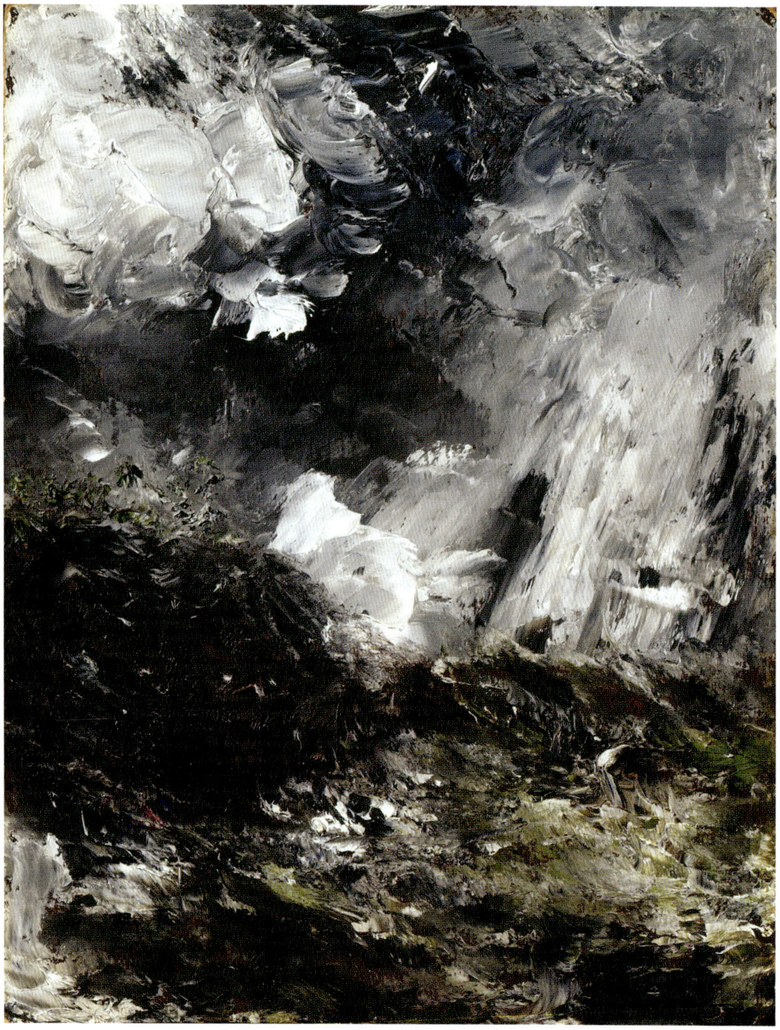

5 August Strindberg, *Night of Jealousy*, 1893. Oil on cardboard, 41 × 32 cm. Strindberg Museum, Stockholm.

This study draws upon the critical art-historical projects centered on definition, contextualization, and thematization, but it seeks provisional answers to questions of a different order. Fundamentally, it asks how symbolist theory could at times be transformed in such a way that, in its antinaturalism, its abstractions, its attentiveness to surface and materiality, and its expressionist distortion, it was liable to be called symbolist, and finds provocative answers in the epistemology of experiment. The argument about symbolism presented here is thus unabashedly partial in both senses of the term. Rather than look at symbolism as a contested whole, it addresses only a small selection of symbolist visual practices, delving into the work of only a few artists at fairly restricted moments in their careers. It does not privilege those artists who were part of symbolism's construction of its own past, such as Moreau or Redon, or those whose works might be labeled primarily symbolist on the basis of their iconography, but considers the work of a handful of artists who, especially in the early to mid-1890s, took particularly self-conscious and intellectual approaches to producing forms at once symbolist and modernist—those, in other words, who searched, and left significant traces of their search, for symbolist form.[44] While my conclusions might be extended to a number of other artists, including Edgar Degas, Paul Cézanne, Paul Gauguin, Paul Sérusier, and Pierre Bonnard, all of whom make cameo appearances, the study concentrates on the work of Denis, Vuillard, Strindberg, and Munch, who all sought to apply symbolist theories to visual practice, to transform naturalist modes into symbolist ones, and often, most evidently in the cases of Denis and Strindberg, to adapt symbolist theories so as to more adequately reflect or constitute their own practices.

In narrowing the focus in this way, the book aims not so much to come to a definitive understanding of symbolist visual form, or even to engage the problem of defining symbolist painting, as to interrogate the epistemological conditions under which these four artists, no doubt four among many others, translated the theories and assertions of symbolism into working practices aimed at producing visual form with the potential to be identified by their contemporaries as symbolist and later as modernist. From the decorative and quasi-abstract early works of Denis and Vuillard, to the "automatisms" of Strindberg, to the protoexpressionism of Munch, the symbolist forms considered here, while never easily subsumed into the dominant narratives of modernism (at least not under the name symbolism), are nevertheless startling in the way they appear to presage twentieth-century modernist currents. But the mechanisms by which symbolism provoked such "modern" and indeed "experimental" works remain obscure.[45] At the same time, the legacy of some of these prototypes has often been questioned in the face of what appear, from within the dominant discourses of modernism, to be regressions (especially when it comes to Denis, Vuillard, and Munch, who all lived well into the twentieth century) into seemingly repetitive naturalisms or impressionisms.

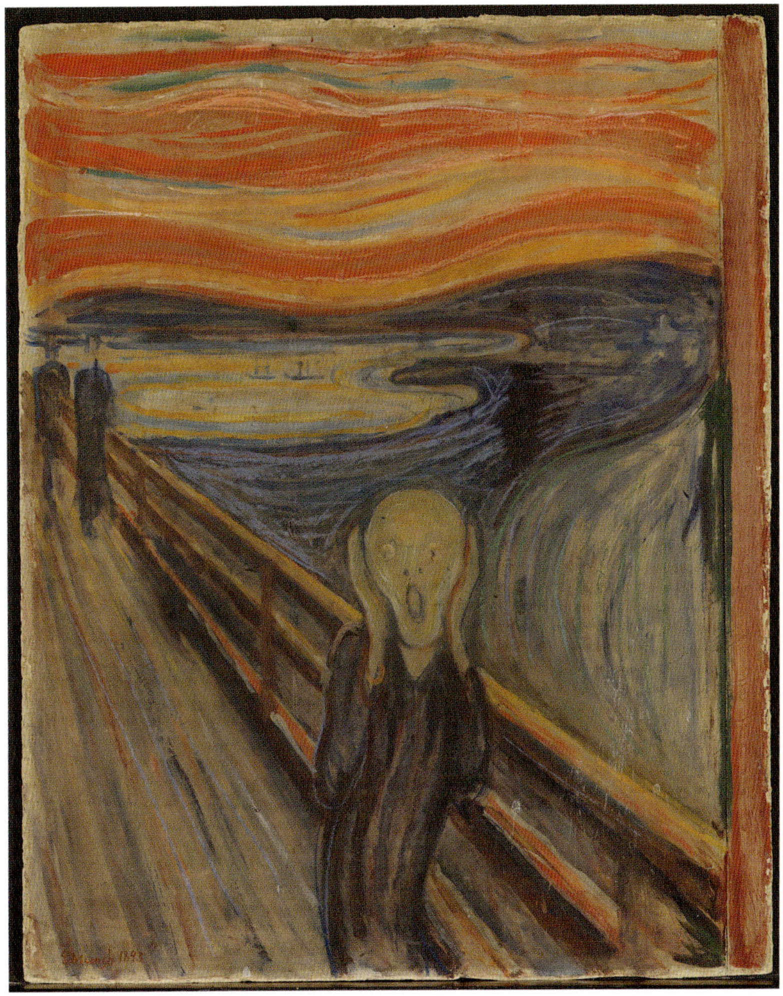

6 Edvard Munch, *The Scream*, 1893. Oil, tempera, and pastel on cardboard, 91 × 73.5 cm. National Museum of Art, Architecture, and Design, Oslo, NG.M.00939.

I argue that the visual and textual practices of Denis, Vuillard, Strindberg, and Munch were enabled and at times disabled by what I am calling nature's experiments, which informed and conditioned these artists' practices and the reception of their works in both productive and counterproductive ways. A number of scholars have demonstrated the extent to which the nineteenth century's promising new field of scientific psychology influenced symbolist theory and practice.[46] And while the new French psychology plays a key role in what follows, my focus is less on the field itself and more on how the field was constituted around the paradigm of nature's experiments. My aim is to investigate how a culture of nature's experiments, as it functioned not only across different scientific disciplines but also within broader contexts both inside and outside France, allowed for and encouraged the cultivation of pathology and otherness to determine ways of producing and receiving symbolist art in the 1890s.

As implied earlier, an art history of nature's experiments, while focused on a small selection of artists, texts, and works of the late nineteenth century, has implications for a discontinuous history of artistic experimentation. Symbolism turns out to provide the specific context in which experiment came to be closely associated with artistic advance. Not at all incompatible with idealism and spiritualism, experimentalism provided a ready-made methodology for fin-de-siècle truth seekers, a mobile doctrine that easily traversed disciplinary boundaries. As a centerpiece of experimentalism, the pathological method proved to be a particularly seductive pathway to something called truth, an epistemological goal that, in the late nineteenth century, squared easily with vanguardism.[47] This study thus questions a transhistorical and transcultural notion of artistic *experimentation* as a persistent, pervasive, and altogether positive avant-garde value, and offers instead an interrogation of a historically specific artistic *experimentalism*, thereby restoring to experimentalism its less attractive values—specifically, its construction of the cultural other as a particularly valuable object of study, the imagined basis on which to establish what was imagined to be universal truth. A scientific revaluing of pathology and otherness thus preceded the emergence of a full-blown avant-garde art, and it is traceable in this study as the residue of the discursive formation of experimental art.

At the outset, I wish to head off one criticism that I suspect will nevertheless stick to certain parts of my text: the charge that in privileging discussions of symbolist visual practice, that is to say, form, I disregard content. The modernist discursive separation of form and content is, in my view, a false construction for much modern art, and especially for the kinds of symbolist practices under discussion here. While this study assumes an unstable notion of symbolist form, emphasizing the search for it rather than an a priori and fixed notion of it, I have nevertheless found helpful, as a point of constant return, Reinhold Heller's analysis of symbolism as not definable solely by subject matter but, at its most characteristic, as instituting a dialectic between form and content at the surface of the work.[48] If I seem to focus more on the problem of form, it is not because I think content can be disregarded (although that claim was made in Denis's protoformalism), but because content often proved less vexing for the symbolist artists considered here. The content, what many symbolists called the "Idea," generally came first. Their most pressing problem became the search for appropriate forms in which to "clothe" or "envelop" that capital-I Idea, forms that would then ideally become completely inextricable from content. The dialectical relationship between form and content proves especially crucial to the analysis of Munch's work that concludes this study. More consistently than perhaps any other symbolist artist, Munch sought out pathological form for pathological content with a view to providing what he and his critical supporters imagined as truthful and universally legible representations of human experience. At the same time, this discursive split is a critical issue for the historiography of modernism, and is relevant especially to my discussion of the writings of Denis in chapter 2, where an expunging of pathological content from the concept of deformation is read as one consequence of an epistemology based upon nature's experiments.

In concerning myself nominally with form, both the form of symbolist visual practice and the rhetorical form of symbolist theory and criticism, I have found it useful to keep in mind the words of T. J. Clark: "It is the form of our statements, and the structure of our visualizations, that truly are our ways of world-making—at any rate the ways that hold us deepest in thrall."[49] I readily admit to being in thrall to much of the symbolist form that I discuss, not because it always or even usually affects me in the way that the artists hoped it would—my twenty-first-century critiques of ideals such as universal truth and experience prevent this—but because the formal solutions to the problems posed by symbolist theory, and especially those put forth under a presumed psychological universalism, strike me as singularly radical and perplexing.

The problem of many symbolists, and it was a problem that would bedevil surrealist artists as well as abstract expressionists, was twofold. Their challenge was to create form that could be read as corresponding to an individual's, usually the artist's, interior emotive world, and to render that form objectively understandable, even truthful, *without* using what were understood to be or caricatured as traditional, academic, naturalist modes of art making.[50] Having ostensibly rejected one system of representation, more often than not identified as naturalism, symbolist artists actively sought another. But another system of representation was not to be plucked from thin air. It still had to function within existing ideological structures. This study therefore concerns itself with both production *and* reception, or, more precisely, production *with a view to* reception. For Denis, Vuillard, Strindberg, and Munch, along with many in their symbolist cohorts, operated with a keen awareness of and attention to the stakes involved in the critical reception of their work. It is, moreover, precisely in those journals that sought to delimit symbolism, above all in *La revue blanche* and the *Mercure de France*, that we can glean a more precise understanding of the relevance of experimentalism, the pathological method, and the work of Taine and Ribot to symbolist audiences. An obscure journalist and devotee of French experimental psychology named Léon Bélugou will be our guide.

French Experimental Psychology, Symbolism, and the *Petites Revues*

Symbolist writers, critics, and artists organized themselves around the numerous and often short-lived *petites revues* that proliferated in Paris in the late 1880s and 1890s.[51] Denis and Vuillard, along with other artists who dubbed themselves the Nabis, were especially

7 Pierre Bonnard, *La revue blanche*, 1894. Lithograph poster in black, gray, and cream, 78.2 × 59 cm. Yale University Art Gallery, New Haven, Connecticut, University Purchase, 1951.9.19.

8 Maurice Denis, *Four Women* or *The Crying Women*, frontispiece for *La revue blanche* 5 (October 1893). Bibliothèque nationale de France.

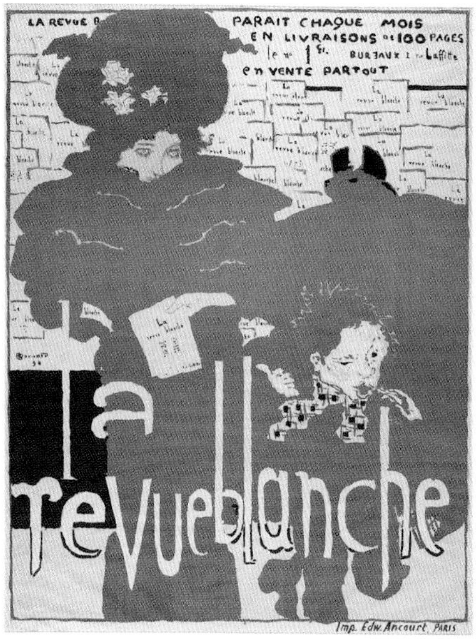

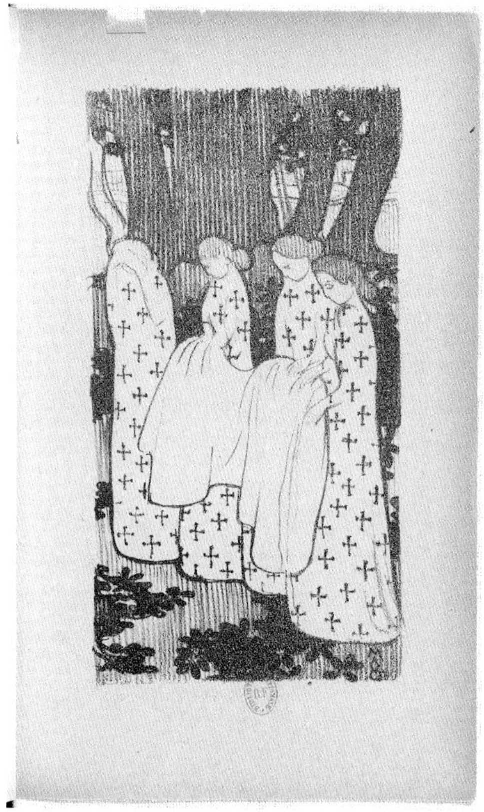

closely associated with *La revue blanche*, the intellectually and artistically ambitious journal founded by the three Natanson brothers in 1889 (fig. 7).[52] The Natansons, and especially Thadée Natanson, were major patrons of the Nabis, reviewed their exhibitions, and frequently solicited contributions from the artists to enliven the journal's pages (figs. 8 and 9). But *La revue blanche* also actively looked beyond Paris. In 1895, Thadée Natanson visited Kristiania (present-day Oslo), where he met Munch and saw a large solo exhibition of his works. Later that year, Natanson published an appreciative article on the Norwegian artist in *La revue blanche*, and he must also have been instrumental in having Munch's lithograph version of *The Scream* published in a subsequent number of the journal (fig. 10).[53] When Munch, no doubt encouraged by Natanson's support, exhibited in Paris in 1896, Strindberg, who had known the artist since Munch arrived in Berlin in late 1892, contributed to the journal an idiosyncratic reflection on Munch's work, one of a number of articles Strindberg published in *La revue blanche* during his time in Paris, between 1894 and 1898 (fig. 11).[54] Essays on and by Strindberg also appeared in the *Mercure de France*, a journal founded in 1890 that published major statements on symbolist visual practice, including Aurier's defining articles on Vincent van Gogh in 1890 and on Paul Gauguin in 1891.[55]

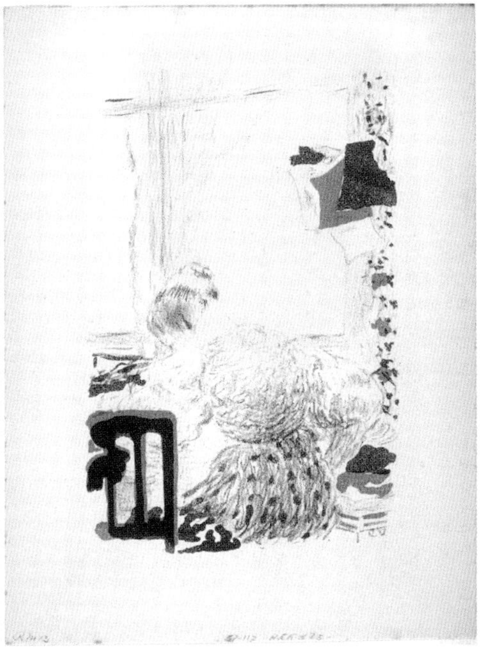

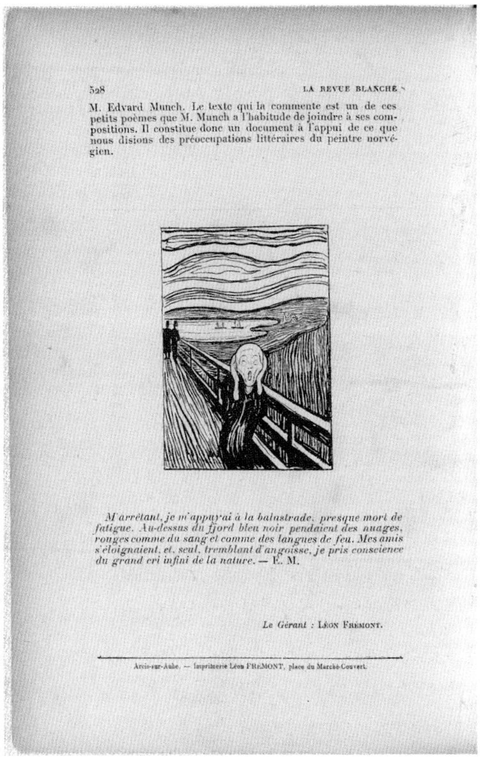

528 LA REVUE BLANCHE

M. Edvard Munch. Le texte qui la commente est un de ces petits poèmes que M. Munch a l'habitude de joindre à ses compositions. Il constitue donc un document à l'appui de ce que nous disions des préoccupations littéraires du peintre norvégien.

M'arrêtant, je m'appuyai à la balustrade, presque mort de fatigue. Au-dessus du fjord bleu noir pendaient des nuages, rouges comme du sang et comme des langues de feu. Mes amis s'éloignaient, et, seul, tremblant d'angoisse, je pris conscience du grand cri infini de la nature. — E. M.

Le Gérant : Léon Frémont.

Arcis-sur-Aube. — Imprimerie Léon FRÉMONT, place du Marché-Couvert.

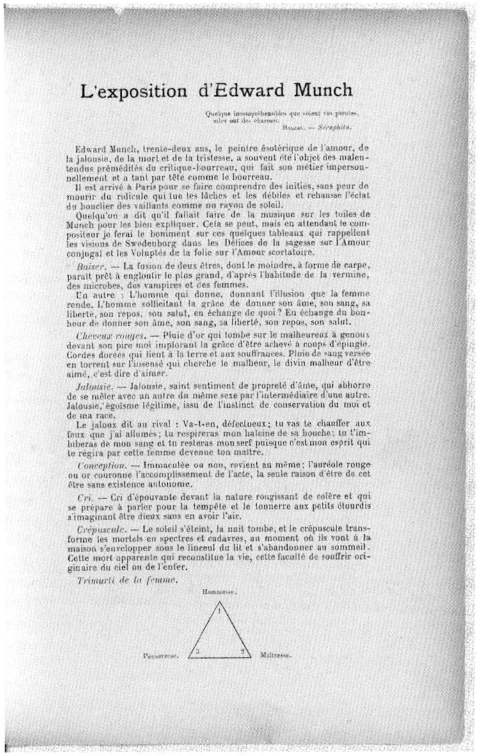

L'exposition d'Edward Munch

Quelque incompréhensibles que soient vos paroles, elles ont des charmes.
Balzac. — *Séraphita.*

Edward Munch, trente-deux ans, le peintre ésotérique de l'amour, de la jalousie, de la mort et de la tristesse, a souvent été l'objet des malentendus prémédités du critique-bourreau, qui fait son métier impersonnellement et a tant par tête comme le bourreau.

Il est arrivé à Paris pour se faire comprendre des initiés, sans peur de mourir du ridicule qui tue les lâches et les débiles et rehausse l'éclat du bouclier des vaillants comme un rayon de soleil.

Quelqu'un a dit qu'il fallait faire de la musique sur les toiles de Munch pour les bien expliquer. Cela se peut, mais en attendant le compositeur je ferai le boniment sur ces quelques tableaux qui rappellent les visions de Swedenborg dans les Délices de la sagesse sur l'Amour conjugal et les Voluptés de la folie sur l'Amour scortatoire.

Baiser. — La fusion de deux êtres, dont le moindre, à forme de carpe, paraît prêt à engloutir le plus grand, d'après l'habitude de la vermine, des microbes, des vampires et des femmes.

Un autre : L'homme qui donne, donnant l'illusion que la femme rende. L'homme sollicitant la grâce de donner son âme, son sang, sa liberté, son repos, son salut, en échange de quoi ? En échange du bonheur de donner son âme, son sang, sa liberté, son repos, son salut.

Cheveux rouges. — Pluie d'or qui tombe sur le malheureux à genoux devant son pire moi implorant la grâce d'être achevé à coups d'épingle. Cordes dorées qui lient à la terre et aux souffrances. Pluie de sang versée en torrent sur l'insensé qui cherche le malheur, le divin malheur d'être aimé, c'est dire d'aimer.

Jalousie. — Jalousie, saint sentiment de propreté d'âme, qui abhorre de se mêler avec un autre du même sexe par l'intermédiaire d'une autre. Jalousie, égoïsme légitime, issu de l'instinct de conservation du moi et de ma race.

Le jaloux dit au rival : Va-t-en, défectueux; tu vas te chauffer aux feux que j'ai allumés; tu respireras mon haleine de sa bouche; tu t'imbiberas de mon sang et tu resteras mon serf puisque c'est mon esprit qui te régira par cette femme devenue ton maître.

Conception. — Immaculée ou non, revient au même; l'auréole rouge ou or couronne l'accomplissement de l'acte, la seule raison d'être de cet être sans existence autonome.

Cri. — Cri d'épouvante devant la nature rougissant de colère et qui se prépare à parler pour la tempête et le tonnerre aux petits étourdis s'imaginant être dieux sans en avoir l'air.

Crépuscule. — Le soleil s'éteint, la nuit tombe, et le crépuscule transforme les mortels en spectres et cadavres, au moment où ils vont à la maison s'envelopper sous le linceul du lit et s'abandonner au sommeil. Cette mort apparente qui reconstitue la vie, cette faculté de souffrir originaire du ciel ou de l'enfer.

Trimurti de la femme.

Hommesse.
1
3 2
Pécheresse. Maîtresse.

9 Édouard Vuillard, *The Seamstress* (*La Couturière*), frontispiece for *La revue blanche* 6 (January 1894). Color lithograph on paper, 32.1 × 24.8 cm. University of Michigan Museum of Art, Museum Purchase, 1959/1.106.

10 Edvard Munch, *The Scream*, 1895. Lithograph, reproduced in *La revue blanche* 9 (1 December 1895): 528. Bibliothèque nationale de France.

11 August Strindberg, "L'Exposition d'Edward Munch," *La revue blanche* 10 (1 June 1896): 525. Bibliothèque nationale de France.

Both of these journals cultivated highly educated contributors and readers, many of whom would have attended one of the elite French lycées and passed their philosophy baccalaureate. By the late 1880s, the lycée philosophy class was producing candidates who were intimately familiar with the most up-to-date developments in experimental psychology and steeped in the pathological method.[56] These students included Denis and Sérusier, who both attended, along with Vuillard, the Lycée Condorcet, a prestigious Parisian institution earlier attended by Bergson at which the symbolist poet Stéphane Mallarmé had taught in the 1870s and early 1880s.[57]

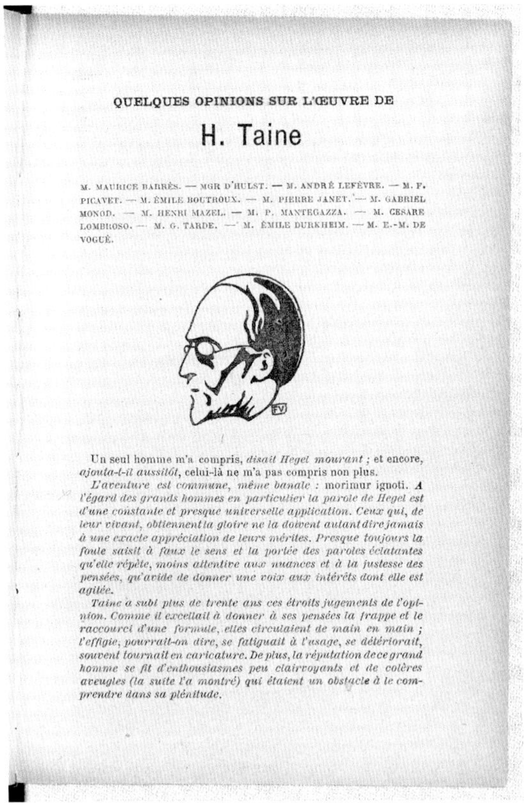

QUELQUES OPINIONS SUR L'ŒUVRE DE

H. Taine

M. MAURICE BARRÈS. — MGR D'HULST. — M. ANDRÉ LEFÈVRE. — M. F. PICAVET. — M. ÉMILE BOUTROUX. — M. PIERRE JANET. — M. GABRIEL MONOD. — M. HENRI MAZEL. — M. P. MANTEGAZZA. — M. CESARE LOMBROSO. — M. G. TARDE. — M. ÉMILE DURKHEIM. — M. E.-M. DE VOGUÉ.

Un seul homme m'a compris, *disait Hegel mourant ;* et encore, *ajouta-t-il aussitôt*, celui-là ne m'a pas compris non plus.

L'aventure est commune, même banale : morimur ignoti. *A l'égard des grands hommes en particulier la parole de Hegel est d'une constante et presque universelle application. Ceux qui, de leur vivant, obtiennent la gloire ne la doivent autant dire jamais à une exacte appréciation de leurs mérites. Presque toujours la foule saisit à faux le sens et la portée des paroles éclatantes qu'elle répète, moins attentive aux nuances et à la justesse des pensées, qu'avide de donner une voix aux intérêts dont elle est agitée.*

Taine a subi plus de trente ans ces étroits jugements de l'opinion. Comme il excellait à donner à ses pensées la frappe et le raccourci d'une formule, elles circulaient de main en main ; l'effigie, pourrait-on dire, se fatiguait à l'usage, se détériorait, souvent tournait en caricature. De plus, la réputation de ce grand homme se fit d'enthousiasmes peu clairvoyants et de colères aveugles (la suite l'a montré) qui étaient un obstacle à le comprendre dans sa plénitude.

12 "Quelques opinions sur l'oeuvre de H. Taine," with vignette by Félix Vallotton, *La revue blanche* 13 (15 August 1897): 263. Bibliothèque nationale de France.

Both *La revue blanche* and the *Mercure de France* used two journalistic formulae designed to enhance their claims to intellectual seriousness: *enquêtes*, quasi-scientific surveys on questions of the day, and *chroniques*, regular overviews of the state of research in various fields designed to keep readers aware of contemporary developments in a wide range of disciplines.[58] In 1897, four years after his death, *La revue blanche* published a survey on Taine, accompanied by a woodcut portrait by the Nabi artist Félix Vallotton (fig. 12). The survey's instigator, Bélugou, invited philosophers, journalists, novelists, historians, and scientists of all philosophical stripes to reflect upon how Taine had influenced them. The respondents' near unanimous appreciation reflects the high esteem for Taine within the *Revue blanche* community, and a special admiration for *On Intelligence* as a formative work. Maurice Barrès, author of the *Cult of the Self* trilogy (1888–91), much admired by Denis and his fellow Nabis, expressed his indebtedness to Taine's methods of observation and experimentation.[59] The Catholic prelate and writer Maurice Le Sage d'Hauteroche d'Hulst praised Taine as a "metaphysician by race" and suggested that Tainean positivism could be easily reconciled with Catholicism, since Taine's method was "at once positive and very a priori, despite the contradiction that those two terms might imply."[60] In a similar vein, Émile Boutroux, the spiritualist philosopher and an important mentor to both Bergson and Émile Durkheim, stressed not only Taine's impact on psychology but also his more inadvertent contributions to a resurgence of idealism.[61]

More predictable were the responses solicited from fin-de-siècle men of science. The psychologist Pierre Janet, who would eventually take up Ribot's chair at the Collège de France, confessed that *On Intelligence* had been on his bedside table when he was a student, and Cesare Lombroso proclaimed Taine "the only master I had after Darwin." The philosopher and Kant scholar François Picavet reminded readers of Taine's importance for anyone claiming to be an intellectual; in Picavet's view, it was simply inconceivable that a well-educated man would not have read his Taine. Concluding the survey, Bélugou confessed his own admiration for Taine's humble morality, absolute conviction, and methodological rigor. And he reminded *Revue blanche* readers that Taine sought what he proclaimed they were all after: the truth.[62]

A journalist and private tutor for a number of wealthy French families, Bélugou, it turns out, is emblematic of the admiration within symbolist communities for experimentalism, experimental psychology, and the pathological method.[63] From 1894, Bélugou's columns for *La revue blanche* and the *Mercure de France* were dominated by his interest in experimental psychology, his admiration for Taine, and his perhaps even greater regard for Ribot. He was especially explicit on the importance of the pathological method for the great strides being made in scientific psychology. In a column published in November 1894, Bélugou discussed recent studies on multiple personality and other "lucid" cases of insanity, citing Ribot and Pierre Janet and praising the work of psychologists who "employed . . . the documents of mental pathology for the solution to psychological problems."[64] Bélugou had in fact been faithfully attending Ribot's courses since 1887, first at the Sorbonne and then at the Collège de France, and would continue to do so throughout the 1890s.[65] Calling Ribot the foremost psychologist of his day, Bélugou recommended to readers Ribot's 1894–95 course on diseases of the emotions, which promised to cover "character anomalies, perversions of moral, social, religious feeling, [and] aberrational notions of the beautiful." For the cultivated audience of *La revue blanche*, Bélugou took pains to clarify that Ribot's approach had nothing to do with decadent curiosity and everything to do with rigorous scientific method: "If he takes his subjects from the asylums, among the insane, the hysterics, and the degenerates, it is so that he can all the better disassemble a normal soul, it is never with the ulterior motive of satisfying some unhealthy curiosity."[66]

When Bélugou reviewed Ribot's *Psychology of Emotions* (1896) and *The Evolution of General Ideas* (1897) for *La revue blanche* in 1897, he once again emphasized the importance of Ribot's work for his aesthetically minded audience. In attempting to "establish some fixity in the undulating morass of emotional life," Ribot had broached territory until then examined only in poetry, novels, and drama. In doing so, Bélugou insisted, Ribot had "made a voyage of discovery," accomplishing for the study of emotions what Taine had achieved for the mind. In particular, Bélugou wrote, *The Evolution of General Ideas* responded directly to the interests and needs of an elite audience of symbolists: "The main goal of this book . . . is to study how the mind abstracts and generalizes, to demonstrate that these two operations come into play at the highest stages of evolution, which is to say that they already exist in perception, and that, in a series of gradual advances that can be determined, they attain the most elevated form, pure symbolism, which can be accessed only by a small number of individuals."[67] To readers who imagined themselves highly refined personalities, members of an intellectual and cultural elite, Bélugou pointed to Ribot as offering the undoubtedly attractive confirmation, gained through the pathological method, that the aesthetic goals of many symbolists—abstraction, generalization, even "pure symbolism"—could be the products of only the most highly evolved minds.

According to Bélugou, Ribot gave a lecture in 1898 on the conscious and the unconscious, inaugurating his Collège de France course that year with a discussion of the beginnings of psychological life in barely sentient creatures such as coral and starfish, from which he proceeded to discuss consciousness in the child.[68] The following year, Bélugou informed *Mercure de France* readers that Ribot's topic was genius, no doubt offering a preview of his *Essay on the Creative Imagination*, to be published in 1900.[69] When Ribot ceded his chair to Janet in 1901, Bélugou reported it as the "university event of the season," and paid homage to Ribot's style of teaching: "The students who came to hear him did much more than take in a doctrine, they assimilated a method, and in the most powerful and lively manner, witnessed his groping and his progress, and watched him work."[70] Throughout the 1890s, that method, the psychological method that Bélugou assimilated and distilled for readers of *La revue blanche* and the *Mercure de France*, was the pathological method.

What appears to be Bélugou's final article for a symbolist journal attempted to shore up the reputation of the pathological method at the precise moment that its influence began to wane, doing so in a creative way that satirized both the increasing reliance on quantitative methods and avant-garde advance. A 1904 article titled "The Power of the Imagination in Children: Experimental Research" purported to be a contribution to "the new experimental psychology or, as one says today, neopsychology." Signed with the names of Gustave Flaubert's misguided clerks, "Bouvard and Pécuchet," but directing the reader to Bélugou "for a certified true copy," the article describes an experiment in which young children are asked to touch first a hot gas lamp and then a cold one.[71] The intrepid investigators propose to measure the power of the children's imaginations by timing how quickly they remove their hands from the cold lamp—in other words, how hot they imagine the cold lamp to be. "Bouvard and Pécuchet" placed their work under the auspices of Alfred Binet, once a proponent of Taine's *petits faits* methods but by the early twentieth century better known for his work with children and his reliance on questionnaires and quantitative measures. Binet supposedly inspired the investigators to ask the children, "What do you feel?" and to perform various "rites of anthropometry," but the results, sadly, "remained obscure." In the end, "Bouvard and Pécuchet" consoled themselves by reminding readers that Binet himself had remarked upon the difficulties of a psychology that aimed to be at once scientific, physiological, and experimental: "It is difficult to perform vivisection, even psychological vivisection," they had Binet say, "on a living being."[72]

Bélugou's last word on experimental psychology for a symbolist journal obliquely reminded knowing readers that Ribot rather than Binet was the true master, and managed to do so by satirizing the avant-garde just as the symbolist avant-gardes were under threat of being eclipsed. "Neopsychology," in Bélugou's view, was not truly new or revolutionary but merely a fad. Quantitative measures of the body were silly, as was asking people questions and tallying up their responses. The only way to truth, Bélugou implied, was the systematic study of pathology, the method by which, year after year, in course after course and text after text, Ribot had succeeded in dissecting the mental functioning of the complex human mind. While psychological vivisection might be difficult for the lesser Binet, relying on quantitative methods, for Ribot, "tireless worker of most glorious renown," it was entirely feasible with a steadfast belief in and reliance upon nature's experiments.[73] As the following chapters reveal, symbolist avant-gardes were just as invested as Taine, Ribot, and Bélugou in pathology as a pathway to knowledge and to what they imagined to be something like objective truth.

Chapter 1 of this study, "Toward an Experimental Symbolism: Ideas and Ideals," places the caricature of Zola flinging shit from a chamber pot at the center of an interrogation of the uses of experiment in artistic discourse before and during the period in which symbolist theories and practices emerged. A rereading of Zola's "The Experimental Novel" situates the essay in relation to its source in Bernard's *Introduction to the Study of Experimental Medicine* (hereafter usually referred to as the *Introduction*), ultimately arguing that Zola used Bernard's text not so much to confer scientific authority on the naturalist novel as to shore up the roles of idea, genius, and temperament in the creation of the naturalist work of art. Focusing on impressionist criticism and on the writings of Jules Laforgue, Gustave Kahn, Émile Hennequin, and others, the chapter proceeds to analyze experimentalism in relation to discursive constructions of impressionism and symbolism, concluding that experimentalism functioned especially effectively within symbolist discourses as a modern methodological ideal. Experimentalism could be, and was, easily detached from the discourses of materialist science and, when necessary, naturalism, in order to be promoted as a method for universalizing individual ideas.

The next two chapters are organized around different symbolist problems to which the central Nabi protagonists of each chapter offered unique solutions. Chapter 2, "Defending Deformation: Maurice Denis's Positivist Modernism," addresses modernist anxieties over deformation as a visual strategy. It analyzes the practices and theorizations of deformation primarily in the work of Denis, but with reference to works by Denis's fellow Nabi artists, Vuillard and Bonnard. Using a lost painting, the enigmatically titled *Décor* (1891), the chapter demonstrates how Denis, after producing a highly distorted female nude that he called a "study in subjective deformation," positioned his writing in opposition to initially tempting, but in the end hostile, conceptions of symbolism, discursively emptying deformation of pathological content in order to make it synonymous with form. The closely linked chapter 3, "Édouard Vuillard's Experimental Arabesques," considers Vuillard's self-conscious development of linear forms that could be read as symbolist. It focuses on iterations of what I call the experimental arabesque in Vuillard's paintings of sleeping figures and the portrait of Lugné-Poë, an image of intense mental concentration. The artist's early notebooks, considered in relation to Bélugou's notes from Ribot's psychology courses, reveal how Vuillard's experimentation on his own automatisms and altered states of mind was ultimately figured through a highly labored use of the arabesque.

The final two chapters introduce the reader to "At the Black Piglet" (Zum schwarzen Ferkel), a Berlin-based but Paris-oriented avant-garde community named after a wine cellar frequented by Strindberg and Munch, among others. Chapter 4, "August Strindberg's Naturalistic Symbolism," explores how the late nineteenth century's most energetic entrepreneur of experiment cultivated altered states in a bid to transform naturalist practices into symbolist ones. Using material published in the Leipzig-based journal *Sphinx*, the chapter suggests the ways in which Strindberg sought out new artistic methods of painting and writing by performing, eventually on himself, psychological experiments on automatism, hallucination, hypnosis, and other forms of abnormal perception. The final chapter, "Madness as Method: The Pathological Experiments of Edvard Munch," traces Munch's deliberate and methodical engagement with different forms of madness as an artistic strategy that enabled him to slough off naturalism and to constitute form liable to be read as symbolist. Analyzing the artist's roulette paintings and writings, produced during and following a time when he made repeated visits to the gambling rooms of Monte Carlo, the iconic painting *The Scream*, and a corpus of works set in Parisian hospitals, I consider the artist's ongoing search for appropriate pathological form to communicate and universalize pathological content. The book's conclusion makes explicit what is elsewhere implicit: the underlying naturalist precepts of symbolist modernism despite symbolism's claims to antinaturalism, as well as symbolism's gendered selves. I also make clear in the conclusion that this book should be read as a critique of a monographic approach to alterity that has resulted in overly teleological narratives of the so-called discoveries of the art of the primitive, the child, and the insane.

Nature's Experiments is one alternative to those narratives, an examination of the epistemological conditions that made otherness in myriad forms available for artistic borrowing under the sign of avant-gardism.[74] Engaging in self-consciously constitutive practices of symbolism, Denis, Vuillard, Strindberg, and Munch rejected existing aesthetic methods and turned to the still emerging methods of experimentalism, above all the belief that an investigation of the pathological could lead to what was imagined to be a form of truth. In exploring the culture of nature's experiments, this study forges a different kind of art history, excavating the epistemological foundations of some of the most formally radical, but thoroughly historical, visual practices of the late nineteenth century.

A final note: my approach to symbolist visual form is reflected in my preference for lowercase in terms such as symbolism, impressionism, and naturalism, a choice that happily adheres to the editorial style of the Press but is also meant to underline all three as discourses (practices) in flux, never able to be entirely pinned down, and involving different stakes and different stakeholders at any given moment.[75] The lowercase downplays the notion of a definable "Symbolism" and more easily allows for multiple symbolisms.[76] It also signals an attempt to speak not from within what Michel Foucault called a "dubious unity" but from without, making use of historical groupings such as symbolism only as starting points, and not always convenient ones, for interrogation.[77]

Lire le 15 Octobre

Lire
dans le Numéro où paraitra
ce Roman, destiné
à produire la plus grande sensation,
une
ÉTUDE LITTÉRAIRE
TRÈS IMPORTANTE
DE
M. ZOLA
intitulée
LE
Roman Expérimental

Na

Suite de

Par Em

1

TOWARD AN EXPERIMENTAL SYMBOLISM

Ideas and Ideals

Some nineteenth-century readers understood Claude Bernard's *Introduction to the Study of Experimental Medicine* as a masterpiece of positivist philosophy. Others, such as Elme Marie Caro, read it as an encouraging sign of science's impending rapprochement with spiritualist views.[1] These divergent readings centered on the role that Bernard had assigned in the *Introduction* to the a priori idea. Tracing the remarkable persistence of the experimental idea in the last two decades of the nineteenth century, this chapter reveals how experimentalism could be held up across a range of belief systems and aesthetic positions, including symbolism, as a modern epistemological ideal. From Émile Zola's controversial theory of the experimental novel to lone voices in impressionist criticism, from the impressionist-symbolist aesthetics of Jules Laforgue to Émile Hennequin's "esthopsychologie," from Gustave Kahn's psychophysical symbolism to fin-de-siècle esotericisms—the authority of the experimental method remained virtually unquestioned, even as debates over the bankruptcy of science and critiques of positivism grew louder.[2] Members of vanguard literary and artistic circles of the 1880s and 1890s ascribed a surprisingly consistent authority to Bernard, insisting, in an almost monotonous refrain, on experimentalism as the most advanced mode of knowledge production, a methodological panacea for transforming a priori ideas into scientific truths. As for impressionism, that artistic practice so often characterized in art histories of modernism as experimental, this chapter argues that the discursive conjunction of impressionism, experimentation, and vanguardism arose, especially in the criticism of Edmond Duranty, but that it was co-opted and used more extensively in discourses around symbolism. As a way of describing artistic practices, "experimental" would, for the most part, have to wait until after the emergence of symbolism in the second half of the 1880s, when French experimental psychology found itself, with the aid of nature's experiments, on surer institutional ground.

Bernard, Zola, and "The Experimental Novel"

In the same year that his *Introduction* appeared, Bernard argued that the heart, and in particular the heart's relationship to the mind, should come under the investigative purview of physiology. In doing so, he addressed those critics who might be tempted to believe that positivism instituted a mechanistic view of mankind and could thus have no truck with matters of the heart, including artistic inspiration. Pleading with his readers to endure physiology's growing pains, Bernard implored them to look toward a future when "the poet, the philosopher, and the physiologist will all understand each other," and to unite in a common search for truth under the ultimate synthesis of experimental method.[3] In 1879, a year after Bernard's death, Zola probably thought he was heeding that call when he published "The Experimental Novel," which appeared in France simultaneously with Zola's follow-up to *L'assommoir*, the novel *Nana* (fig. 13).[4] One might expect the essay to have instituted, or at least to have smoothed the way for, an art understood as experimental, but at least initially, it did not. Thoroughly misconstrued, virulently attacked by Zola's detractors, the essay was greeted largely with silence from even his most loyal supporters.

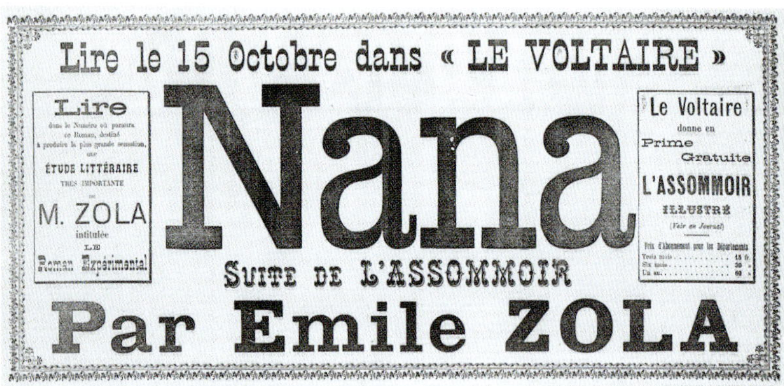

13 Advertisement for Émile Zola, *Nana* and "Le roman expérimental," *Le Voltaire*, no. 467 (15 October 1879): 4. Bibliothèque nationale de France.

The contemporary novelist, Zola claimed at the outset of the essay, needed only to replace the word *doctor* with *novelist* in order to apply Bernard's method to the naturalist novel. Zola situated the experimental novel both as the next step in the ongoing evolution of human thought, which had most recently given rise to experimental physiology, and as the most appropriate literature for a new scientific age. He characterized its emergence as part of an inevitable process of detaching the pursuit of knowledge and truth from belief in the irrational and the supernatural, an evolution that was in turn part of a larger movement from the unknown to the known and from the indeterminate to the determinate. Naturalism, Zola maintained, was superior to idealist forms of writing, above all because it provided for more truthful representations of human beings. Recently established as the model for investigating all organic phenomena, including the most complex, experimentalism thus provided a ready-made and up-to-date method for naturalist forms of truth making.

In establishing experimental physiology as the basis for experimental medicine, Bernard had hoped not just to obtain causal knowledge but to become a master of causes so that disease could be treated, a goal that underpinned Bernard's reliance on the pathological method. Similarly, Zola stressed that the experimental novelist was not simply an author-creator but a "moralist-experimenter," determining the root causes of humanity's diseased intellectual and passionate life with a view to rendering it healthier.[5] In his novels, so often concerned with the physiological, psychological, and social effects of hereditary degeneration, Zola claimed to reveal the underlying causes of individual and social pathologies, accomplishing that first step in knowing and mastering those causes before undertaking treatment.[6]

The immediate responses to "The Experimental Novel" set the terms for its uncertain place in literary history and for nearly a century of disquiet about the essay's intentions and effects. Avowed critics of Zola's naturalism roundly condemned it for crudely applying the methods of positivist science to the subtle art of the novel. Ferdinand Brunetière denounced what he saw as Zola's mechanistic and fatalistic view of humanity, and accused him of stripping the novel of one of its key elements, characters who could exercise their own wills. In his view, Zola's theory rivaled his novels in vulgarity. Worse yet, Brunetière accused the novelist of failing even to understand what it meant to experiment. Otherwise, the critic wrote, Zola would have known that "the novelist, like the poet . . . can only experiment on the self, not on others."[7] Another critic willingly admitted that all novels were experimental in the sense that they were experiential, but judged the conjunction of the "filthy" naturalist novel with the "pure renown" of Bernard an act of "vanity" and "self-infatuation."[8] Bernard's acolytes joined this chorus, accusing Zola of so grossly misusing the theories set forth in the *Introduction* that they felt the need to defend their master from this opportunistic appeal to his authority, all the more offensive in appearing so soon after his death. One called the essay a "monstrous promiscuity," wondered if it was not some kind of "fantastic hallucination," and judged Zola's implied assimilation of the experimental practice of vivisection to a stylistic procedure to be nothing less than a symptom of mental illness.[9] But critics chafed above all at the text's seeming promotion of a scientific literature for a scientific age, bristling at what they saw as the resulting discounting of the roles of genius, imagination, and even, it seemed, temperament in the creative process. Brunetière, who had long castigated naturalism, pointed gleefully to inflammatory citations from the essay that apparently evidenced Zola's pessimistic view of humanity, and he found ample grist for his mill. "Here we find the practical utility," Zola had written, "and the high morality of our naturalist works . . . that take apart and put back together piece by piece the human machine, in order to make it function under the influence of different environments."[10]

Few if any of Zola's supporters spoke out publicly in support of the text, and a number expressed their criticisms privately, albeit pointedly.[11] Henry Céard, who later claimed to have lent Zola the *Introduction*, reproached his friend for extending Bernard's method into the psychological realm of the novel.[12] The day after the last installment of the essay appeared, Gustave Flaubert, who admired Bernard enormously, complained to Guy de Maupassant of Zola's vapid theorizing: "Do not talk to me about realism, naturalism, or the experimental! . . . What empty nonsense!"[13] The dearth of positive critical reception among Zola's near and dear probably stems in part from the waning influence of naturalism as a collective movement in and around 1880.[14] But perhaps critics and allies alike were also conserving their energies for either vigorous attacks on or forceful defenses of *Nana*, published at exactly the same time. In the face of predictable critiques from the likes of Brunetière, and, much worse, the silence of his supporters, Zola refrained from pursuing the argument in subsequent writings and would thereafter shy away from public theorizing.[15]

But structuralist analyses have set aside the charges of opportunism and naïveté to interrogate the inordinate role of Bernard's a priori idea in "The Experimental Novel," and the discursive force of Zola's appeal, not so much to the prestige of science as to the scientific gaze of the experimenter.[16] In channeling Bernard, sometimes nearly to the point of silencing his own voice, Zola, the reader begins to suspect, was trying to reiterate something that Bernard had said that Zola felt had gone unheard. More than a decade earlier, in his art criticism, Zola had mitigated naturalism's unavoidable scientism by ascribing a crucial role to man's imagination, derived from his temperament. In his review of the 1866 Salon, Zola had rejected the mantle of the doctrinaire positivist in order to insist upon the primacy of individual temperament *despite* a "spirit of the age" that he claimed inevitably drove artists toward positivism and a minute study of reality.[17]

"The Experimental Novel" in fact used Bernard's *Introduction* to reinforce the notion that the *idea*, construed as both a product of an individual temperament and a generating force for experiment, remained paramount in Zola's theory of naturalism, as it did in Bernard's text. "In order to show the mechanism of events," Zola insisted, "it is necessary that we produce and direct phenomena; there is our part in invention, the genius in the work. . . . An experiment . . . is always based on an idea." To provide the ultimate approbation, Zola quoted from Bernard directly: "The appearance of the experimental idea . . . is entirely spontaneous, and its nature is entirely individual; it is a particular feeling [*sentiment*], a *quid proprium*, that constitutes the originality, the invention or the genius of each experiment."[18] Spontaneity, individuality, feeling, originality, invention, and genius: clearly, the experimenter was also an artist.

Both Bernard and Zola emphasized that, after the a priori idea sets an experiment in motion, the experimentalist must maintain a stance of nescience regarding the experiment's outcome, continuing to doubt the root causes of the given phenomenon under investigation. This stance would provoke the exhaustive process of testing that would eventually lead to an objective determination of causes. Far from being a constraint, the doubt of the experimenting novelist, Zola made clear, allowed him to realize the full powers of his creative and rational faculties, both his subjectivity and his objectivity. The originating idea would in the end become more meaningful and more universally significant when proved objectively true: "What becomes of genius for the experimental novelist? It remains genius, the a priori idea, only it is controlled by experiment. Naturally, experiment cannot destroy genius. On the contrary, it confirms it. Take a poet, for instance; is it necessary for him to possess genius, that his feeling, his a priori idea is false? No, clearly, because a man's genius will be even greater once experiment has further proven the truth of his personal idea."[19] Far from "subjectifying the objective," then, which would stand as Kahn's pithy version of Zola's familiar definition "art is a corner of nature seen through a temperament," experimental method, in my reading, would in fact serve to "objectify the subjective."[20] And far from governing the uses of experimentalism within impressionist discourses, "The Experimental Novel" might better be understood as prefiguring, possibly licensing, and certainly, in one instance—the case of August Strindberg discussed in chapter 4—determining the uses of experiment within discursive formations of symbolism.

There is, then, an extraordinary irony in the critical and scholarly reception of Zola's essay.[21] Well before 1879, naturalism had been derided as too scientific, too photographic, too mechanistic, critiques all explicitly referenced in "The Experimental Novel." But Zola seems to have tried to use Bernard's *Introduction* to respond to such critiques, above all choosing passages from the text that proclaimed the importance of the idea in experimental method. "The Experimental Novel" thus tried to reintroduce the imagination into a discussion of the naturalist novel via Bernard's text, attempting but ultimately failing to refute readings of both naturalism and experimentalism as overly scientific, to the detriment of inspiration. Bernard was not to be held up as a doctrinaire positivist but as the philosophical author of the "Study on the Physiology of the Heart," the humble man of science who judged inspiration to be eternal and simply aided, at the current stage of human development, by experimental method. And Zola, it appears, hoped that he would be judged similarly. As he put it, performing modesty and genius at the same time, "the method is only a tool; it is the idea that the worker provides that makes the masterpiece."[22]

I offer one final point about "The Experimental Novel" before turning to the historiography and criticism of impressionism. Zola must have known that he was deluding himself in setting up the experimental novel as the next step in a natural evolution of the sciences that had most recently given rise to experimental physiology. By the late 1870s, when he published the essay, experimental psychology had a much better claim to that territory, particularly through the efforts of Hippolyte Taine and Théodule Ribot, of which Zola was certainly aware. Even allowing for a more or less feigned naïveté, the fact that Zola could assert the experimental novel in 1879 as, in his words, "a kind of scientific psychology," without mentioning the actual emergence of the field, signals, if nothing else, the still embryonic state of the discipline. A decade later, however, with Ribot ensconced as chair of "experimental and comparative psychology" at the Collège de France, experimental psychology, rather than the experimental novel, had finally arrived.

Impressionism, Experimentalism, and Art History

In 1898, a series of scathing and at times anti-Semitic caricatures, each representing one of Zola's works, attempted to capitalize on the author's notoriety at the time of the Dreyfus Affair. Under the title *The Experimental Novel*, a typically scatological illustration pictured the author standing in front of an easel flinging fecal matter from a chamber pot onto a canvas (fig. 1). Other images in the series had a more obvious relationship to the text depicted. *Doctor Pascal* showed Zola as both himself and his eponymous character, hunched over his manuscript at a desk on which appeared a jar with a preserved fetus and popularizing books with titles such as *Science for All* (fig. 14).[23] But what was the rationale for representing the author of "The Experimental Novel" as a painter? Zola's work as an art critic and erstwhile defender of the impressionists was certainly well known, as was his distancing from impressionism around 1880 and what was often seen as his betrayal of impressionism with the publication of the novel *The Masterpiece* in 1886. But precisely what connection did the image assert between painting and Zola's infamous co-optation of Bernard's *Introduction*? Depicting an act of haphazardly flinging material onto a canvas undoubtedly mocked impressionism, calling to mind, for instance, Louis Leroy's satirical review of the First Impressionist Exhibition, in which a Monsieur Vincent, standing in front of Claude Monet's *Boulevard des Capucines* (fig. 15), exclaims, "But these spots were done with the same method used for whitewashing stone: Pif! paf! v'li! v'lan."[24] Lebourgeois's illustration, it seems, was meant to ridicule both "The Experimental Novel" and impressionist painting for their lack of skill and method, for being overly materialist and frankly crude.

Art-historical discourse generally affirms impressionism as experimental, privileged for having been new in its time, radical, and vanguard.[25] But scholars have asserted that impressionism was experimental not only in this transhistorical sense but also in the historical sense, experimental insofar as it purportedly aimed to produce knowledge through an engagement with the natural world comparable to, or even directly influenced by, nineteenth-century scientific experimentation.[26] Given Zola's close relationships with Édouard Manet and Paul Cézanne, and his critical support for impressionism, "The Experimental Novel" would appear to reinforce that claim, although its near-complete absence from the art-historical literature signals, perhaps, that it does so only nominally.[27] While a view of impressionism as straightforwardly objective and/or scientific has long been discounted,[28] many historians refer to impressionism as experimental for good reasons, while also bracketing out, more or less, the scientific connotations of the term. But it also seems useful to distinguish between nineteenth-century meanings of *experiment* and our own. In other words, it is useful to be clear when we are talking of experimentalism, as in the experimental method, when we are talking about the influence on impressionism of specific experimental sciences such as physics and chemistry,[29] and when we are speaking more generally of heuristics.

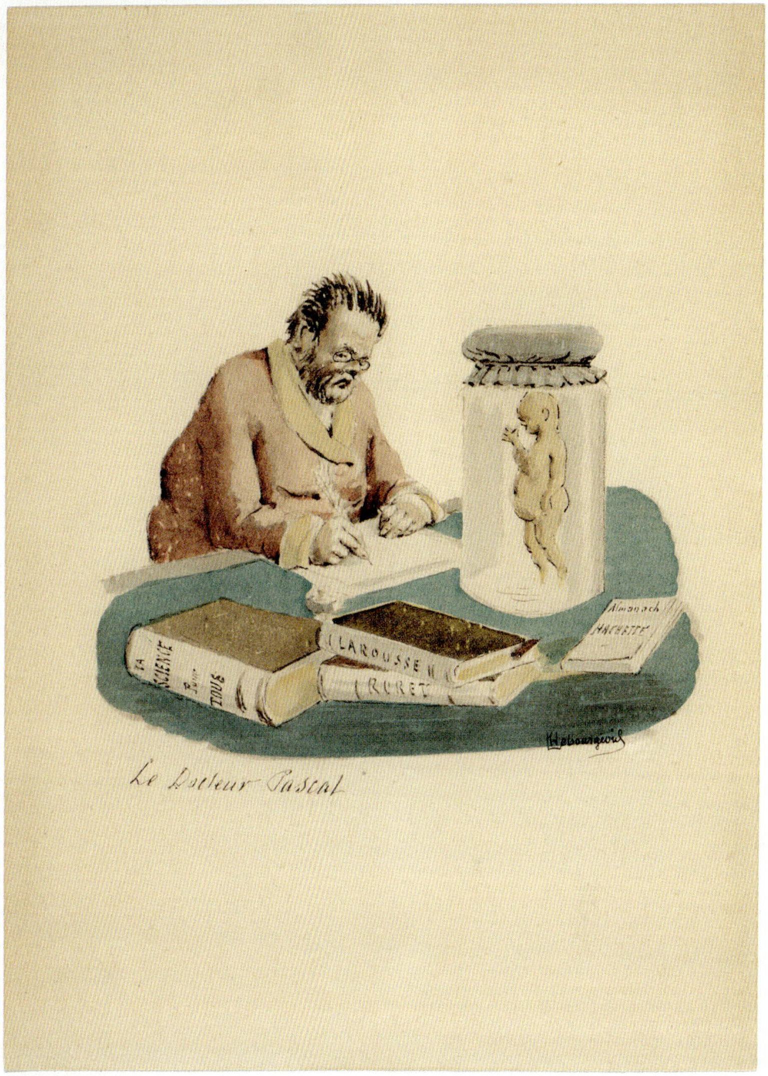

14 H. Lebourgeois, *Doctor Pascal*, from *L'oeuvre de Zola*, 1899–1902. The Thomas Fisher Rare Book Library, University of Toronto.

In 1978, Ernst Gombrich outlined a history of experimentation in the arts in which he traced the heuristic processes of "negative feedback" that enabled artists to "objectively test" the success of any attempt at naturalistic representation. Gombrich deemed Apelles, Leonardo da Vinci, and John Constable experimenters par excellence, artists who used experiment as an instrument to aid their creative processes. Constable, with his well-documented interest in meteorology, provides a particularly salient example. "Painting is a science and should be pursued as an inquiry into the laws of Nature," the British landscapist wrote. "Why, then, may not landscape painting be considered as a branch of natural philosophy, of which pictures are but experiments?"[30] As Gombrich well knew, by 1836, when Constable set down those words, he was swimming against the tide in a Europe in which painting and science were increasingly assigned to separate domains rather than united under the umbrella of natural philosophy.[31] In Gombrich's view, the subsequent history of artistic experimentation was largely and lamentably one of co-opting experimentation as a value, vitiating its heuristic function as self-criticism by appropriating science's prestige and connotations of progress in a mad pursuit of alternatives to traditional (read: naturalistic or illusionistic) artistic modes. Gombrich concluded that while experimentation and artistic creation had frequently had a cozy, instrumental relationship early in the nineteenth century, modern understandings of artistic experimentation, in thrall to the allure of scientific progress rather than to science itself, generally impoverished our understanding of the creative process.[32]

Zola's "Experimental Novel" proves extremely awkward for Gombrich's narrative. He no sooner mentions it than ushers it off stage to refer obliquely to impressionist "experiments in art," and to discuss Claude Lantier's musings on science in *The Masterpiece*, concluding only that Zola's writing as a whole indexes the contemporary authority of science. Impressionism then becomes a tipping point, still putatively scientific, but heralding the avant-garde's dismantling of the instrumental relationship between art and experiment. Gombrich's is a curiously split text, divided between wanting to establish experimentation as a transhistorical value and wanting to tie the processes of "negative feedback" to specific scientific cultures, seeing "true" artistic experimentation only at those specific moments when artists took an avid interest in the science of their day. The view that impressionism constituted a moment of profound artistic experimentation, in the sense of being in the vanguard, and that this artistic experimentation was somehow allied with or related to the scientific cultures, and specifically to the experimentalism, of mid-nineteenth-century France, was commonplace for much of the twentieth century. But the awkward assimilation of Zola's essay into Gombrich's attempt to trace the history of artistic experimentation reveals a fundamental incoherence.[33]

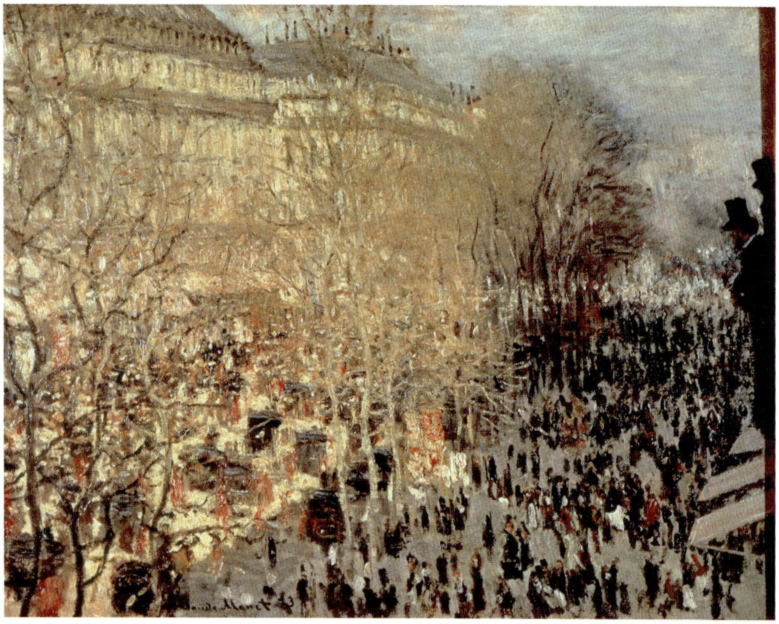

15 Claude Monet, *Boulevard des Capucines*, 1873. Oil on canvas, 61 × 80 cm. Pushkin Museum, Moscow.

The Anglo-American formalist tradition is especially rife with the heroic, value-laden language of experiment, rhetoric that long served to insert impressionism smoothly into a masculinizing narrative of modernist abstraction.[34] As early as 1894, Roger Fry, albeit in a text that went unpublished at the time, made both "experience" and an experimental approach to nature central to his understanding of impressionism.[35] Later, Fry would implore his readers to think of the works on display at the Grafton Gallery, including those of Cézanne, as "serious experiments—. . . not always successful experiments—but still serious experiments made in perfectly good faith towards . . . discovery."[36] More recently, an exhibition invited audiences to participate in an "experiment in looking," to set aside the easy to digest and "pretty" aspects of impressionism in order to attend first and foremost to the radical nature of the impressionist surface, so as to grasp it still as "contingent, experimental, and risky."[37]

The will to experiment seems to be corroborated by numerous contemporary documents that ostensibly confirm the discursive conjunction of impressionism, experimentalism, vanguardism, and, at times by implication or association, science. Edgar Degas apparently exhibited a number of his prints at the Fifth Impressionist Exhibition as "experiments and plates in different states."[38] Edmond Duranty supposedly claimed that artists were, in 1870, "experimenting with the knife and would try out the spoon if it seemed promising."[39] Cézanne, thanking the critic Gustave Geffroy in 1894, allegedly wrote, "I read the long article you devoted to shedding light on the experiments I have been making in painting."[40] Cézanne also apparently assured the dealer Ambroise Vollard, "I am continuing my experiments and I will inform you of the final result as soon as my study has yielded some satisfaction."[41] In 1891, the symbolist critic Adolphe Retté supposedly argued that in "[Émile] Bernard we find an experimenter."[42]

In each of these cases, however, it is English translation that has rendered the relationship between experiment and impressionism seemingly natural, whereas the original French belies such ease and reflects rather the period's shifting understandings of *expérience* analyzed in my introduction. Degas, in fact, exhibited "Essais et états de planches."[43] Duranty wrote that up-and-coming artists "essaient du couteau, ils essaieraient de la cuillère si elle s'y prêtait."[44] Cézanne thanked Geffroy for "mettre en lumière les tentatives que j'ai faites en peinture,"[45] and assured Vollard that "Je poursuis donc mes recherches et vous ferai part du résultat acquis."[46] Retté in fact spoke of Bernard as a "chercheur."[47] *Essais, tentatives, recherches:* all three words abut contemporary understandings of avant-garde experimentation and appear frequently in the Anglo-American literature translated as "experiment." But their meanings for French speakers of the nineteenth century nevertheless remained somewhat different from the range of meanings available at the time for *expérience.*

Indeed, impressionist criticism reveals a remarkable discursive lack of *expérience* with either scientific or avant-garde connotations, although this should be less surprising after the etymological analysis offered in my introduction, in which I demonstrated the extent to which the denotations and connotations of *expérience* were in flux in the second half of the nineteenth century. In the criticism surrounding the eight impressionist exhibitions, we find an entire vocabulary that relates to our own understanding of artistic experimentation—*essais, tentatives, recherches*. These words are used over and over again to illuminate, in both a positive and a negative light, the impressionists' striving for a painting that was "vrai," or, less often, "juste." And certainly, such "recherches" were in dialogue with scientific advances in optics, chemistry, color theory, and, increasingly, psychology. But just as Zola's "Experimental Novel" failed at first to find many sympathetic readers, it appears that critics could call such research "experimental" only with some difficulty.

The word *tentative,* in both singular and plural forms, appears more than twice as often as *expérience* and its cognates.[48] Nearly two-thirds of these latter instances relate to the more workaday definition of *expérience* as "knowledge of things, acquired through use over a long period" (the second definition of both the French Academy's and Littré's dictionaries, in use at the time), connoting expertise or skill. More often, critics were remarking upon a lack thereof. Elie de Mont wrote of Henri Rouart, "His paintings indicate a remarkable inexperience [*inexpérience*], I would almost say a remarkable clumsiness in the use of color."[49] Four reviews, including Stéphane Mallarmé's article of 1876, published in English, tie experience much more closely to an involuntary subjectivity, as in Paul Adam's 1886 reference to "the always uncertain data of experience."[50] Three begin to link ideas of vanguardism with a generalized notion of experiment as new and progressive, although usually in a negative sense, as in Philippe Burty's assessment of 1879: "Today the experiment is over and we can now judge it more severely." This leaves only six separate instances of *expérience* with scientific connotations, more than half of them from 1879 or 1880, around the time that "The Experimental Novel" appeared. Joris-Karl Huysmans referred in 1880 to specific experiments done by Jean-Martin Charcot on the perception of color. Henry Fouquier recoiled in horror at Degas's dancers, which he likened to corpses submitted to electricity experiments, while Émile Blavet, in 1876, wrote positively of the new "realists," who needed "great freedom to experiment and a laboratory of their own."[51]

But it was the criticism of Duranty that most closely linked impressionist painting as a new, progressive, and tentative form of art making with scientific notions of experiment. Duranty's 1876 pamphlet *The New Painting* had used the language of science to propose a physiological basis for impressionist "discoveries," but there Duranty had avoided the term *expérience* in favor of *tentative*.[52] In an April 1879 review, however, Duranty cautiously likened impressionist painting not only to *tentatives* and *essais* but also to the experiments of chemists and physicists, those exemplary modes of experimentation generally cited prior to Bernard's *Introduction*: "If one has followed painting for quite some time, has been associated with people who are seeking new outcomes in art, the attempts [*tentatives*] of this group are interesting in a way that they cannot be for those who, wanting to have immediate and near-certain values shown in their galleries, concern themselves only with definitive results and are more inclined to scoff than to welcome laborious attempts [*essais*] that resemble the experiments of the chemist or the physicist."[53] Duranty's review, however, fell on fallow ground, at least in France.[54] Henry Havard found Duranty's analogy between attempts in painting and scientific experiments novel and enticing, but he was uncertain whether such *essais* (Havard himself eschewed the term *expériences*) should be exhibited.[55] Duranty, who was undoubtedly drawing upon his conversations with Edgar Degas, Édouard Manet, and Zola for this novel understanding of impressionism as both new and acutely experimental, had little opportunity to pursue this line of argument before his sudden death in April 1880. And in the criticism surrounding the 1881, 1882, and 1886 impressionist exhibitions, there are just three instances of the term *expérience*, and only one, that of Émile Hennequin, that connotes either vanguardism or science.

Duranty wrote at a moment when the connection between the naturalist novel and impressionism, between Zola and the "Zolas of oil," seemed tightest.[56] More often than not, however, as Lebourgeois's caricature suggests, the comparison was mobilized for the purpose of heaping scorn on both. Already in a review of 1877, a critic had called both impressionism and *L'assommoir* forms of "premeditated cynicism."[57] And in November 1879, in a cunning move, Brunetière attempted to co-opt impressionism for the novel, likening Alphonse Daudet to an impressionist painter and attempting to shunt Zola off to the side of both naturalist and impressionist aesthetics.[58] But the ties that bound naturalism and impressionism together had already begun to loosen even before "The Experimental Novel" appeared. Zola's own review of the 1879 impressionist exhibition caused a stir among the artists Zola had previously professed to support. Acknowledging their role as pioneers, he nevertheless criticized the group, singling out Monet, for being all too easily satisfied with middling efforts.[59] After 1879, with the exception of a short article in 1884, Zola ceased to take up his pen in strong support of his impressionist friends. Some, Cézanne perhaps among them, saw the 1886 novel *The Masterpiece* as Zola's final statement on the impressionists' inability to achieve *tableaux*.[60]

While impressionism was no doubt experimental in many senses of the term, and artists were surely thinking experimentally without actually articulating their practices as such, the discursive conjunction of impressionism, experimentation, and vanguardism is less a product of nineteenth-century art criticism than a strategy of twentieth-century Anglo-American art history. Although Duranty, and to a lesser extent Blavet, conceived of impressionist paintings as something like scientific experiments, this novel idea, despite Zola's essay (or, more probably, because of it), failed to take hold in contemporary writing, a hypothesis confirmed by the complete absence of the adjective *expérimental* in the criticism of the eight impressionist exhibitions. Certainly, positivist science was at the height of its authority in the first two decades of the Third Republic, but it would take the experimental *human* sciences, and above all experimental psychology, as it was taken up in emerging theories of symbolism, to solidify the discursive conjunction between modern forms of art making and experimentalism.

Experimental Idealism: The Impressionist-Symbolist Aesthetics of Jules Laforgue

Zola's insistent emphasis on the central roles of idea, inspiration, and genius in Bernard's experimental method initially failed to find many sympathetic readers in France. However, among the juvenilia of the poet and aesthetic theorist Jules Laforgue (fig. 16), a series of short articles for the Toulouse-based journal *La guêpe* conveys Laforgue's deep admiration for Zola's work and especially for the much derided essay. Less than two weeks after "The Experimental Novel" appeared in France, the nineteen-year-old Laforgue carefully summarized Zola's theory for his provincial audience, and defended it as a "beautiful and noble dream."[61] What might be dismissed as teenage adulation is significant, however, because in his more mature writings, Laforgue put experimentalism at the center of his idealist aesthetics. Most of his texts, including his 1883 notes on impressionism, remained unpublished during his short lifetime but became available within symbolist circles—a number were published in *La revue blanche*—via Laforgue's friendships with Gustave Kahn, Félix Fénéon, and Téodor de Wyzewa. Laforgue's aesthetics significantly expanded upon Duranty's physiological justifications of impressionism, picked up on Zola's promotion of Bernard's a priori idea as a means to temper a view of both naturalism and positivism as vulgarly materialist, and presented experimentalism and the pathological method as fundamental for both aesthetic theory and artistic practice.

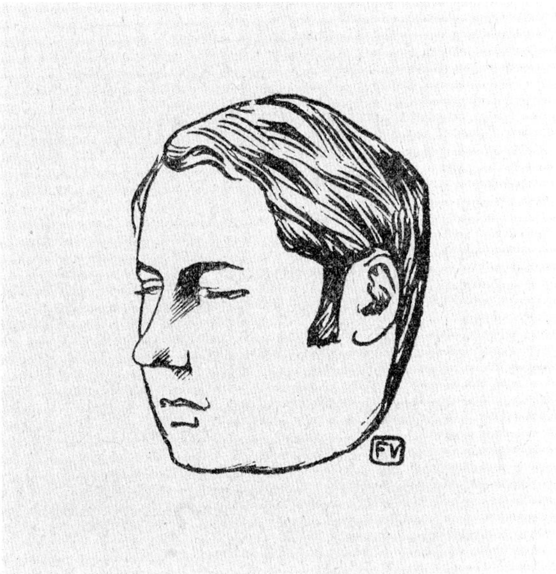

16 Félix Vallotton, *Jules Laforgue*, in Remy de Gourmont, *Le livre des masques* (Paris: Mercure de France, 1896), 204. Bibliothèque nationale de France.

Laforgue's draft notes on impressionism, intended for a German audience, propose a theory of impressionist painting informed by recent research in the fields of psychology, philosophy, physics, and aesthetics, especially the work of Taine, Eduard von Hartmann, Hermann von Helmholtz, and Charles Henry, the latter a mutual friend of Laforgue's and Kahn's.[62] Laforgue wrote jokingly to Henry that the German public, after reading what he had to say about impressionism, would surely believe that "impressionism—with its madness—was born in Germany out of Fechner's law."[63] Indeed, Laforgue linked his theory of impressionism directly to psychophysical experiments on how the eye responds to the stimulation of light in order to produce a physiological sensation of color.[64] In Laforgue's view, the impressionist painter was a supranormal personality with an eye physiologically capable of seeing "naturally" or "primitively." The "impressionist eye," however, was to be understood not as atavistic but as more highly evolved than the "academic eye," because it worked in concert with a mind and a hand able to forget any previous artistic training, able to slough off "childish" conventions such as modeling, perspective, light, and chiaroscuro in order to paint "naïvely." Participating in an inevitable evolution toward increasingly modern forms of painting, the more highly evolved eye would also be most at home in its own body, most in tune with its own temperament. When the individual eye reigned, Laforgue concluded, there would be no need for schools, juries, medals, or state patronage, and art would exist in the context and be the product of individual liberty, "nihilistic dilettantism," and indeed anarchism.[65] Laforgue's radically individualist stance led him to posit the Hartmannian unconscious

(Hartmann had insisted on capitalizing the word), seat of the individual, as his core aesthetic principle.[66] Hartmann had argued that the unconscious, made up of illogical will checked by reason (or idea), was, when actualized, the fundamental generative force of all life.[67] In notes on aesthetics published posthumously in *La revue blanche* in 1896, Laforgue explicitly aligned Hartmann's unconscious with genius, instinct, and, most important for the purposes of this study, "sacred *hypothesis*, fertile mother of all determinism," emphasizing their mutually reinforcing roles in both scientific and artistic pursuits.[68]

The notion of an evolutionary drive toward an ever more physiologically perfect eye derived largely from Laforgue's readings of Herbert Spencer, which he would have known in Ribot's French translation.[69] Laforgue's evolved eye had achieved a high degree of specialization by repeatedly performing the same tasks. It had perfected itself in order to work more efficiently with the other organs, including the mind and the hand, contributing to an ever more improved human organism, which in turn produced ever more improved cultures and societies. At the end point of this increasing specialization and improvement lay an ideal state of equilibrium, in which all parts (of the body, of society, etc.) corresponded to one another harmoniously.[70] Uniting Spencerian laws of progress with Hartmann's unconscious, Laforgue presented art as but one indicator of the unconscious's part in the drive toward this future ideal state. Sketching the relationship between art, the ideal of any given moment, and that ultimate state, which he called the capital-I "Ideal," Laforgue wrote, "This ideal must order the Ideal that is the object of research in Aesthetics (my article): The ideal in art—increasing specialization of the organ used—and as for choice—it happens by itself—through the natural selection of the Unconscious."[71]

Laforgue's idealism was not, therefore, based on static or unchanging notions of tradition or classicism but on an analysis of how specialized the relevant organs that produced a certain art had become, how far along the pathway of progress they had traveled. Lowercase ideals, in other words, were to be found not in the artwork produced but primarily in the mind and body of its creator. This aesthetics would thus need to base itself on psychological and physiological investigations. In a letter to his former employer and friend Charles Ephrussi, written in late 1883, Laforgue explained that he had been rereading different aesthetic theories and had begun to write his own, which "accords with Hartmann's Unconscious, Darwin's transformism, the works of Helmholtz." "In any case," he went on, "it is very new, and touches on the latest problems of human thought and is not in disagreement with either the modern physiology of optics, or the most advanced psychology."[72]

This drive toward both ideals and the Ideal, Laforgue reminded his readers, occurred not just in art but in all domains of human activity, including science, where scientists had just as much recourse as artists did to the unconscious: "In science, as in all human paths toward the Ideal, there is instinct and thought [*réflexion*], the Unconscious and knowledge. As if Claude Bernard . . . had nothing to do with instinct, with the Unconscious!"[73] Against Ernest Renan's argument that art, like science, would become more and more dependent on thought, Laforgue mobilized Bernard, in whose laboratory his good friend Henry had once worked as an assistant, to argue that instinct and reason were always united, whether in art or in science, by way of the unconscious. They were not the basis of successive states, as Auguste Comte had asserted, but existed simultaneously, as Bernard had emphasized. As in the case of Zola's "Experimental Novel," Bernard was again called upon to play the part of a unifying figure who underlined the important role of instinct in experimental method, just as in all strivings toward the Ideal.[74] Laforgue similarly enlisted Leonardo da Vinci, later held up by Joséphin Péladan as the preeminent practitioner of "experimental idealism," in the positivist-idealist cause.[75]

Laforgue presented his own emerging ideas on aesthetics as a corrective to Taine, whose work, he argued, contained internal contradictions. Although Taine, whose lectures Laforgue had attended at the École des beaux-arts in 1880–81, claimed to be working toward a positivist analysis of cultural production solely as a product of race, environment, and moment, he had nevertheless continued to uphold traditional notions of classical ideals, arguing that classicism remained the highest and healthiest form of artistic practice. But if, as Laforgue held, the Hartmannian unconscious was the sine qua non of all human activity, then it simply *was;* it could be judged neither healthy nor diseased. And this, Laforgue reasoned, would shatter the link between health and beauty to which Taine had clung. Laforgue tied this part of his argument directly to individualist artistic practices; decadence, as an aesthetic, could therefore in no meaningful way be criticized as pathological.[76]

Although Laforgue argued that impressionism was an ideal of its time, it had not yet attained the Ideal. Indeed, Laforgue anticipated and hoped for further evolution. His clearest statements on aesthetics appeared posthumously in *La revue blanche* in 1895. There, Laforgue explained carefully that his aesthetics was an attempt to unify positivism and idealism and to bring the results of an "experimental metaphysics" to bear on an aesthetics that would both define and show the way to the Ideal. Expressing absolute fealty to experimentalism, Laforgue argued that the "inductive metaphysics of the experimental life sciences" had demonstrated the unconscious as an essential law, "the unique, constant, guiding force forever evolving toward pure consciousness through the inevitable selection produced by the flux of conflicting desiring forces of love, religion, language, sciences, arts, social apostolate, new mysticism, etc."[77] Although he still took issue with Taine, it was thanks to Taine's insistence on a pathological approach that experimentalism had revealed the underlying law of aesthetics to be the unconscious. Elsewhere, Laforgue echoed Taine even more closely, revealing that the pathological method had a specific place in this new aesthetics. In making aesthetic judgments, he wrote, "the ordinary naked eye has no more authority than ordinary consciousness in the old psychology. . . . As is done today for the cerebral center . . . there is a series of specialized sciences that need to be organized, a psychology of the eye, psychology of the ear, of the palate, of the olfactory nerve, and from an artistic point of view—a psychology of *artistic insanity* [*aliénation*]."[78]

In notes related to these aesthetic principles, Laforgue went even further, suggesting not only that experimentalism had furnished his aesthetics with its underlying principles but also that the elusive Ideal groped its way toward an objective understanding of itself through "experimental" art making: "We sense above us an Ideal. . . . It objectifies itself dynamically, floundering around with withering genres and with contradictory works that go from one excess to another. Its disillusionments and its new glimmers, however, give birth to geniuses. The Ideal inevitably continues making its experiments according to an obscure law that says to itself: you were wrong last century, let us take another path." By bringing the truths discovered through experimental method to bear on his approach, and by aligning experimentalism with the progressive march of the Ideal, Laforgue envisioned a "radiant and certain" aesthetics that would "respond to both Platonists and empiricists." In a letter to Kahn, he added "determinists" and "pure Hegelians" to the list of those whose aesthetics he was superseding: "Yes, idealists, we are idealists as well."[79]

Like the chemist Marcellin Berthelot, who had argued that the factor common to "positivist science" and "idealist science" was none other than experimental method,[80] like spiritualists such as Caro, who had seen Bernard's experimentalism as a potential helpmate for approaching metaphysical questions, and indeed like Zola, who had mobilized Bernard's a priori idea to counter the view of naturalism as overly positivist, Laforgue placed his unswerving faith in experiment. He believed in its power to reveal fundamental laws, above all the foundational principle of the unconscious; he envisioned its bringing humankind to an ideal state of equilibrium; and he emphasized experimentalism's almost alchemical power to yoke together seemingly incompatible beliefs.

Toward an Experimental Symbolism: "Esthopsychologie," Psychophysical Aesthetics, and Experimental Spiritualism

Until his premature death in 1887, Laforgue maintained close ties with neoimpressionist painters, especially with Georges Seurat, who was among the few mourners at Laforgue's funeral; with the critics who initially championed neoimpressionism, especially Kahn and Fénéon; and with Henry, the idiosyncratic devotee of psychophysics championed by a number of neoimpressionists, whom one art historian has called a "scientific dilettante."[81] Many scholars, over the past twenty years, have productively countered the view of neoimpressionism as simplistically scientific, and have demonstrated the shifting and at times contradictory engagements with scientific and pseudoscientific theories, practices, and rhetoric among neoimpressionist artists and the critics who championed them.[82] More recent analyses of neoimpressionism emphasize that certain artists and critics subscribed to a vision of science as a kind of modern ideal that might point the way toward universal modes of communication, a view not at all incompatible with idealist pursuits, even Wagnerian aesthetics.[83] Science was a tool to be used at times in the manner of utopic fantasy, as in Seurat's interest in Henry's theories of dynamogeny, and at times in a more allegorical fashion, as in Fénéon's scientist rhetoric.[84] To these critical accounts I would simply add that neither science writ large nor any particular scientific field could confidently and indisputably claim to be a modern ideal. Experimental method, however, could. Scientists, writers, critics, and artists alike might have questioned the motivations of different scientific practitioners, their more or less materialist or spiritualist biases, or their choices of objects for research, but experimentalism reigned supreme as a modern methodological ideal for producing truth.

In the year after Laforgue's sparsely attended funeral, Kahn would mourn another victim of an early death, the critic Émile Hennequin, who had sought to apply experimental methods to art and literary criticism in an attempt to create a field of research he called "esthopsychologie." In 1882, Hennequin had approached impressionism through the lens of the positivist aesthetics of Eugène Véron, and in 1885 had considered Wagnerianism from a Spencerian perspective.[85] But his volume *Scientific Criticism* (*La critique scientifique*), published just before his death in 1888, outlined his scientific methods of criticism in the most comprehensive terms. Highlighting Hennequin's absolute rigor in applying only the experimental method to his criticism, Kahn expressed his appreciation for how Hennequin had tried to counter the usual critical methods by rejecting those overly concerned with traditionalism, hagiography, and the relative age of an idea. By contrast, Kahn emphasized that Hennequin's approach was admirably experimental, comparative, and based on the pathological method, proceeding as it did from "slight aberrations that approach abnormality."[86]

Kahn (fig. 17), who had promoted psychophysical research as a crucial tool for the symbolist search for universal modes of communication, and who, unlike Hennequin, had viewed neoimpressionism as compatible with symbolist theory, had himself demonstrated that the "rational principle of polychromy" would form the basis for a future aesthetics. For this, he had cited the work of Henry, which "proved . . . experimentally" the physiological effects of both color and linear direction. "Pathological observations," Kahn wrote, were especially revealing, demonstrating, by comparison, that normal individuals were naturally more stimulated by different colors and linear shapes.[87] In his September 1886 manifesto, the "Response of the Symbolists," Kahn reiterated not only that Henry's scientific aesthetic principles offered guidelines for symbolist practice, but that experimentalism and idealism could be conjoined to produce symbolist forms of representation.[88]

The anarchist Kahn would continue to hold up science as the great hope for society's improvement. Once liberated from bourgeois state institutions, science would be free, he argued, to manage its own discoveries for the social good.[89] Responding to Jules Huret's questions in 1891 about the "future of positivism in literature," Kahn argued that science had finally moved beyond positivism, replacing "little positivist experimentations" with a grander form of experimentation guided by the "theoretical intuition" of "new scientists" such as Henry. That even some "positivist scientists" were now basing their experimentation on the work of such "intuitives" was, Kahn implied, to be especially noted by the symbolist generation.[90]

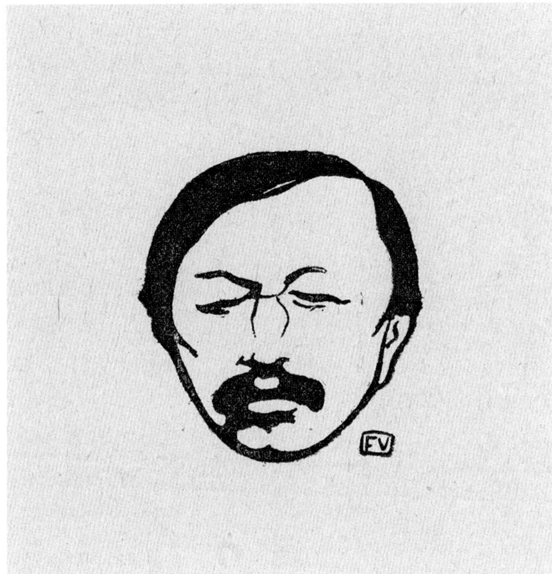

17 Félix Vallotton, *Gustave Kahn*, in Remy de Gourmont, *Le livre des masques* (Paris: Mercure de France, 1896), 242. Bibliothèque nationale de France.

While the overall prestige of positivist science had begun to wane by the late 1880s, Bernard's alleged idealism, along with the kinds of pathological subjects studied by experimental psychology, enabled experimentalism to retain its authority easily throughout the 1890s, perhaps most intriguingly in those domains that might have seemed most inhospitable to positivism, namely, spiritualism and spiritism. Many observers saw the great promise of the fin de siècle as the reintegration of science and religion, a new relationship between nature and art, and a new unity of the spiritual and the material. Among many writers, critics, and artists there remained an overwhelming faith and confidence in scientific methods, and above all in experimentalism, which, in its application to psychology, now seemed poised to fulfill the hopes of positivists and spiritualists alike. The institutional success of experimental psychology indicated to some that an all too restrictive Comtean positivism had finally been superseded by a new generation of thinkers who would willingly join with philosophers and poets to fulfill the dreams of Bernard.

Bernard continued to be enlisted in spiritualist and idealist causes in the 1890s, often in the proliferating symbolist journals. Between 1890 and 1892, René Caillié, a follower of the occultist Eliphas Levi, published a series of articles in the esoteric journal *L'étoile*. Under the title "Spiritualisme expérimental," the articles provided a history of spiritism to the present, expressing a sincere and overriding optimism about the ongoing reorientation of science away from an "atheistic Materialism" and toward a study of spiritism, which Caillié argued would in turn reinvigorate spiritualism as a philosophic endeavor.[91] In the pages of *L'étoile*, Caillié quoted Bernard at length, in a citation "proving" that this "high authority" was far from a straightforward materialist:

> The human body is composed of matter that is constantly being renewed. Each day you lose a part of your physical being, and you replace what you have lost with food, so efficiently that, in the space of eight years, your flesh and bones are replaced by new flesh and bones. . . . Your skull is no longer occupied by the same cerebral matter as it was eight years ago. But . . . how is it possible that you perfectly remember things that you heard, or learned eight years ago? If these things are lodged, as some physiologists argue, in the lobes of your brain, how is it possible that they survive the complete disappearance of those lobes? . . . It must be that there is something immaterial, permanent, always present, independent of matter. This something else is the soul.[92]

Caillié followed this long passage from Bernard with an admonition to those who had ignored the metaphysical side of the great physiologist, almost apologizing for having to enlist such a big name to support his claim: "Certainly, these words come from the mouth of an authority, a high authority. But none are so deaf as those who refuse to hear, nor none so blind as those who refuse to see." The pleading and defensive tone was not altogether different from Zola's in "The Experimental Novel."

An 1892 open letter to the psychologist and occultist Jules Bois, published in another short-lived symbolist journal called *Le chimère*, provided its readers with a similarly nuanced view of the latest developments in positivism. Its author, Pierre Dévoluy, criticized the narrow presentation of positivism in the recent work of Charles Richet, who would himself go on to seek a unity of science and spiritualism, performing extensive experiments with mediums under the banner of what he called metapsychics (*la métapsychique*). Positivism, Dévoluy argued, should not be conflated with materialism, nor should it be simplistically opposed to idealism: "Positivism is sometimes purposely confused with materialism: but this is the crudest of errors (one that naturalism bumps up against, for example). But they are like day and night. . . . To oppose idealism to positivism is another great heresy. I fervently believe that the youth of tomorrow will be positivist in this larger sense of the word."[93] Naturalism may at times have confused positivism with materialism, but the greatest positivists (Dévoluy cited Bacon, Descartes, and Darwin alongside Comte, Strada, and Spencer), he insisted, were scientists of subjectivity, faith, and the unknown.

Édouard Schuré's *The Great Initiates*, published in 1889, dramatically demonstrates the continuing currency of idealist strains of positivism at the fin de siècle, and it reflects the ongoing hagiographic role ascribed to Bernard on the basis of his experimental method, in particular its recent application to psychology via the pathological method. Schuré's book, widely read by the Nabis and owned by both August Strindberg and Edvard Munch, describes how the "great initiates" throughout history, including Rama, Isis, Moses, Plato, and Jesus, sought to understand the universal truths of humankind.[94] Schuré introduced his text by deploring a tragic lack of spiritual understanding in contemporary society. He lamented the fractured relationships among religion, science, and philosophy. Religion had enclosed itself in dogma and become powerless in the face of the impressive advances made by science, while science had become arrogant in the wake of its recent successes. Philosophy, for its part, had relinquished its responsibility of mediating between religion and science and had retreated into skepticism. The only way forward, in Schuré's view, was the reconciliation of religion and science, with philosophy as helpmate.

At the outset of his text, Schuré offered an epigraph from Bernard that will by now be familiar to the reader: "I am persuaded that a day will come, when the physiologist, the poet, and the philosopher will all speak the same language and will all understand each other."[95] Mobilizing Bernard once again, not just as an authority but as *the* authority to guide this imminent reconciliation, Schuré put his text under the sign of experimental method. The epigraph set the stage for Schuré's avowal of experimental science as a "great intellectual force" that nevertheless had been so "dazzled" by its power in the physical realm that it had lost sight of the "psychic and intellectual worlds." Fortunately, though, Schuré sensed a growing tendency to associate matter with the idea of a force, which, just as Caillié also hoped, would lead to a renewal of spiritualism. Schuré further argued that experimental psychology, the most promising of the new experimental sciences, supported this shift in human consciousness. An outgrowth of physiology, the new psychology, despite its initial injunction against metaphysics, was thus poised to provide knowledge of another world, that of the soul. Its investigations of somnambulism, automatism, and hypnosis, in particular, had already dealt a severe blow to vulgar materialism. Schuré made clear that science's methods were entirely sound; science simply needed to follow the lead of experimental psychology by continuing to widen its scope.

In his introduction, Schuré spoke directly to contemporary artists, encouraging them to see themselves as playing a role in this development. Viewing the impending transition as a conversion of base naturalism into an exalted new relationship with "Nature," Schuré heralded the "sanctification" of nature of which Maurice Denis would write in his "Définition du néo-traditionnisme" the following year. While both realism and decadence had debased nature, Schuré argued, "systematically negating the soul and the intellect," he predicted that a new generation of artists would soon return to an "unconscious esotericism."[96] Schuré's text is one of the most useful guides to understanding how an artist, seeking to produce work to be read as symbolist, could simultaneously adhere to positivism, certain strains of naturalism, and even a form of materialism, as well as idealism, spiritualism, spiritism, and esotericism. It was a complex and often contradictory cocktail of belief systems held together by a seemingly unshakeable belief in the expediency and efficacy of experimental method. What greater proof of experimentalism as a fundamental good, as an essentially right way to reach something called the truth, than Bernard's own hopes for a future reconciliation between physiologists, philosophers, and poets?

Obviously, the antipositivist position of G.-Albert Aurier laid out in my introduction should not be taken to indicate a wholesale antipositivism among symbolists.[97] But even Aurier's position was not nearly as dogmatic as is often assumed. His most polemical text, the 1892 article "The Symbolists," is a little too insistent on oversimplified oppositions between art that is "exclusively materialist, experimental, and immediate" and a "new, idealist, and mystical art." In the same essay, Aurier implied that his views were less strident than they might at first seem, when he referred to "false science," in the clear implication that there might actually be a "true science."[98] In another text, Aurier gave a sense of what that "true science" might be—a science based, paradoxically, on nescience: "the cherished naïveté of the Primitives who translated, as far as they were capable, with such a touching ignorance . . . that I call THE TRUE SCIENCE, so many great and deep-seated psychologies."[99] Aurier was not about to embrace experimental psychology or an "experimental art," but the terms under which he could imagine a "true science" included unexpected echoes of the language with which experimental psychology would be approached by other artists and critics active within symbolist circles.

Indeed, much is made, in symbolist theory and criticism, of "false science," "banal and petty naturalism," "worn-out positivism," and "vulgar materialism."[100] But such qualifiers leave much room for these concepts and constructions to be reappropriated and regenerated. The problem was not science, or even necessarily positivism, but how science and positivism had been used, the kinds of questions their adherents had asked, and the kinds of problems to which they had been applied. The tools of science, however, such as experimentalism and the pathological method, remained authoritative and available for higher moral and aesthetic purposes. Although Zola had opposed naturalism to idealism in literature, he had nevertheless held the experimental method up as a new modern ideal, a method capable of verifying and objectifying individual ideas to the point of universal truths. Experimental method converted ideas into knowledge. It weeded out the untruthful ideas and transformed the best ideas of geniuses into indisputable scientific facts. The allure of this idealizing view of experiment, and the concomitant idealization of experimentalism for critical and aesthetic discourse, intensified with the institutionalization of experimental psychology. In the writings of Laforgue, Kahn, and Hennequin, as well as in the spiritualist, idealist, and esoteric discourses exemplified by Caillié, Dévoluy, and Schuré, experimentalism provided an unassailable way forward toward new aesthetic goals, including those of the emerging symbolist avant-gardes.

Experimentalism, as incarnated by Bernard, is less a red thread winding its way through my narrative and more of an arrow. For the writers, critics, and theorists discussed, its aim was true. Its authority as a method for determining human truths virtually unquestioned from the time of the *Introduction*, it guided late nineteenth-century writers of different aesthetic positions toward an imagined destiny, the discovery and determination of universal truths about humankind. Its authority in relation to both aesthetics and art making was established by the second half of the 1880s, not as a metaphor for the new, or as an allegorical appeal to the allure of science, but as a real, workable approach to "objectifying the subjective," modeled above all on experimental psychology and nature's experiments.

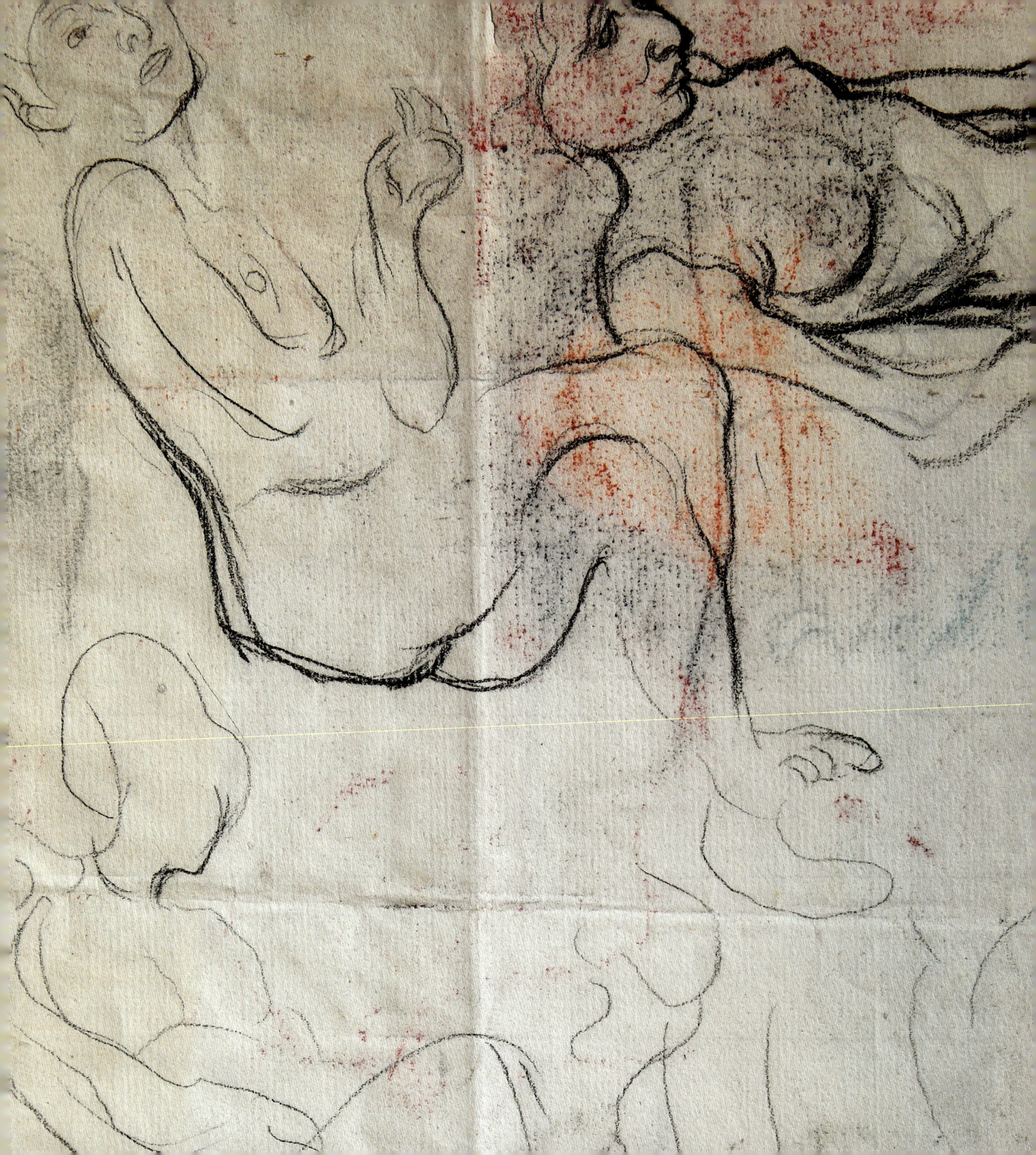

2

DEFENDING DEFORMATION

Maurice Denis's Positivist Modernism

What astonishes us most, in Cézanne's work, is his research for form, or to be more exact, for deformation.

—MAURICE DENIS, 1907, TRANSLATED BY ROGER FRY, 1910

When Maurice Denis published his protoformalist manifesto, the "Définition du néo-traditionnisme," in August 1890, he strategically deployed the rhetorical conventions of what Martha Ward has called the "aesthetics of normality" to devalue academic and naturalist practices by branding them pathological in comparison to what he coined as "néo-traditionnisme."[1] Using the words *hallucination*, *perturbation*, and *anomaly* to describe the minutely detailed work of trompe l'oeil specialists such as Ernest Meissonier, Denis reserved the words *pure*, *true*, and even *normal* for the painting of Pierre Puvis de Chavannes. In the "Définition," the word *déformation* identified naturalism as a false and diseased doctrine. Meissonier had embarked upon a "deformation" of the Dutch tradition, while a mania for mimesis had "deformed the eye of the Academy's teachers."[2]

But one year later, Denis made deformation central to his symbolist visual practice, and soon, as the epigraph to this chapter suggests, to his modernist theories.[3] At the 1891 Salon des Indépendants, his début as a symbolist painter, Denis courted the attention of symbolist critics by exhibiting, among other works, a highly distorted and brightly painted female nude, a painting, now lost, that Denis enigmatically titled *Décor*. He positioned this life-size nude to rival female nudes by Paul Cézanne, Paul Gauguin, and Émile Bernard, and to function as a corollary to the "Définition," demonstrating that a painting, "before being a battle horse, a female nude, or some other anecdote, is essentially a flat surface covered in colors arranged in a certain order."[4] In a letter to a critic that he surely anticipated would be published, Denis proclaimed the work a "study in subjective deformation."[5] But while the "Définition" proved a success, an apparent license for modernist abstraction, *Décor* failed to find many admirers. Denis withdrew it from the public realm—it remains untraced—and almost immediately tamed the distortions of his painted female bodies.[6] When Denis salvaged the idea of deformation for his symbolist theory, he argued that symbolist painting was ideally a balance between the extreme "subjective deformations" exemplified by Vincent van Gogh and the "objective deformations" manifested by Gauguin. This maneuver, underpinned by Denis's education in positivist philosophy and the epistemological paradigm of nature's experiments, revalued the pathological connotations of deformation, providing deformation with a scientific cast, which in turn eased modernism's discursive tendency to separate form from content.

Conserving as it does some chimerical notion of right, correct, or true form, deformation has been identified as something of a false problem for modernism.[7] Richard Shiff, approaching what he called this "inherently evasive subject," argued that *déformation*, which he translated as the less bodily "distortion," was a widely deployed critical and visual strategy of the late nineteenth century that functioned largely as a guarantee of artistic sincerity.[8] His argument deflates critics of modern art who denigrated modernism by comparing its visual language to physiological deformity. But in normalizing modernist deformation as a sign of authenticity, Shiff also neutralized it, eliding the conjunctions at play between deformation, pathology, and the body.[9] Alongside concerns over sociocultural degeneration, those associations hardened at the fin de siècle, just as critics and artists were increasingly debating deformation as an artistic strategy.[10]

"Deformation," as Gilles Deleuze argued, "is always bodily," and it "subordinates the abstract to the Figure," by which Deleuze meant not the figurative but rather the "sensible form" of a painter like Cézanne.[11] The critical dispute exacerbated by the 1891 exhibition of *Décor* was essentially a moral one, centering on whether deformation of the human body, and especially the female nude, could ever constitute the basis for a viable symbolist art practice, an argument that was but a cipher for the arguments over abstraction.[12] Many critics, including Alphonse Germain and Yvanhoé Rambosson, said that it could not, arguing that deformation could never be anything but a dangerous sign of pathology. Others avoided the issue, perhaps aware that deformation could all too easily be mobilized to denigrate representations of a body unpalatable to particular tastes and ideologies. Denis offered, I contend, the first modernist solution to this problem by placing deformation at the very center of his understanding, if not his execution, of symbolist visual form, but only after revaluing pathology and thus diminishing its pejorative connotations. He did this by recourse to positivist methods of induction derived from Claude Bernard and the new experimental psychology, instituting a continuum between subjective and objective deformation effectively modeled on nineteenth-century assumptions about the essential, continuous, and useful relationship between the normal and the pathological. Denis's discursive reliance upon nature's experiments, however, would create a problem for the role of the will in art making, a problem that Denis would debate with his fellow Nabi artist Édouard Vuillard and that would be one of his legacies to formalist modernism.

Realist, Naturalist, and Symbolist Deformations

When Denis referred to *Décor* in 1891 as a "study in subjective deformation," he picked up on a resolutely new vocabulary. For much of the nineteenth century, the substantive *déformation* was largely restricted to medical texts, but its use increased in the second half of the century alongside the new scientific impetus given to pathology as nature's own privileged experiment. At this time, the term *deformation* came into more general usage, the earliest figurative uses appearing in the 1880s. Jules Lemaître in 1885, echoing Émile Zola, used the soothing term "even," signaling the terminology as relatively new and surprising. "In the final analysis," he wrote, "what is interesting about a work of art is the transformation and even the deformation of the real by a mind [*esprit*]."[13] Twenty years earlier, Zola had used the language of deformation in private correspondence, employing the metaphor of a window screen to convey the distorting role of the artist in representing nature: "Exact reality is . . . impossible in a work of art. We say that we denigrate or idealize a subject. Same thing in the end. There is a deformation of what exists. There is a lie. It matters little whether this lie is beautiful or ugly. I repeat, deformation, the lie that is created . . . depends on the nature of the Screen."[14] Placing deformation, denigration, idealization, the ugly, and the beautiful all on the same plane of difference from nature, Zola devalued the conventional idealizing project of art, negated a discourse of the beautiful, and insisted upon man as the key, transformative element in any creative endeavor. He relied upon his correspondent to be struck by the pejorative connotations of deformation, using a shock tactic by then familiar in the increasingly competitive arenas of criticism. But while Zola toyed with the idea of deformation in private, he avoided the term in public discourse and soon abandoned the metaphor of man's temperament as a screen. By comparison, the well-known definition "a work of art is a corner of nature seen through a temperament" is neutral, even rather bland. Although Zola admired Léon Bonnat, he was hardly praising the artist when he wrote, in 1878, that his "brush deforms nature."[15]

The criticism of impressionism includes very few instances of *déformer* or its cognates, especially before 1880. This no doubt stems in part from the dominant role of landscape painting within impressionism; to speak of a *paysage déformé* would have made little sense. Deformation only really assumed currency in relation to the work of Edgar Degas, in which critics perceived that a bodily norm had been transgressed not by the artist but as a result of the labor involved in the physical activity depicted. One sympathetic critic, for example, remarked upon how Degas's dancers "furnish him with types singularly deformed by work."[16] Degas's deformations were thus recognized not so much as products of his temperament but as signs of his fidelity to the real deformities of his subjects' bodies, a mark less of his modernism, in other words, than of his allegiance to realism.

The scientifically minded critic Émile Hennequin, proponent of the "esthopsychologie" discussed in chapter 1, deployed deformation in ways dependent on the scientific aesthetics of Eugène Véron as well as on Zola's notion of a deforming temperament. In an 1882 review, Hennequin argued that a key innovation of impressionism had been its recognition of "the experimental fact" that in natural light, "the contours of objects are deformed."[17] Véron's *Aesthetics* of 1878 recommended that painters be aware of two kinds of deformations, those resulting from movement and those resulting from light.[18] Although he carefully avoided explicitly recommending deformation as an artistic strategy, Véron nonetheless implied that these natural deformations might be used to counter fixed, academic ideas of form. The artist, he argued, should take into account scientifically observable deformations of bodies in movement, which demonstrated the falsehood of static, classicizing beauty. At the same time, Véron criticized the conventions of perspective as entirely "subjective" deformations, arguing that while academicism promoted such conventions as objective ways of painting, these conventions were in fact the very antithesis of objective. Véron, an ardent republican, thus pathologized conventional formal norms as subjective deformations, while normalizing an alternate and, in his view, more truthful form of painting that would take into account only those deformations that had been observed objectively.[19] In his review of the 1886 impressionist exhibition, Hennequin identified the aim of truth to nature as fallacious, arguing that realism and art in fact contradicted each other. Artists who sought to render nature with as little deformation as possible, such as the neoimpressionists, only negated individuality. True painters "deform nature," Hennequin claimed, a position that allied him at once with Zola and with an emerging symbolism.[20]

In the theories of decadence and symbolism, the concept of deformation came to be used in more consistently polarizing terms. Jean Moréas, whose symbolist manifesto was written as a response to decadent aesthetics, justified and promoted the symbolist novel on the basis of its examination of environments deformed by individual temperaments. "In this *deformation*," Moréas claimed, "lies the only *reality*. . . . The symbolic [*sic*] novel will be edified by *subjective deformation*." Moréas acknowledged that his conception of subjective deformation might be imagined to be aligned with Zola's definition, but he argued that subjective deformations were more real than any purportedly objective reality forming the basis of naturalism. "In objectivity," he insisted, "art can only find a simple and extremely limited point of departure."[21] Kahn's symbolism, which, unlike Hennequin's, would encompass neoimpressionism, was much less concerned with formal deformation as an index of individual temperament. While Moréas had produced a list of weird, archaic, and surprising linguistic devices, deformations of which the symbolist author might avail himself, Kahn promoted an objective analysis of the self and, perhaps consequently, left deformation aside.

Véron's, Moréas's, and Kahn's strategic distinctions between the subjective and the objective reflect wider uses of what Lorraine Daston and Peter Galison have identified as the relatively new "epistemic virtue" of objectivity. As Daston and Galison argue, a reordering of values within scientific discourse around the mid-nineteenth century saw passive objectivity held up as a necessary check on a fundamentally active subjectivity. A starker opposition emerged between the artistic self, the romantic vision of the abundantly subjective artist, and the scientific self that operated by subduing the active self, functioning as far as possible as an automatic recording device. While the artistic self courted a "willed willfulness," the scientific self proceeded, paradoxically, by "willing the self to will-lessness."[22] Zola's promotion of individual temperament and Moréas's adherence to subjective deformation over and against objectivity prolonged the romantic paradigm, but Kahn's more scientific-sounding formula, the formula that Denis eventually took up, contained within it fundamental challenges. How, from a practical standpoint, was an individual poet or artist supposed to "objectify the subjective" in ways that were understandable, or at least legible? To put it in Daston and Galison's terms, how might an artist, in the service of objectivity, suppress the active self but at the same time make the active self the very foundation, the raison d'être, of the artwork? Denis would eventually propose a continuum between subjective and objective deformations, arguing, unlike Véron, that the two were connected and that objective deformation could act as a check on subjective deformation for the purposes of symbolist visual form. But Denis's continuum structure, which he modeled on the paradigmatic continuum between the pathological and the normal, would threaten the starker distinction that scientific discourse had made between active willful subjectivity and passive unwilled objectivity.

Décor and Bodily Deformation

Denis's ten submissions to the 1891 Salon des Indépendants—ten was the maximum number allowed—testify to the ambitions of the twenty-year-old artist (fig. 18). With eight paintings, one fan design, and a suite of drawings inspired by Paul Verlaine's *Sagesse*, Denis contrived to attract the attention of symbolist critics, especially with *Décor*, under whose title in the catalogue he included a suggestive line from Charles Baudelaire's translations of Edgar Allan Poe.[23] In his letter to the symbolist critic Adolphe Retté, excerpted in Retté's review of the Indépendants, Denis justified *Décor* as a "study in subjective deformation," echoing less the language of Aurier, whose essay on Paul Gauguin had only just appeared, than that of Moréas, who had placed subjective deformation at the center of his symbolist theory.[24] Unlike the rest of Denis's submissions, all smaller and mostly of overtly religious or spiritual motifs, *Décor* presented viewers with a life-size, odd-looking female nude lying on a deep red sofa against a bucolic landscape. The title referred not to a specific decorative scheme but inserted the work into emerging debates over decoration and symbolism.[25] It was not to be seen primarily as a female nude but literally as a *décor,* "a flat surface covered with colors arranged in a certain order."

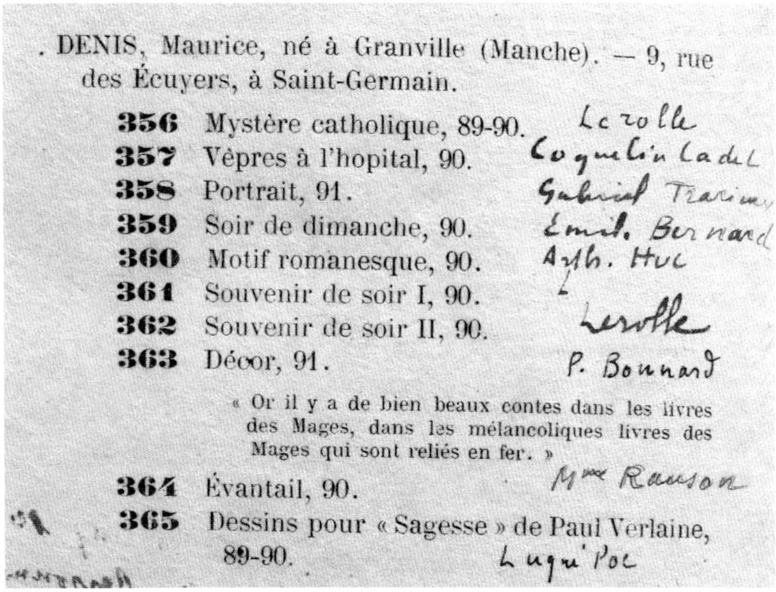
. DENIS, Maurice, né à Granville (Manche). — 9, rue des Écuyers, à Saint-Germain.

356 Mystère catholique, 89-90. Lerolle
357 Vêpres à l'hopital, 90. Coquelin Cadet
358 Portrait, 91. Gabriel Trarieux
359 Soir de dimanche, 90. Emil. Bernard
360 Motif romanesque, 90. Arth. Huc
361 Souvenir de soir I, 90.
362 Souvenir de soir II, 90. Lerolle
363 Décor, 91. P. Bonnard

« Or il y a de bien beaux contes dans les livres des Mages, dans les mélancoliques livres des Mages qui sont reliés en fer. »

364 Évantail, 90.
365 Dessins pour « Sagesse » de Paul Verlaine, 89-90. Lugné Poe

18 Page from the catalogue for the Salon des Indépendants, 1891, with notations in the hand of Maurice Denis. Musée départemental Maurice Denis, Saint-Germain-en-Laye.

Despite the anti-academic rhetoric of the "Définition," Denis had continued to study at the École des beaux-arts through the fall of 1890, regularly submitting drawings to the various juried competitions. In November 1890, he received his third and final mention for a study after a classical cast (fig. 19).[26] Flipping over this prize-winning drawing, Denis subsequently produced a large study for *Décor,* thereby enacting the break with academic and naturalist modes that he had called for in the "Définition" (fig. 20).[27] In doing so, he reprised an earlier moment in his artistic training, recounted in the "Définition," when he had tried to "embellish, idealize, and synthetize a nude figure," only to have his teacher remark, "that's not nature, you would never sleep with that woman!"[28]

Individualized features coupled with a woman's name, "Anna Thibout," and an address on the sheet suggest that Denis hired a model for the study, possibly together with Bonnard and Vuillard, with whom he was then sharing a studio.[29] At the time, he told a friend, the future theater director Aurélien Lugné-Poë, that he was working on a "mystical nude woman," which he subsequently gave a name and a rationale: "My reclining woman . . . is named Nahasch, the snake, symbol of universal attraction."[30] A Hebrew word for serpent, Nahasch not only invites all the eroto-symbolic possibilities of a snake woman, licensing the figure's serpentine attenuations; it also identifies the work with the Nabi group's early years, when obscure terminology helped bond the brotherhood.[31] The point of departure was thoroughly conventional, but as Denis searched for effective means and sites through which to study deformation, his preparatory works moved, first, away from the conventional veiled eroticism of nineteenth-century portrayals of the female nude toward a more frank eroticism, and then away from this more frank eroticism toward antinaturalist abstraction.

While the outstretched pose is a familiar one, Denis depicted the model not so much reclining as lying down, her head crooked at an awkward angle against the mere indication of a sofa arm. The model's left arm is arranged naturalistically alongside her body, but the crooked right arm hangs heavily into an undefined space, ending in a bloated hand. Using stylized contour, Denis nevertheless searched for ways to mitigate the figure's deformations, appealing to line as a source of both bodily distortion and pleasure.[32] There are no usual signifiers of female modesty, however, such as an artfully arranged sheet or a hand covering the mons pubis or breasts. Denis assayed different attitudes for the face, eventually producing a figure seemingly unaware of the beholder, caught absorbed in a somnolent or semisomnolent state. Denis's recording of his teacher's comment, "you would never sleep with that woman," exposed the hypocrisy of the transcendent stance taken by studio masters and their male pupils drawing from the female nude.[33] At the same time, it carved out an artistically and morally superior position not only for Denis's antinaturalism but also for more overt forms of eroticism.[34] The large drawing presents a lithe, supine body exposed—ready, frankly, to be mounted. The worked contour of the lower back, which indents slightly to suggest a tilted pelvis, and the breasts are especially eroticized. The left nipple stands erect, nearly "touching" the contour of the left arm, an area that Denis labored over in smaller sketches on the sheet, while the right nipple confronts the viewer directly, an eye staring straight out of the drawing.

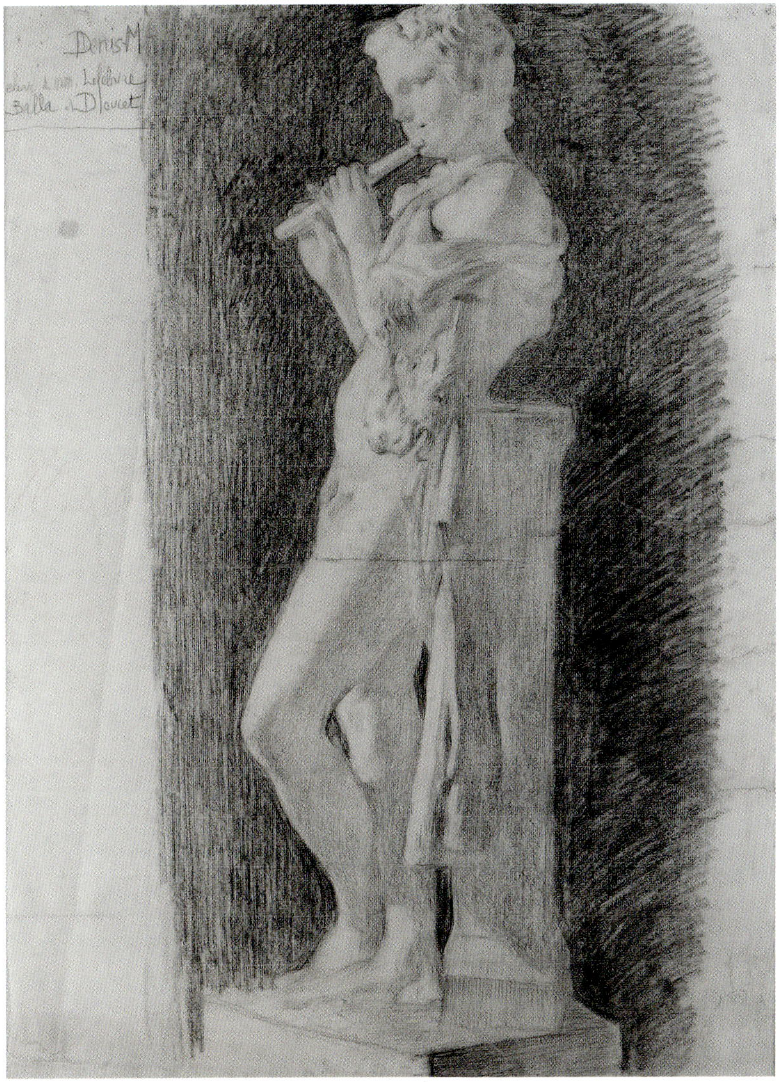

19 Maurice Denis, study after a cast of a faun, 1890. Charcoal on paper, 62.5 × 47.5 cm. Musée départemental Maurice Denis, Saint-Germain-en-Laye, Gift of the Family of Maurice Denis, 1976, PMD 976.1.17.

In subsequent sketches, Denis worked through different degrees of and different sites for deformation, introduced color, eventually rejected the use of contour to consolidate the figure's distortions, and de-eroticized the portrayal in favor of an antinaturalist emphasis on medium.[35] In a partial version of the pose in the upper right corner of a sheet of sketches, rapid, seemingly unpremeditated execution dictates the extreme distortion of the arms and hands (fig. 21). The joints of the left arm are nonexistent, while the right arm moves from a skeletal shoulder joint, recalling anatomical studies, to the right forearm and hand, which appear as highly abstracted, rapidly performed arabesques. In another sketch, Denis changed his position to establish an awkward point of view from above the model, a perspective from which

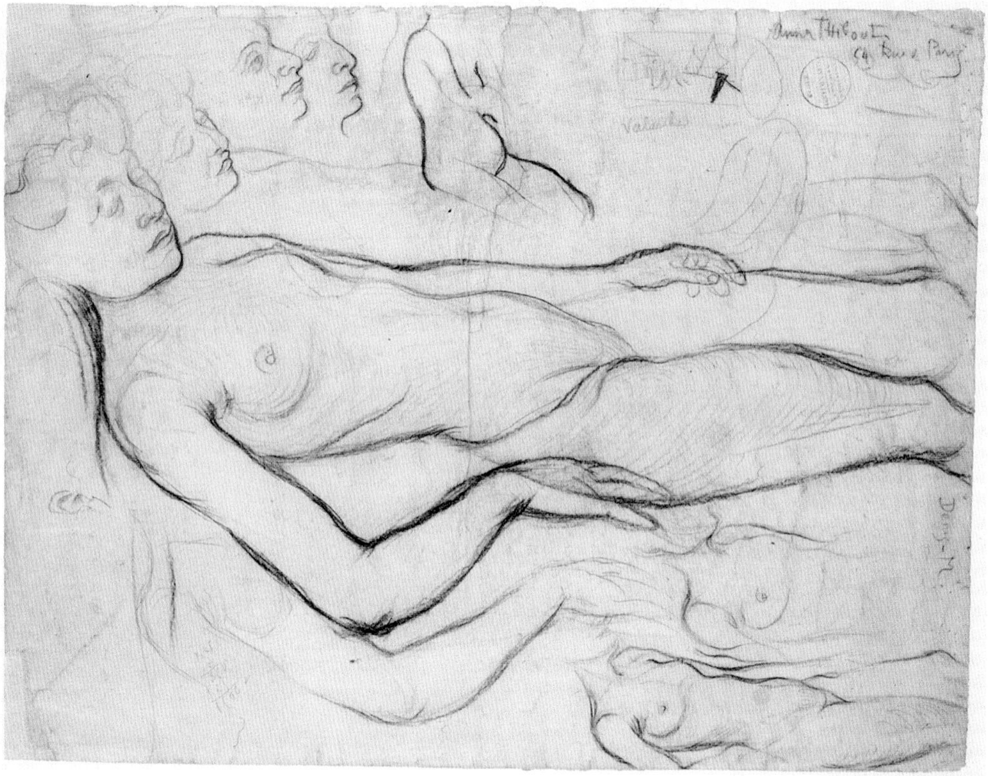

20 Maurice Denis, *Symbolist Nude* (study for *Décor*), ca. 1890–91. Charcoal on paper, 47.5 × 62.5 cm; verso of fig. 19. Musée départemental Maurice Denis, Saint-Germain-en-Laye, Gift of the Family of Maurice Denis, 1976, PMD 976.1.17.

Thibout's sinewy curves become ungainly and bulbous, as if her flesh had been pushed by gravity against a surface (fig. 22). Another pair of drawings, including one with color pastel, tries out a third point of view from behind the model's head (fig. 23), a composition that Denis would later use to illustrate André Gide's *The Voyage of Urien* (*Le voyage d'Urien*).[36] The resulting foreshortening changes the possibilities for deformation, as the figure's shoulder, breasts, ribcage (or hips), and knees form a series of hills receding into space.

Another pastel and charcoal sketch rejects this strategic use of unusual viewpoints to provoke distortion and returns to the composition of the large study (fig. 24). Sinuous contour attenuates and flattens the body in a schematic drawing approaching caricature.[37] If pastel is a medium particularly suited, in the words of Jean-David Jumeau-Lafond, to "clothing the Idea in sensible form," and thus particularly apt for symbolism, it is also a medium capable of suggesting intimate relationships between vision and touch.[38] Heavily applied red pastel evokes velvety upholstery, while other tactile signifiers recall, in an almost satirical way, the overt eroticism of the initial drawing. The few lines of orange pastel at the mons flout academic conventions of decorous eroticism, and the curved line beyond the left hand suggests masturbatory movement.

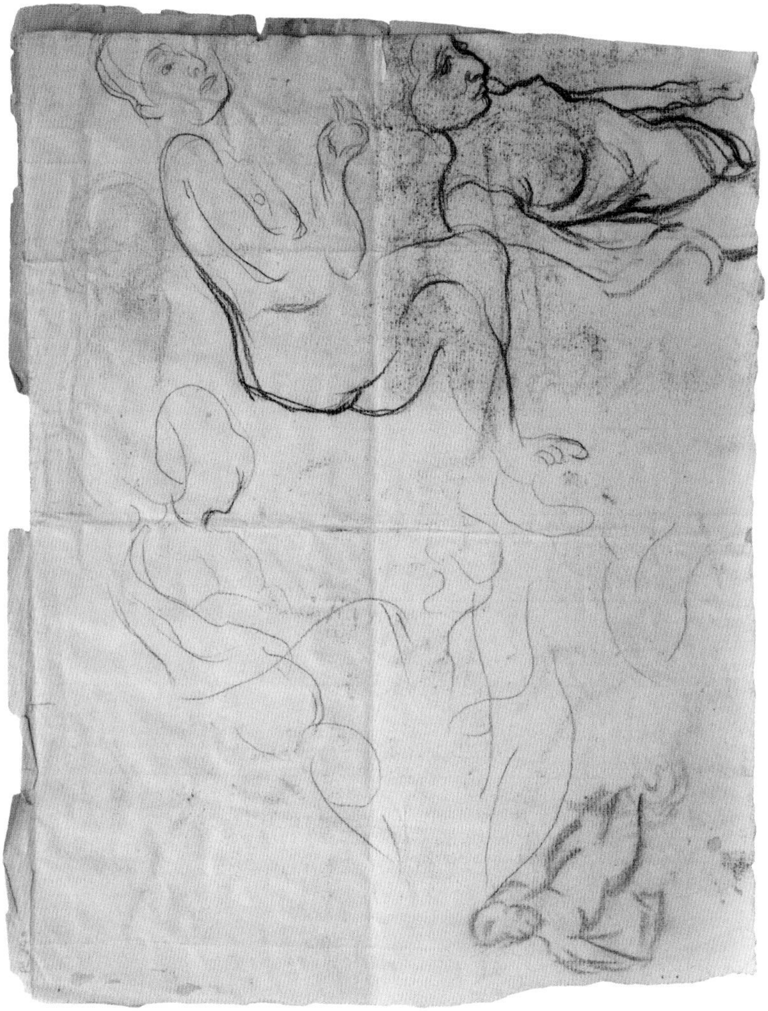

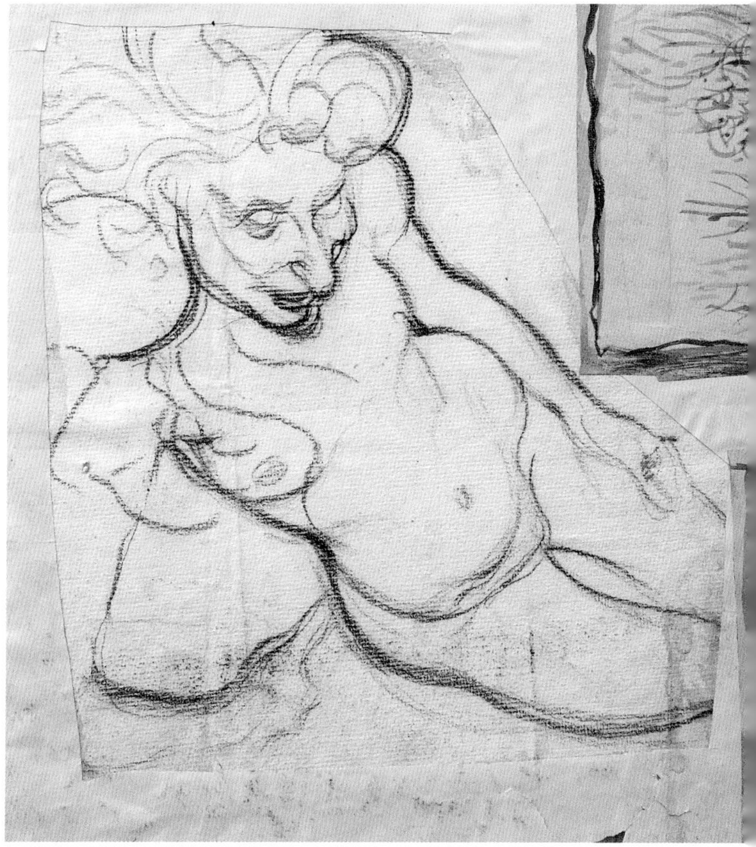

21 Maurice Denis, study for *Décor,* ca. 1890–91. Charcoal on paper, 31.2 × 24.7 cm. Private collection.

22 Maurice Denis, study for *Décor,* ca. 1890–91. Charcoal on paper, 21.5 × 20.5 cm. Private collection.

Denis settled on this composition for two small oil sketches in preparation for the final work. In one, he removed the figure to the lower left to concentrate on elaborating the landscape (fig. 25). In the other, in which virtually all detail and contour are eliminated from the body, the figure now appears as a radically simplified arrangement of broad, pale yellow brushstrokes and dabs distributed horizontally across the center of the small panel (fig. 3). Only narrow strokes of thinned turquoise provide any indication of three dimensions, offering a modicum of form to the right breast, torso, mons, and legs. As he moved toward the painting to be exhibited, Denis further de-individualized and de-eroticized the body, removing nearly all identifying features, save for the model's red/orange hair, and nearly all formal invitations to take pleasure in the female form.

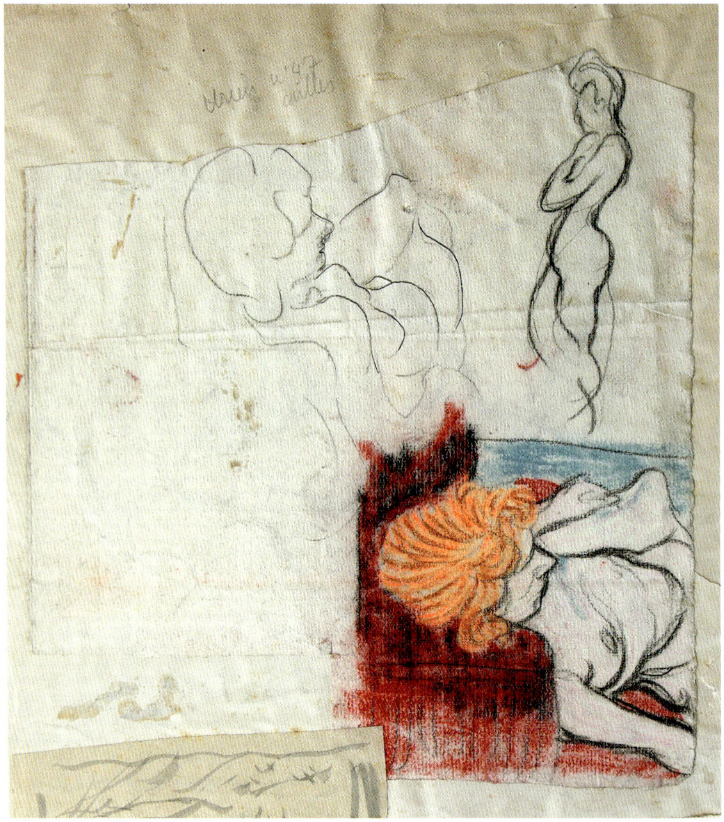

23 Maurice Denis, study for *Décor,* ca. 1890–91. Charcoal, ink, and pastel on paper, 25.5 × 24 cm. Private collection.

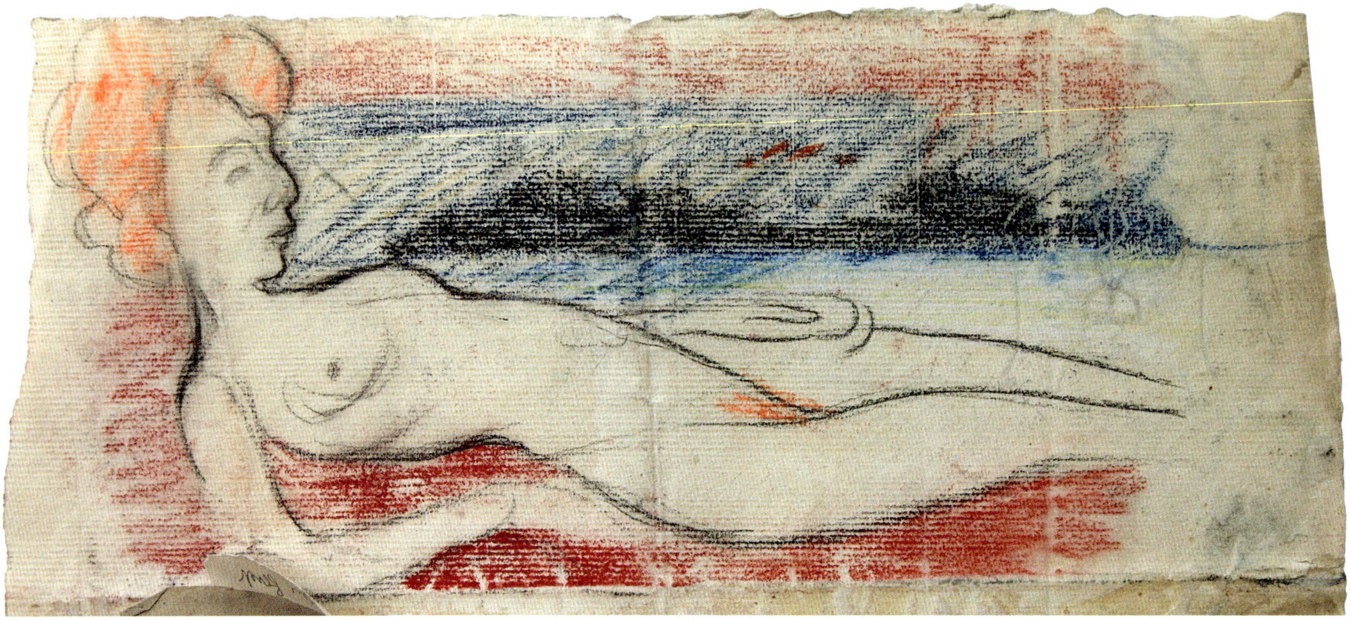

24 Maurice Denis, study for *Décor,* ca. 1890–91. Charcoal and pastel on paper, 31.2 × 24.7 cm; size of sheet, verso of fig. 21. Private collection.

The abstracted composition makes the body just one of a series of horizontal swaths of high-contrast red, pale yellow, olive green, and sky blue, broken only by the pathetically withered right arm, which itself is cannibalized by a red triangle of paint. Rejecting the use of an unusual point of view, a strategy reminiscent of Degas's bathing women, or sinuous contour to license or tame the figure's distortions, Denis offered up a different kind of painterly fantasy. The deliberate lack of modeling and the nonnaturalistic palette adamantly declared the figure's deformations to be entirely the result of the artist's setting brush to support.

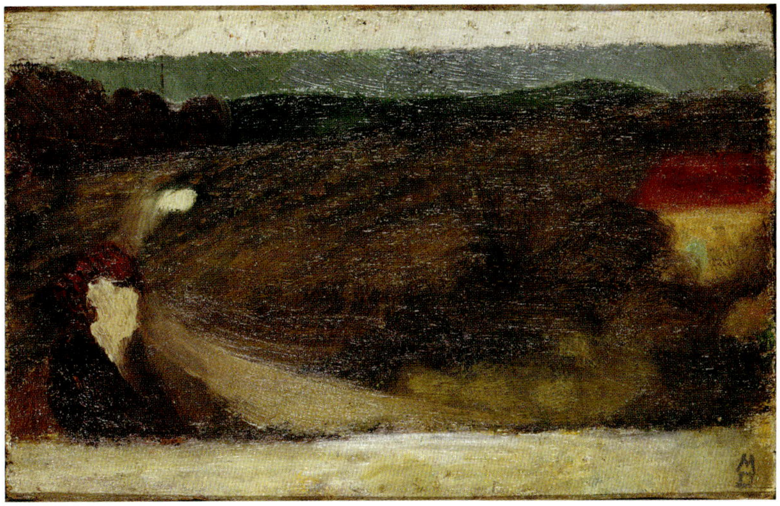

25 Maurice Denis, study for *Décor*, ca. 1890–91. Oil on wood panel, 16 × 15 cm. Private collection.

In 1892, Denis gave *Décor* to Bonnard, who in 1916 depicted it in the background of his painting *The Mantelpiece*, reflected in a large mirror behind a standing female nude (fig. 26).[39] Bonnard certainly enacted his own deformations on *Décor*, providing, for example, a more blocklike figure than in any of the sketches, but his representation nonetheless maintains elements of Denis's preparatory works, including the red sofa, the red/orange hair, and the landscape setting.[40] One of the most arresting features of Bonnard's rendering is how turquoise paint simultaneously suggests and defies contour. Turquoise describes the elongated mons just as in one of Denis's oil sketches, as well as the torso and arms, but also pools at the chest, both flattening the figure and describing the odd, islandlike breasts that threaten to drift away from the midline of the torso. Beyond the knees, an area that Denis steadfastly refused to develop in his sketches, turquoise contour merges with the orange-yellow painted flesh to become an unresolvable melding of indiscernible form. And a hard-to-read string of small orange-yellow circles, more visible in the oil study for the work (fig. 27), further unsettles the depiction, as if the figure were dribbling out beyond arbitrary boundaries.[41] Both Denis's final oil sketch, with its horizontal bands of color, and Bonnard's version of *Décor* rebuff the erotic address of some of Denis's earlier sketches, largely by way of deformations mitigated by neither color nor contour. Bonnard's inclusion of *Décor* in what Charles Harrison has called a meditation on representation and the gendered self marks Denis's painting as an outsized talisman of the early Nabi years, perhaps the first attempt to produce an "icon" on a grand scale.[42] With *Décor*, Denis elaborated the possibilities of deformation for symbolist visual practice, performed a rejection of academic and naturalistic modes, and reminded viewers that this particular female nude was first and foremost a painting.

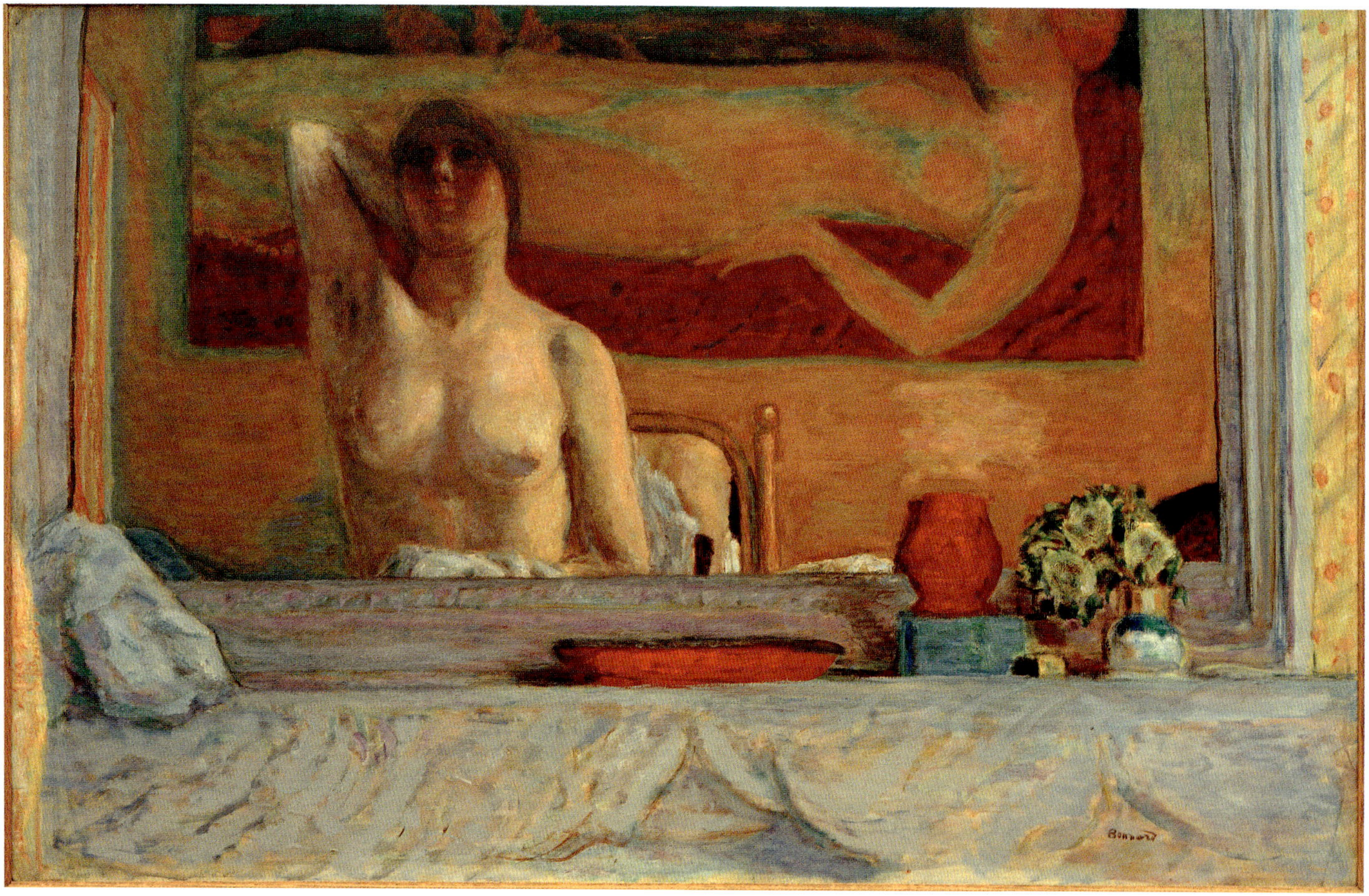

26 Pierre Bonnard, *The Mantelpiece,* 1916. Oil on canvas, 80.7 × 126.7 cm. Private collection.

27 Pierre Bonnard, *Young Woman at Her Toilette* (study for *The Mantelpiece*), 1916. Oil on canvas, 66 × 89 cm. Margaret and Eugene McDermott Collection, Dallas, Texas.

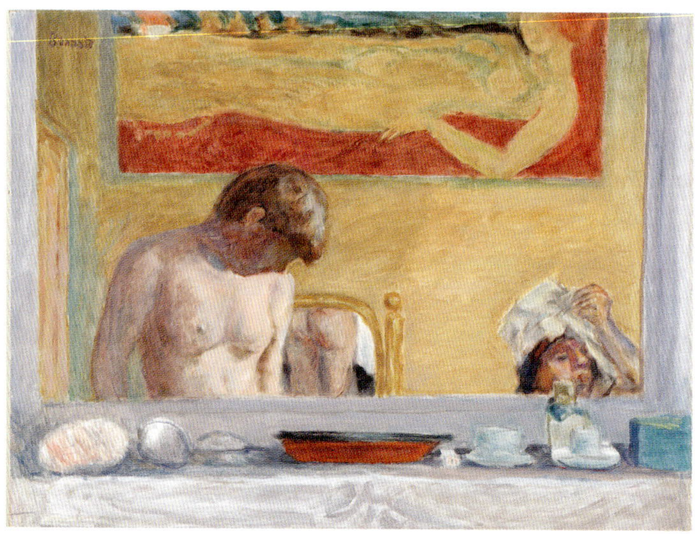

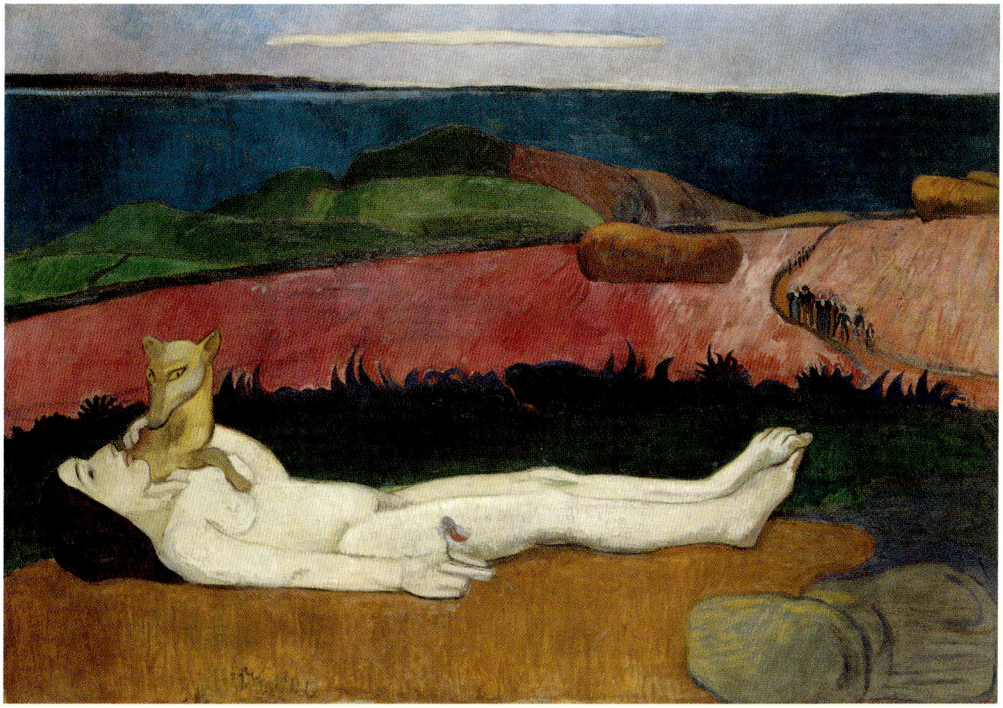

28 Paul Gauguin, *The Loss of Virginity*, 1890–91. Oil on canvas, 90.2 × 130.2 cm. Chrysler Museum of Art, Norfolk, Virginia, Gift of Walter P. Chrysler Jr., 71.510.

But of course it mattered a great deal that Denis chose a life-size reclining female nude as a major statement on subjective deformation. Denis's *Décor* engaged in a familiar gambit, referencing the already established tradition of painted nude women as battlegrounds for modern painting.[43] Manet's *Olympia* had just returned to the spotlight, on display at the 1889 Universal Exhibition, after which it entered the Musée du Luxembourg, where Gauguin could be found making a copy in February 1891.[44] That same winter, while Denis was painting *Décor*, Gauguin completed *The Loss of Virginity* (fig. 28), and although no written source confirms that one or both were aware of the other's painting, the stiff supine poses, the impassive expressions, and the landscapes sloping from right to left are too similar to ignore.

For those in the know, *Décor* also referenced a notoriously distorted female nude by Cézanne. Now lost, the nude, sometimes called the *Wife of the Garbage Collector* (*Femme du vidangeur*), had been in Gauguin's collection and was, by early 1891, housed at Père Tanguy's shop along with Cézanne's portrait of the painter Achille Emperaire (fig. 29).[45] Both rejected from the 1870 Salon (fig. 30), the paintings together posed the problem of deformation for modern painting, raising questions about the artist's motives, skill, and intentions. To what extent had Cézanne, in a quest for realism-not-idealism, sought to paint physiological deformity as it existed in "real" life, declaiming his realism by choosing

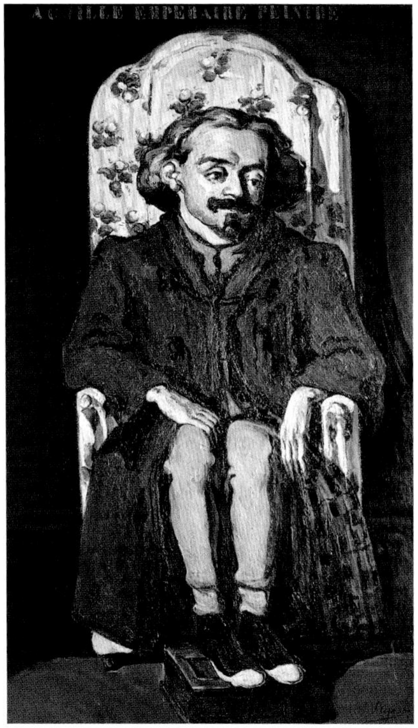

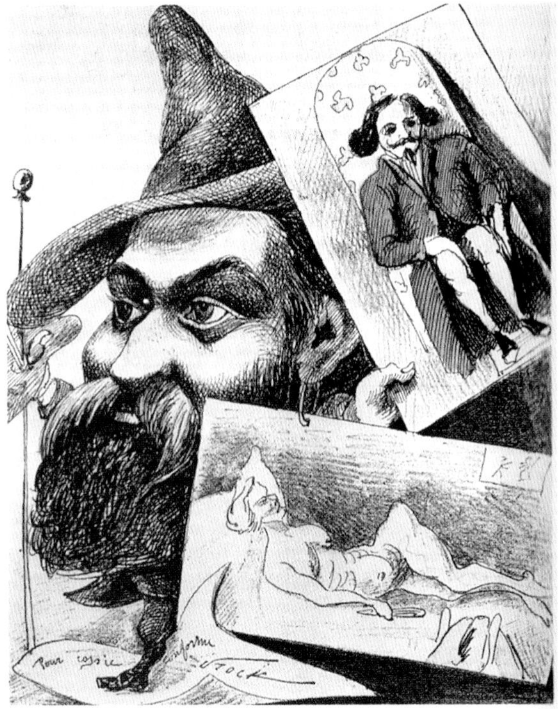

29 Paul Cézanne, *Portrait of the Painter Achille Emperaire*, ca. 1868. Oil on canvas, 200 × 120 cm. Musée d'Orsay, Paris.

30 Caricature of Paul Cézanne with two paintings rejected by the Salon jury of 1870 (nos. 139 and 140), from *Album Stock* (Paris: Dufour et Cie, 1870).

nonideal, even pathological bodies such as that of a known dwarf and hunchback, or the presumably deformed working-class body of the elderly (as she was described) wife of a garbage collector? To what extent were those deformations due not to the models but to the artist's own practice, a supposed lack of skill, a deliberate intention to distort, or, as modernist critics would later have it, a sincere method with unintentional by-products?[46]

Like many aspiring avant-garde artists around 1890, Denis went to Tanguy's shop to "burn . . . the idols of the École des beaux-arts," as Bernard put it.[47] Bernard pronounced Cézanne's nude "very ugly, but nevertheless magisterial," and Denis, much later, recalled his own encounter with the painting, in all certainty while working on *Décor*.[48] He remembered being mystified by Cézanne's work and especially by *Achille Emperaire* and the female nude, the latter a "piece of thickly painted bravura."[49] In retrospect, he would connect the challenge of Cézanne's deformations with three canvases he had seen around 1890 in Bernard's studio, a frieze of nude figures set in a landscape that documents a search for deformations of the female body concurrent with Bernard's attempts to come to grips with Cézanne.[50] And, as scholars have pointed out, echoes of Cézanne's nude, with her jutting hip, appear in all three canvases, most obviously in *Bathers with a Red Cow*, where she can be seen from the rear, in a bulbous representation that includes an act of buttock multiplication predating Cézanne's late bathers (fig. 31).[51] While Cézanne's reclining

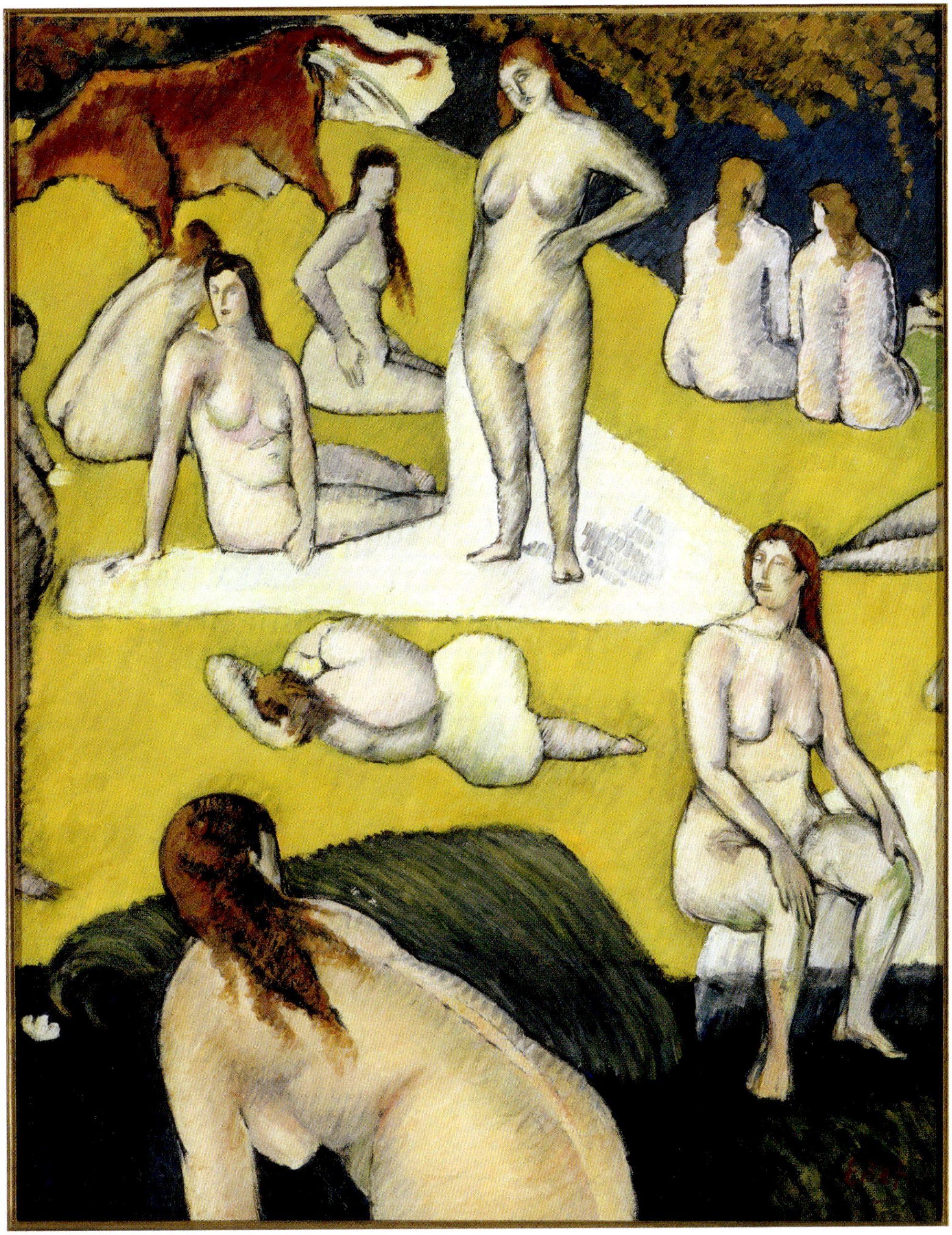

31 Émile Bernard, *Bathers with a Red Cow*, 1887. Oil on canvas, 92.7 × 72.4 cm. Musée d'Orsay, Paris, RF1984–21.

female nude had maintained a realist palette, Bernard deployed antinaturalist coloration, most evidently in the bright yellow of the ground, making clear that the figures were not to be compared to a visible reality, and thus severing any possible connection between artistic deformation and actual physiological deformity.

Absent from the descriptions of Cézanne's nude and, it seems, from the work itself, was any conventional erotic address. I can only guess all the means by which Cézanne conveyed the figure as elderly, but certainly part of its strangeness derived from how signifiers of an older woman combined with the deformations and the thick impasto to rebuff a desiring gaze. Cézanne's nude confounded the expectations of many who saw it in and around 1890, those who still expected from their nudes a modicum of titillation. Although Gauguin never showed it in a major exhibition, *The Loss of Virginity* would still have provided such titillation. *Décor*, it seems, did not.[52] The moment that *Décor* appeared at the 1891 Indépendants was an especially fractious one for emerging symbolist avant-gardes.[53] Van Gogh was dead. Gauguin, despite having just been proclaimed the paradigmatic symbolist painter, was about to leave for Tahiti. Georges Seurat died two weeks into the exhibition. Denis's *Décor*, with its extreme level of deformation, antinaturalistic color, melding of form and contour, and rejection of an eroticizing gaze, offered itself as a rival to Gauguin's Breton Eve, Bernard's frieze, and Cézanne's notorious nude, ushering modernist deformation onto a larger and much more public stage.

Critiquing Deformation

As the competition to define symbolist visual practice intensified in and around 1891, deformation proved ever more contentious. In one of the few reviews of the 1889 Volpini exhibition, Jules Antoine, generally sympathetic toward the Pont-Aven artists, sharply criticized deformation of the human form, proclaiming there to be "an absolute rule." "If one draws a man or a woman," Antoine wrote, "the figure must hold up; if the artist feels the need to deform, he must do so while maintaining certain essential proportions." Antoine took special issue with Gauguin: "M. Gauguin knows how to draw, and there is no aesthetic that justifies arms that are too long, torsos that are too narrow, or a woman's head resembling that of a mouse."[54]

In March 1891, Aurier defended Gauguin's need to "exaggerate, attenuate, deform," but prior to that, Aurier's trio of alteration had been a duo. Previously, Aurier had warranted an artist's right to "exaggerat[e] or attenuat[e]," but in 1891 he insisted upon a distinction between the deformations of realism—Zola's deformations—and those of symbolism.[55] Impressionism, merely an extreme form of realist practice, invited certain deformations in its "rapid subjective synthesis," Aurier argued, allowing for objects to be distorted as they passed through the artist's "soul." But the symbolist could distort

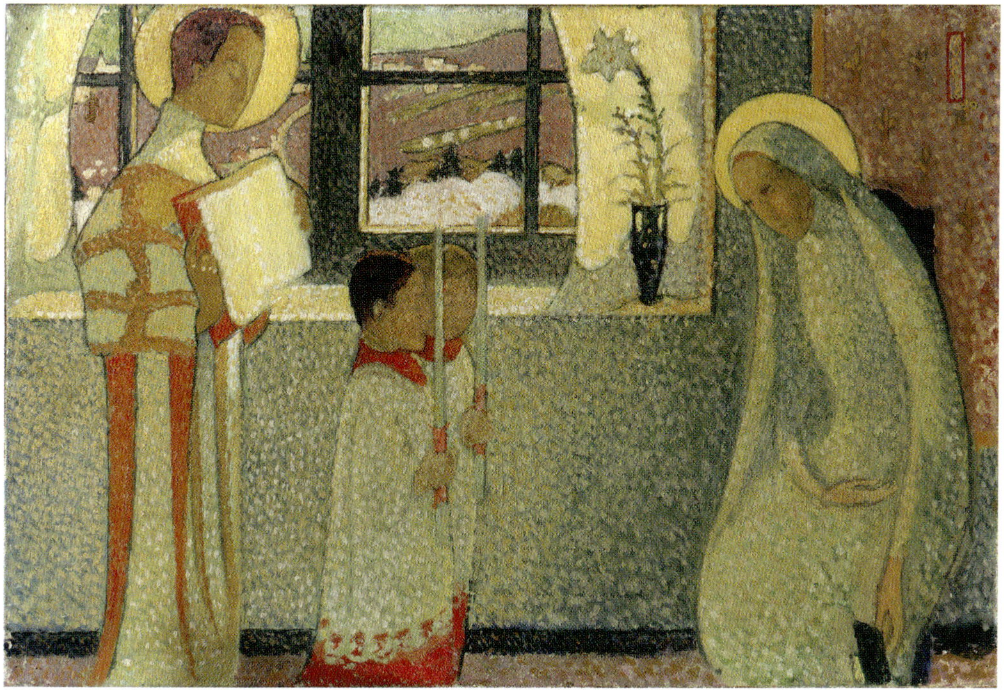

32 Maurice Denis, *Catholic Mystery*, 1890. Oil on canvas, 57 × 77 cm. Private collection.

twice over: "the artist will always have the right to exaggerate, attenuate, deform . . . not only according to his individual vision, following the molds of his subjective personality (within which he can even arrive at realist art), but to exaggerate, attenuate, deform . . . again, according to the needs of the Idea to be expressed."[56] Realism, to which Antoine still apparently clung, thus implied limits dictated by visible reality. Symbolism removed those limits, licensing more radical deformations based on the artist's "Idea." But Aurier was careful to temper the pejorative connotations of deformation strategically, not only by situating it in relation to realist modes but also by grouping it with attenuation and exaggeration, terms that lacked the same bodily and pathological associations.

When Retté, preparing his review of the Indépendants, wrote to Denis seeking clarification on *Décor*, Denis not only insisted that the painting was a "study in subjective deformation" but also defended the work's extreme deformations by distinguishing between two types: "Deformation . . . does not imply ugliness. . . . Here is a woman, elongated, without hips, and genuine; a landscape in which the hills are like ephebic stomachs (first series of deformations). Then the choice of the painting, the dimensions of the frame in which the subject is presented. Other necessities: such as the search for an appropriate support, the agreement of the whole ensemble, the orchestration of the colors chosen (second series of deformations)."[57] In this, the first inkling of his theory of the two deformations, Denis decoupled deformation from ugliness and implied that *Décor* defied any heteronormative erotic address by mentioning the figure's prepubescent proportions and the landscape's "ephebic" hills. Oddly, for an 1891 statement on subjective deformation, he also obfuscated any connection between deformation and temperament in favor of a rationalized explication of *Décor*'s supposed three levels of signification, from base, to spiritual, to intellectual: "For most people," Denis wrote to Retté, "it will be a woman of serpentine lines—for the more spiritual: the fearsome symbol of seductive flesh—for the wise: a representation of the material force in its circulatory manifestations," after which Denis obliquely referred Retté to Poe's prose poem *Eureka* of 1848.[58]

Retté, who had arranged for Denis to show his *Sagesse* drawings to Verlaine in 1890, was much more enamored with another of Denis's submissions to the Indépendants, *Catholic Mystery* (fig. 32).[59] In this annunciation scene, Denis mitigated the attenuated arms and hands of the Virgin with delicate contour and a pointillist pattern that unified the surface of the work. But *Décor*, its typological counterpart, seemed to repulse Retté. Footnoting the word *déhanchée* from Denis's letter with the comment "too much; and it is ugly," Retté went on, "This reclining woman, whose torso is too long, is unpleasant; it is stiff, hard; the right arm, insufficiently drawn, does not work; the face lacks expression; finally the red of the drapery is disagreeable."[60] No serpentine line, pleasing colors, or legible expression on the figure's face countered the deformations that Denis had enacted on the figure.

No other critic reviewing the 1891 Indépendants mentioned *Décor* at any length. Émile Verhaeren read the painting as a decoration for Poe's poem, and thus mentioned what he imagined were the work's mystical and occult preoccupations.[61] Julien Leclercq admired the *Sagesse* drawings but complained that *Décor* lacked "harmony between the color that vibrates too much and the line which is and should remain silent."[62] Roger Marx, Gustave Geffroy, Charles Saunier, and Jules Antoine all praised Denis at length, singling out *Catholic Mystery* and the *Sagesse* works, but scrupulously ignored Denis's most ambitious submission.[63]

Despite Retté's view of *Décor*, the mutual admiration between Denis and the critic continued through 1891, with Retté contributing an admiring profile of Denis to *La plume*'s special issue on painting, published in September of that year. For his part, Denis addressed some of Retté's concerns that summer by painting *On the Theme of Trinitarian Evening by Ad. Retté: Ornamental Painting* (fig. 33), a work inspired by one of Retté's poems, which tamed the deformations of *Décor* through line and pointillist pattern in a manner similar to *Catholic Mystery*. Studies for the two nudes in the foreground attest to Denis's concern with fine-tuning the figures' deformations.[64] Sketches of the foremost figure, a reworking of J.-A.-D. Ingres's *Valpinçon Bather* (1808), demonstrate a careful calibration of the distortions of torso, right arm, and hand, thickenings and thinnings that in one sketch produce an awkward opening through which patches of flesh and ground are visible (figs. 34 and 35). In the finished painting, torso and arm are in close proportion, with obvious distortion remaining only in the flattened hand relegated to the bottom right corner of the canvas.

Retté's profile of Denis for *La plume* avoided the issue of deformation altogether, probably in deference to Alphonse Germain, the symbolist critic who, in the same issue, engaged in a full-scale attack on artists he called the "deformers."[65] Germain had initially been impressed by Denis's "Définition," which he believed responded to his own theories of symbolist painting.[66] In an open letter to Denis, Germain even appeared to encourage deformation as a sign of a striving temperament at work: "Feel your way, wander, be incompetent!" he wrote with passion. "Nothing more human than failing!—but be Yourself and create. Then you will make art, even without beautiful forms, without charming color, like the Egyptians and the primitives."[67] In November 1890, the exchange between Germain and Denis reverted to private correspondence, the critic writing to the artist, "I fervently wish to make your acquaintance, a writing, aestheticizing painter such as yourself is so rare!"[68] Later that fall, before he had met Denis or probably seen any of his work, Germain assured Denis of his continuing support for the artist's theoretical endeavors.[69] Finally, in late February 1891, the two met when the critic traveled to Denis's home in Saint-Germain-en-Laye. As he prepared for the Salon des Indépendants, feverishly, as he wrote to Lugné-Poë, *Décor* probably remained in the Paris studio.[70] Impressed both by their conversation and by the works he saw in Saint-Germain-en-Laye, Germain subsequently renewed his pledge to help disseminate Denis's theories, but he requested clarification: "your paintings of such sweet mysticism transported me into the beyond," he mused, so much so that he feared he had missed some crucial aspect of Denis's "research."[71]

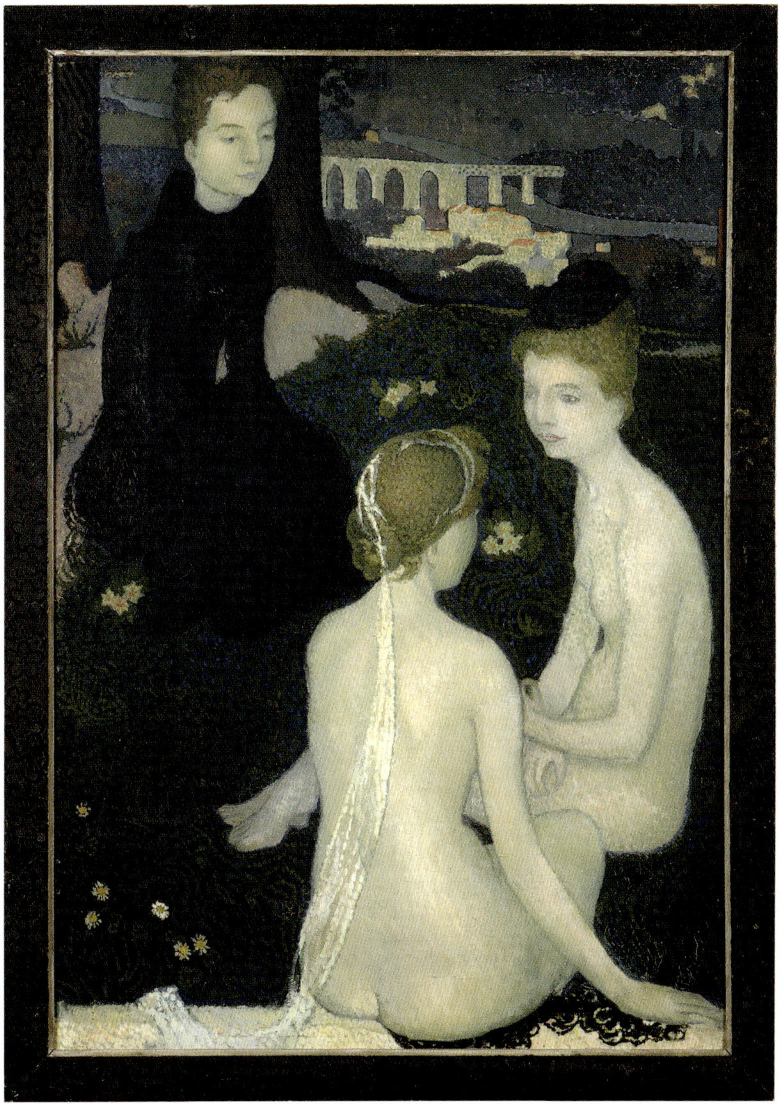

33 Maurice Denis, *On the Theme of Trinitarian Evening by Ad. Retté: Ornamental Painting*, 1891. Oil on canvas, 105 × 72 cm. Private collection.

While Germain's article "Modernism and the Beautiful," published the day before the Indépendants opened, echoed the antinaturalist rhetoric of Denis's "Définition," his review of the Indépendants registered a sudden and much diminished enthusiasm

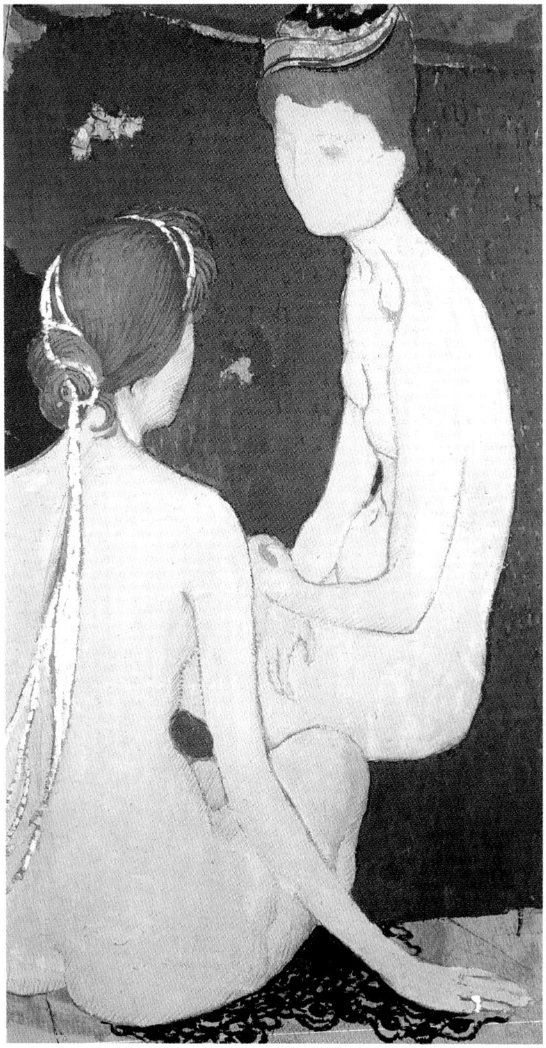

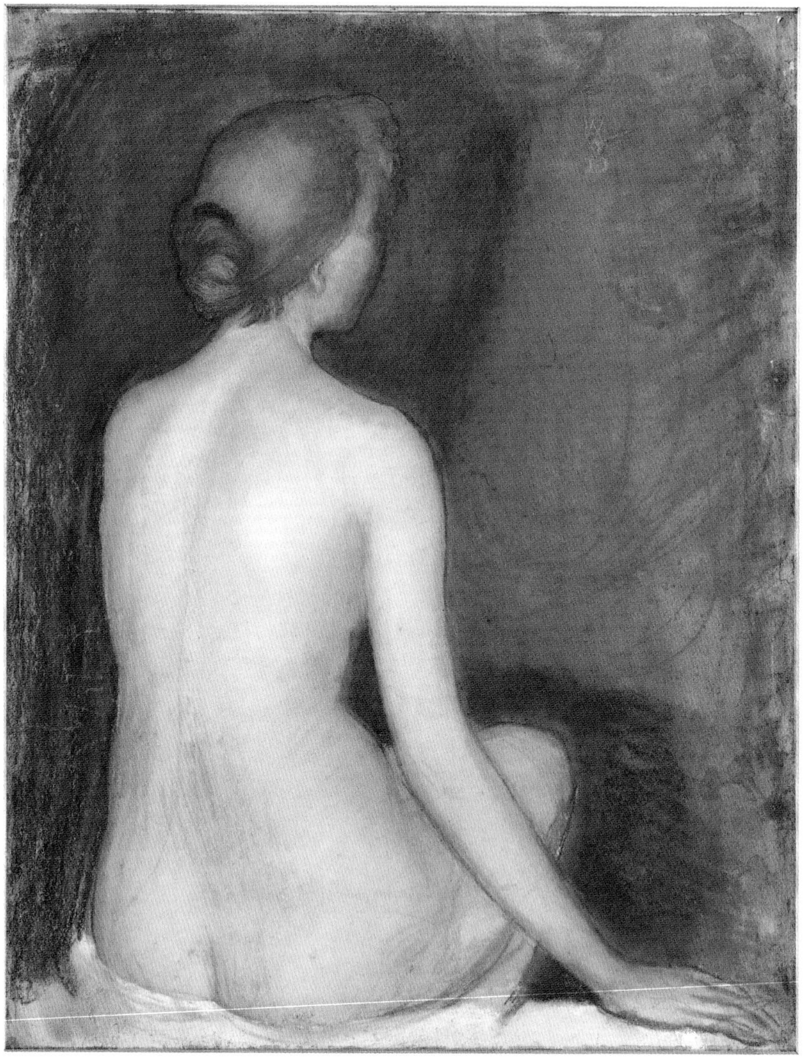

34 Maurice Denis, study for *Trinitarian Evening*. Glue-based paint on cardboard, 81 × 44 cm. Private collection.

35 Maurice Denis, *Nude Woman, Seen from Behind, Turned to the Right* (study for *Trinitarian Evening*), 1891. Pastel and charcoal on paper, 58.2 × 74.5 cm. Musée d'Orsay, Paris, RF14915.

for Denis's work, which I suspect resulted from the critic's first encounter with *Décor*.[72] Germain noted the "superiority" of Denis's "icons" but lamented an oversimplification and lack of modeling, and he made a number of pointed remarks about the dangers of such "neoprimitive" techniques. "Neither naïveté nor simplicity excludes knowledge," Germain wrote. "Return to babbling! . . . To no longer respect form because art is no longer reality! Would all masterpieces since the Renaissance be only magnificent aberrations?"[73] Nevertheless, Denis escaped Germain's most vituperative criticism, which was reserved for Bonnard, Louis Anquetin, J. F. Willumsen, and Gauguin, all of whose excessive deformations Germain characterized as monstrous, savage, and puerile.

This did not bode well for *La plume*'s special issue on painting, which Germain was coordinating and which provoked anxiety among the Nabis in the summer of 1891. In a long letter to Denis, Vuillard indicated that he had recently seen Germain in order to pass along Denis's drawings for the issue. Intimating that Germain was no longer on their side, Vuillard expressed concern over this loss of critical support and wondered, in the absence of a public moniker, what *La plume* would call them: "When he let loose the name *néo-traditionnistes* as a title for the article . . . I grimaced and asked him to do nothing until you had given your opinion. . . . He also asked me to ask you if you had any changes to make in your theories since last year. . . . You were telling me recently that it was likely that you wouldn't be quite so categorical if you had to do it over again."[74] Given Germain's earlier promise of support, that they were now out of touch and that the reticent Vuillard was acting as a go-between suggests a falling-out over practice, theory, or probably both. Judging by Vuillard's summing up of their recent conversations, it also sounds as if Germain's criticisms had hit their mark and that Denis was wavering.

Vuillard had good cause for concern. When the issue appeared, Germain conserved "néo-traditionnisme" in the body of the text in which he discussed the work of the Nabis, but he chose the somewhat surprising title "Theory of the Deformers," and he made clear that he intended to both "explain" and "refute" that theory. Germain reiterated that he admired the "Définition" and expressed a now toned-down appreciation for Denis's painting, but on the whole he claimed that the works of the "deformers" had failed spectacularly in comparison to their theories. In choosing to follow Cézanne, Gauguin, Van Gogh, and the "Japonais," their deformations were nothing short of pathological, the sign of an atavistic return to an unhealthy, dangerous state of ignorance. "So long as they confuse the simple with the deformed, so long as they depart from natural laws, all decorative beauty will escape them," Germain wrote hyperbolically, "and instead of giving birth to an original work, they risk aborting." He criticized their admiration for Gauguin's "diseased mentality," Van Gogh's "hypertrophic vision," and Cézanne's "dysphoric" manner. Their works were the true signs of a "degenerate race" better suited to a "teratological" atlas.[75]

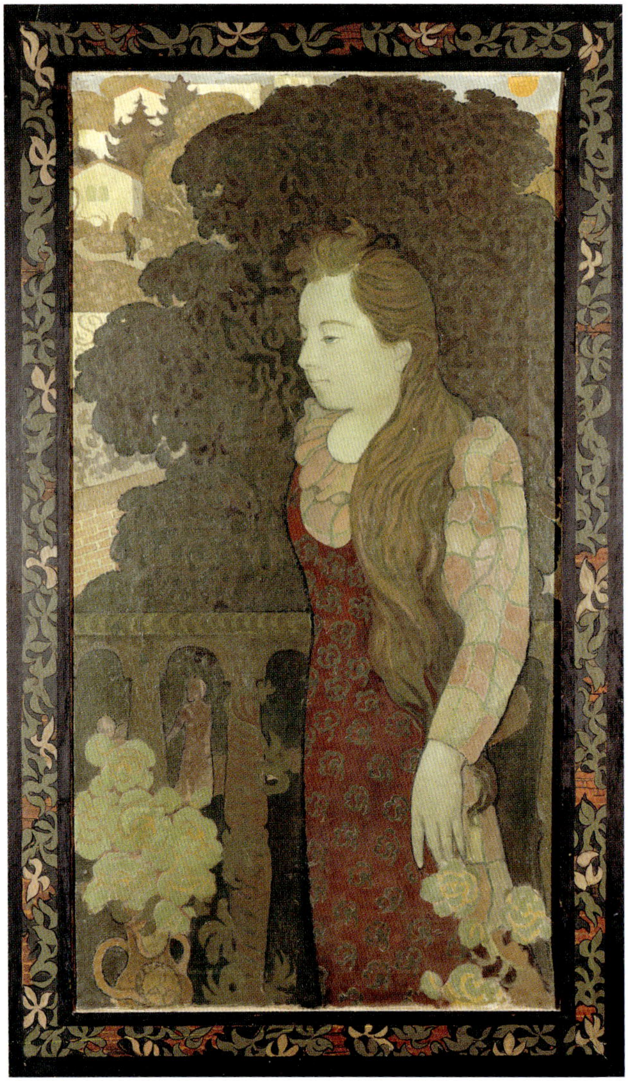

This must have been a blow to Denis, coming not from an anti-avant-garde source but from a symbolist critic who had extended a hand of support, a man with whom he had much in common, including a deep admiration for Puvis de Chavannes, classicism, and a "Latin" aesthetic. The impact could hardly have been lessened by Germain's unenthusiastic article on the "Chromo-luminaristes" or by Retté's much more positive profile of Denis. In the fall of 1891, using the "aesthetics of normality," Germain ensured that the meaning and significance of deformation within discursive constructions of symbolist visual practice were far outside Denis's control and equally far from being a neutral indicator of the artist's sincerity of temperament. Using deformation as a rhetorical weapon against its apparent practitioners, Germain forcefully reminded readers of its bodily and pathological associations, rejecting any attempt to neutralize or normalize its practice.

Yvanhoé Rambosson took Retté's and Germain's criticisms to their logical conclusion by directly questioning the use of a theory of deformation for symbolist art. In a review published in 1892, he expressed doubts about whether "a durable and anything but perverse, precious and fleeting art, can be founded on the principle of deformation." Rambosson found positive things to say about Denis's paintings, which "make the best of his theory's faults," but nevertheless recommended that the artist "return to tradition, not the tradition of the École, but the classical tradition." In fact, he perceived that Denis was already starting to reject the pathological "perversity" of deformation in favor of a normative classicism: "His submissions seem to suggest a return to the norm of the eternal laws of form."[76] Both *Trinitarian Evening* and the painting originally titled *Portrait of a Young Girl in an Evening Décor* validate Rambosson's claim that by 1892 Denis had distanced himself from his earlier "pathological" deformations. A stiff portrait of Denis's new fiancée, Marthe Meurier, posed against the landscape of Saint-Germain-en-Laye, the painting (sometimes known as *Symbolist Marthe*) upends and reverses Anna Thibout's pose while clothing the figure (fig. 36). The image has an air of gothic penance about it, heightened by the painted frame and *hortus conclusus* setting. Placing himself in the work as a tiny caricatured face on the vase in the lower left corner, Denis offered the painting in atonement for the sins of *Décor*, a virginal image of his "bien aimée," whom by early 1892 he knew not just spiritually but also carnally.[77]

36 Maurice Denis, *Portrait of a Young Girl in an Evening Décor* (also called *Symbolist Marthe*), 1892. Oil on canvas, 130 × 71 cm. Niigata Prefectural Museum of Modern Art, Niigata.

Defending Deformation with Positivism

Retté's, Germain's, and Rambosson's critiques all, I believe, contributed to Denis's retreat from the practice of deformation and, at least temporarily, from public theorizing. Between late 1890 and early 1895, Denis published very little, instead consolidating his position as a leading painter.[78] When he returned to making public statements on art, in a March 1895 review of works by the Pont-Aven artist Armand Séguin, he discarded his pseudonym for the first time and immediately prioritized a thoroughgoing theorization of deformation. Addressing Germain directly, he listed the various names under which the symbolist vanguard had been known, identifying Germain as the originator of "Symbolists and Deformers" but simply ignoring the fact that the critic had used "Deformers" so pejoratively. Taking aim at Germain, he went on to criticize "young *littérateurs*" and "learned *collégiens*" who confused discussions of art by speaking of norms and ideals, and accused these pretenders of "invok[ing]" abstractions such as "the Laws of nature and the Norm of Harmony."[79] By insisting that Séguin maintained at once "a very pure and . . . classic sense of form" and "a very correct sense of deformation," Denis belatedly entered into dialogue with the symbolist critics of deformation.[80] Now prepared to counter their claims, Denis offered objective deformation as a positivist corrective to subjective deformation.

In 1896, Denis insisted that the work of the Nabi artists had been the "immediate result of positivist philosophies, then in vogue, and the methods of induction that we held in such great respect."[81] This long-discounted claim, first explored by art historians Filiz Burhan and Jean-Paul Bouillon, is supported by references to positivist thinkers in Denis's journals, published writings, and an unpublished notebook that he used to prepare for his philosophy exams in 1887–88.[82] Denis's notebook reveals the various ideas to which the studious youth was exposed, demonstrating his formative understanding of experimentalism, experimental psychology, and in particular the role of objective methods, including the pathological method, in the search for knowledge.[83] It offers a scientific rationale for rejecting subjective deformation and for deploying objective deformation as a methodological corrective.

At the outset of his notebook, as a kind of exhortation to himself to take his philosophical studies seriously, Denis, citing Francis Bacon, noted that "truth is the goal of thought . . . to think well is to form true ideas . . . an idea is always real, true." A few pages later, in notes taken during a course given by Émile Boirac, Denis recorded "the goal of science" to be "proving in a clear and incontestable manner the truth to everyone" and classified the sciences as either abstract or concrete.[84] The abstract sciences demanded mathematical methods, while the concrete sciences, including the physical, natural, and "moral" sciences, necessitated experimental ones. Denis's introduction to scientific method concluded with Claude Bernard's definition of induction: "Induction is to deduce under specific conditions: it is to make a provisional deduction that demands experimental verification."[85] There follow extensive notes on psychology informed by second-generation positivism, which held up the new experimental psychology as an alternative to previous introspective methods. Distinguishing between subjective and objective methods, Denis diligently recorded a proof refuting the claims of subjective introspection and noted that objective psychological methods were more "modern" ways of attaining knowledge. Among objective methods, he highlighted the pathological method, observing people "other" than the "adult, civilized, scholar," including "original natures" and children, as well as instances of pathology, including blindness, deaf-muteness, somnambulism, hallucination, and madness. As key authorities for the new modern, objective methods of psychology, Denis dutifully noted the names of Charles Darwin, Hippolyte Taine, Alexander Bain, Wilhelm Wundt (misspelled "Wendt"), and Théodule Ribot (initially misspelled "Ribaud" and subsequently corrected).

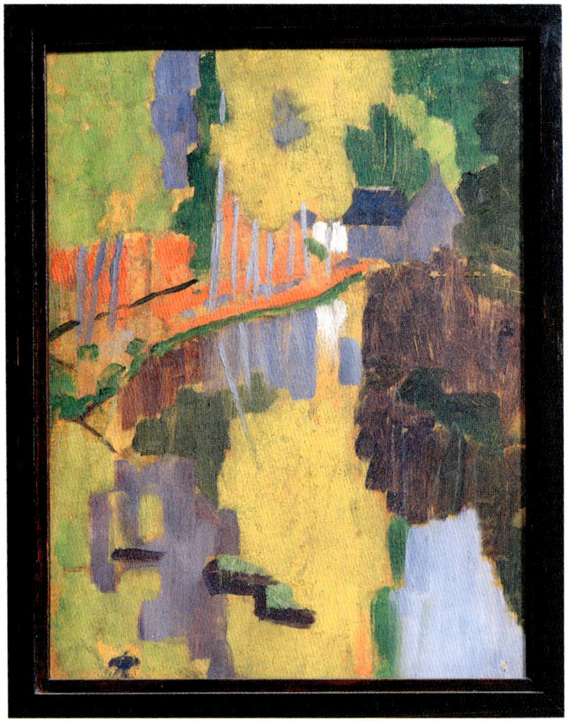

37 Paul Sérusier, *The Talisman, the Aven River at the Bois d'Amour*, 1888. Oil on wood panel, 27 × 21 cm. Musée d'Orsay, Paris.

Years later, Denis recalled studying for his exams on the train into Paris, whereupon, arriving at the Académie Julian, he discussed what he was reading with Paul Sérusier, who had passed both his philosophy and science baccalaureate exams in 1883. Denis saw Sérusier as a mentor, "a mind of superior culture . . . an intellectual and artistic guide."[86] When Sérusier first showed the small painting later called *The Talisman* to fellow Julian students, it was not as an independent artwork but as a demonstration of the initial stages of Gauguin's working method (fig. 37). Perhaps only in retrospect, it provided one answer to the crisis of method faced by artists who sought to break with academic and naturalist methods in the name of symbolism.[87]

Immediately after painting *The Talisman*, Sérusier confessed himself confounded by Gauguin's approach, which he viewed as willfully "illogical" and "childlike." What bothered him, he wrote to Denis in the summer of 1889, was how to judge aesthetic value in the absence of visible nature as a reference point; what Gauguin was producing struck Sérusier as far too removed from nature to advance as a model.[88] But as he continued to work with Gauguin, Sérusier began to see a way to rationalize Gauguin's practice. Taking the form of a philosophical theorem set out in a letter to Denis, Sérusier's elaboration of Gauguin's approach became the basis, as Denis readily acknowledged, for the "Définition" itself. Above all, Sérusier insisted, there were "immutable principles," the universal truths of art, which could be either deduced within the self—so long as the self, in an echo of Jules Laforgue, had not been "spoiled" by too much academic training—or induced by observing past art. While the first method was more direct, it was also more difficult, given the influence of academic traditions. It was more practical to proceed by induction, by studying past works by artists who had presumably been able to deduce the immutable principles from within themselves.[89] Sérusier suggested that it was up to each artist to decide how to attain the immutable principles. At any given moment, an artist could proceed by the more direct method of deduction, exploring the self, or could choose the less direct path of observing the work of other artists. While Denis proceeded more often by induction, his "néo-traditionnisme," fundamentally based on the study of the past and especially the work of the *primitifs*, other Nabi artists such as Vuillard, who grappled with Sérusier's ideas along with Denis, developed more personal methods of deduction.

Claude Bernard's theory of induction, copied into Denis's notebook, argued that induction was merely provisional deduction proved true through experiment. With such a definition in hand, Denis could easily forge a link between Sérusier's two methods of discovering the immutable principles. Experimental method was both the way to prove that the immutable principles deduced from the self were true and also a model for comparing an individual's "results" with those of past masters. This suggests a twin approach to the genesis of an artwork: creation derived from a self imagined as unspoiled, and ongoing comparison against a canon of great works. In other words, Denis could promote an artistic practice based on self-deduction of an internal "other" and also experimental verification of results through comparison to other works. If objective psychology recommended the study of "others" (understood as alterities and/or cases of pathology) as a pathway to truth, an artist's intent to seek the immutable principles would naturally lead to a focus on "others" both within and without, the "unspoiled" "original nature" of the self, as well as on experimental psychology's trio of privileged objects of psychological study, "madmen, primitives, and children," most often construed by the Nabis as *primitifs*—French, Italian, Breton, Tahitian, Egyptian, or Japanese. As Denis claimed just before the close of the century, "Symbolism . . . was the most strictly scientific attempt to make art . . . based on a truth confirmed at once by tradition and experiment."[90]

In the early twentieth century, Denis more explicitly connected the deformations of the generation of 1890 with a scientific approach to creation, linking a "primitive," childlike attitude toward the world with both a scientific stance and the nescient curiosity of an artist like Ingres, whom Denis had come to see, along with *primitifs* such as Fra Angelico, as the exemplar of "correct" deformation. The "primitive" artist, Denis insisted, sought not to represent the appearance of objects but to explore their qualities like a scientist: "he looks at them with a new soul and his curiosity is as much that of a scientist as an artist." Maintaining this childlike curiosity, much as the experimenter maintained doubt or nescience, enabled the artist-scientist to both analyze and synthesize, with the ultimate goal being to arrive at universal truths. Denis came to revel in the paradoxes of this claim for knowledge through ignorance, artistic vision through the vision of a child, playing on the dual meaning of *expérience:* "Their inexperience comes from their hundreds

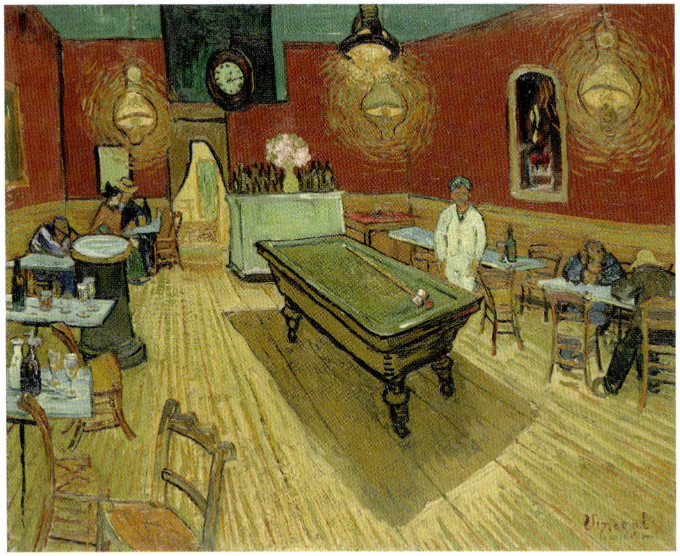

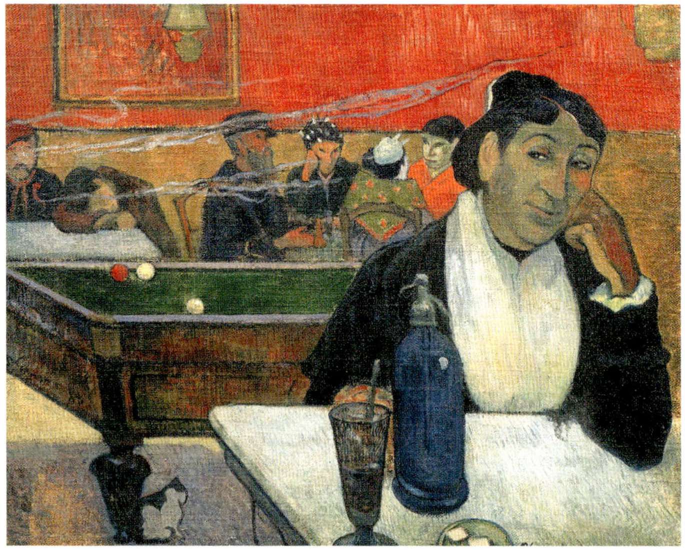

38 Vincent van Gogh, *The Night Café*, 1888. Oil on canvas, 72.4 × 92.1 cm. Yale University Art Gallery, New Haven, Connecticut, Bequest of Stephen Carlton Clark, B.A. 1903, 1961.18.34.

39 Paul Gauguin, *Café at Arles*, 1888. Oil on canvas, 72 × 92 cm. Pushkin Museum, Moscow.

of experiences [or: experiments]; their ignorance from their forever learning."[91] Denis might have added that deformation enabled the formation of truth, that the deformation of nature, even if pathological—indeed, precisely because it was pathological—provided the opportunity to know nature in all its forms.

When Denis introduced his theory of the two deformations in 1895, he situated objective deformation as a methodological corrective to subjective deformation, which he conflated with Aurier's symbolism, Zola's definition of the work of art, and Bacon's formula *Ars est homo, additus naturae.* While subjective deformation was a necessary element, it had, he argued, little claim to truth unless its excesses could be checked by objectivity. The best painters of the generation of 1890 already practiced both subjective and objective deformation, Denis argued, "probably unconsciously . . . like all Great Masters."[92] Definitively formulating the theory of the two deformations in 1909, Denis established a continuum between subjective and objective deformation. At one extreme he placed Van Gogh, an artist whose deformations were almost solely an expression of his own temperament. At the other extreme lay Gauguin, whose excessively objective deformations manifested themselves in an almost academic concern for harmonious decorative arrangements.[93] Van Gogh's *Night Café* and Gauguin's *Café at Arles* provided an ideal illustration of their differences (figs. 38 and 39). In 1909, Denis observed in his journal, "Gauguin's [painting] with the Arlésienne, the bottle and smoke—dark and well established; the other by Vincent, bright and all askew, but so full of feeling and life that the Gauguin seems academic."[94] While Van Gogh embodied the romantic and at times irrational artist, Gauguin, in the decorative and logical rigor of his compositions, demonstrated what Denis saw as a persistent rationalism. Relying on the burgeoning critical tendency to cast Van Gogh as mentally ill, Denis pathologized subjective deformation run amok, shunted it into the past, and condemned what he increasingly saw as a dangerous Baudelairean liberation of imagination that licensed even the most excessive "caricatural transpositions," including, it would seem, his own *Décor.*[95] The 1909 essay also marked the apotheosis of Cézanne in Denis's thinking. While Van Gogh and Gauguin worked at the two poles of deformation, Cézanne worked precisely in the middle, balancing subjective and objective deformation such that his work proposed a new kind of classicism.

Positivist Modernism and the Will

By asserting an ideal synthesis of subjective and objective deformation as a new kind of truth-producing practice counter to the "laws" and "norms" of a previous way of making art, Denis effectively regulated deformation. He layered onto the continuum of the normal and the pathological, form and deform, a continuum between subjectivity and objectivity. If managed correctly, an artist could achieve what Denis called a new, "healthy" classicism, over and against a "neurasthenic" impressionism.[96] But in suggesting that an artist could balance subjective and objective deformations, the analogical structure of a continuum created an intractable problem for the will. Scientific discourse had constructed the objectivity-seeking scientist as already split, as willing himself to "willessness," removing, as far as possible, his active will from his observations and experiments.[97] Denis proposed objective deformation as a corrective to an all too active subjectivity, not, however, through a passive stance, but through an implicitly active, willful rationalism exemplified by Gauguin's quasi-academicism. If an artist were then to actively balance subjective and objective deformations, was not the will once more engaged in this balancing, the artist then doubly split? In making deformation useful for form, in situating deformation as an "experiment prepared by nature," Denis revalued deformation by revaluing pathology, effectively purifying deformation such that it became synonymous with form, but at the same time he created an equivocal position for the will. From the early twentieth century on, Denis's texts would curiously be divided between those that insisted upon the fundamental role of the will alongside a normative classicism, texts often concerned with situating the Italian masters in relation to new forms of sacred art, and those retrospective texts on symbolism that elided the will in order to theorize deformation, an elision that would exacerbate Denis's own ambivalence about modernist distortion as he saw it emerging in the work of such artists as Henri Matisse.

When Roger Fry translated Denis's 1907 essay on Cézanne, he rendered *déformation* not as *distortion* but more literally as *deformation*, conserving the relationship between form and deform crucial to Denis's meaning. But when he came to write his own texts for the Grafton Gallery exhibitions, Fry instead chose *distortion*, and in a telling aside, echoing Zola, addressed his readers' and, I suspect, his own lingering anxieties even over this less bodily term: "If you boggle at the words misrepresent and distort," he wrote, "you may substitute mentally whenever I use them, the consoling word idealise, which comes to exactly the same thing."[98] Predictably, it was Walter Sickert who needled Fry for failing to tackle the issue of deformation head on. Accusing Fry of stealing his ideas from Denis, Sickert insisted that "a law as to deformation or distortion" should be established once and for all, and then proceeded to attempt this, arguing that deformation was an essential quality of art so long as it was the inadvertent result of sincere efforts. Deformation or distortion was allowable, desirable even, so long as the artist had no intention to deform.[99] While Fry dismissed Sickert's critique as ridiculous, he continued to equivocate on the role of the will when it came to formal distortion, elsewhere praising Cézanne for working as the "passive, half-conscious instrument of some directing power."[100] Similarly, Clement Greenberg insisted that Cézanne's distortions were due to "a growing compulsion, more or less unconscious, to adjust the representation in depth to the two-dimensional surface pattern."[101] Fry and Greenberg required their modernist master to be fully in control of his faculties but also to submit himself to the higher power of his unconscious in creating his art. This admittedly difficult position excused artists like Cézanne for their deformations, which were thus situated as the unintended by-products of their at once willful, rational, logical, sincere, and necessary practice. This tension between the willed and the unwilled undergirds what Shiff called Cézanne's "technique of originality," a practice that he argued could be understood paradoxically as "making a find."[102] Despite Denis's rhetorical cunning (or perhaps I should say his rhetorical convolutions) and his strategic recourse to positivism, the specter of the pathological continued to haunt modernist deformation.

From Rome, in 1898, Denis wrote to Vuillard, "What is important in a work of art is the fulfillment of the artist's effort, the power of his will." Vuillard, admitting that his own experience made it impossible for him to abide by this theory of superintentionality, replied, "whether by a weakness in my nature or a curable disease, I really only taste this power of the will rarely, this effort that you like to talk about. . . . In this wild gust of wind, I only have my instinct, pleasure, or more accurately my ready-made satisfaction to guide me." In response, Denis admitted that they had arrived at an impasse: "The problem of the will cannot be posed in the same way for you as for me."[103] I explore this exchange further in the following chapter, but for now, suffice it to say that I imagine Denis and Vuillard had always been on different sides of this issue, even as early as 1889–90, when they encountered Sérusier's ideas, and during the following winter, when they shared the Paris studio with

Bonnard and all three probably sketched the red-haired Anna Thibout.[104] Because the female nude is relatively rare among Vuillard's Nabi works, scholars have argued that nearly all of his early nudes relate somehow to Denis's *Décor.*[105] *Reclining Nude* transposes Thibout to an enormous bright orange couch, deemphasizing the snake-woman as a symbol of universal attraction by placing her in a disturbing domestic setting on an engulfing and unnaturally colored piece of furniture (fig. 40). Like Denis's first sketches, but unlike the final painting, she is overtly eroticized; her smooth curves, gracefully arched lower back, orange pubic hair, left arm held just above rather than resting on her thigh, and the hint of movement suggested by the pentimenti around the lower torso produce an intimate, sensual, yet disconcerting image of autoeroticism.

In another pastel, Vuillard further schematized Denis's horizontal bands of color into flatter and more concise strips of deep red, yellow, and blue pastel, simultaneously eliminating the elongations to present the female form as a mountainous lump of flesh (fig. 41). The pastel relates to a charcoal and watercolor study more evidently done from the living model, which helps to clarify the model's position (fig. 42). While her lower body is tipped parallel to the picture plane, the upper body is torqued perpendicular to the plane, which splays the breasts away from each other and away from the center of the body, effects that Vuillard exaggerated in the pastel and that Bonnard later played with in *The Mantelpiece.* In Vuillard's pastel, the left two-thirds of the body have been radically abstracted; covering the right third of the pastel, the remainder of the work bears no resemblance to a body. The head is immobile and mask-like, while the splayed breasts, enormous red nipples, and stomach roll (the antithesis of an ephebic stomach) are in stark contrast to Denis's representation of a lithe torso and pert breasts. While the left side of the work accords with Denis's insistence on a painting as above all a décor, an arrangement of colors on a flat surface, the right side seems to revel in unidealized fleshiness, a disjuncture in the work that suggests the simultaneous presence of the willed and the unwilled, the conscious and the unconscious. Not without, I suspect, a certain glee, Vuillard satirized the female nude both as an object of titillation and as a "study in subjective deformation," bringing his friend back from the lofty heights of theory down to earthbound material experience.

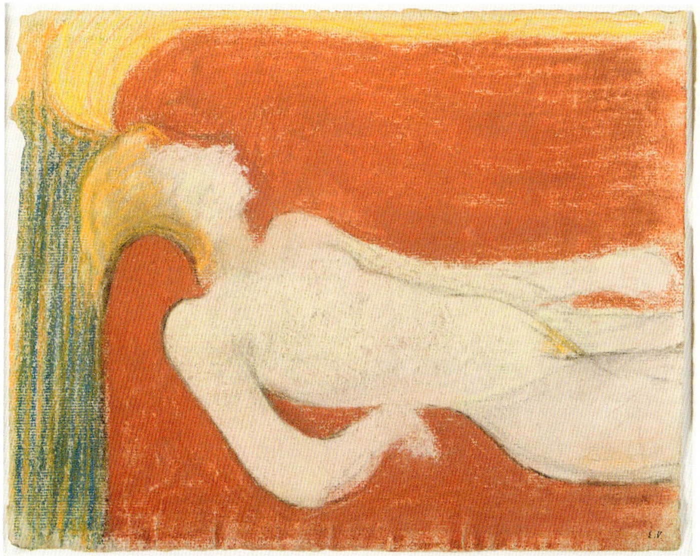

40 Édouard Vuillard, *Reclining Nude,* ca. 1891. Pastel on paper, 24 × 30 cm. Private collection.

Writing in homage to Bonnard toward the end of his own life, Denis hoped that viewers would see more in his friend's works than just their deformations: "Most of his

41 Édouard Vuillard, *Reclining Nude*, ca. 1891. Pastel on paper, 13 × 32 cm. Current whereabouts unknown.

42 Édouard Vuillard, *Reclining Nude*, ca. 1891. Charcoal and watercolor on paper, 11.2 × 34.2 cm. Private collection.

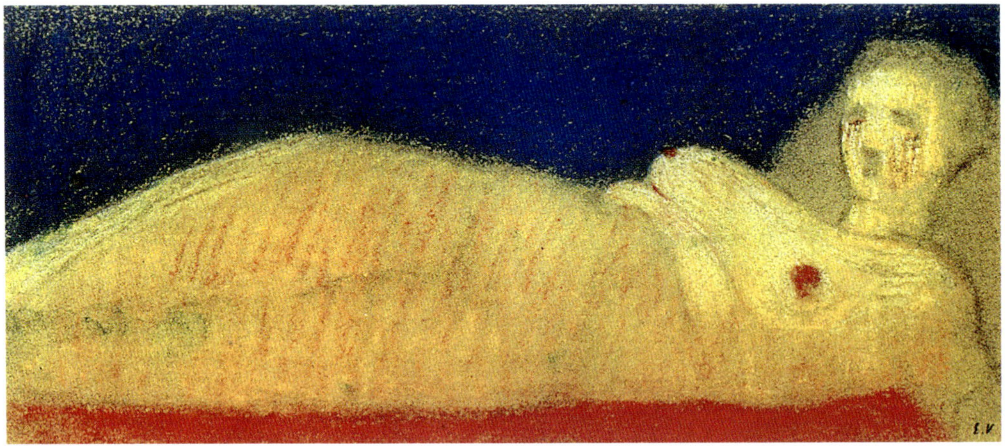

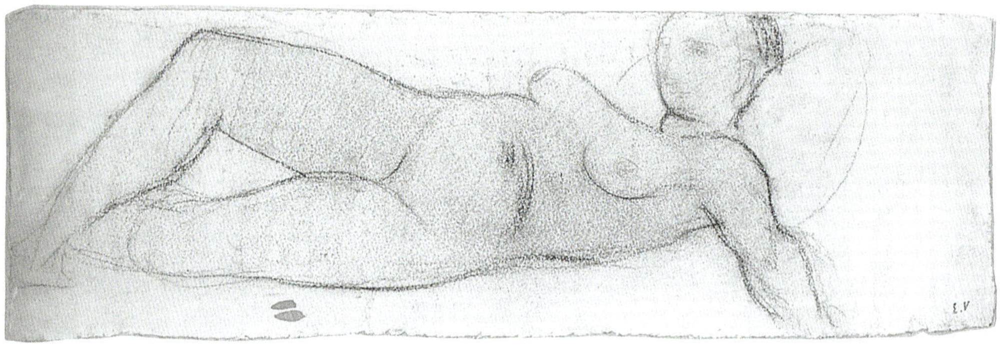

young admirers only appreciate, I fear, the apparent disorder of his composition, the audacity of his dissonances, the extreme fantasy of his imagination. They love the unexpected, the spontaneity of his technique, his deformations, his negligence, more than his profound qualities of color and drawing. Because his system is to not have a system, he easily escapes analysis. . . . His painting . . . is sometimes naïve and at other times complex and reasoning."[106] Denis underlined the contradictions in Bonnard's work, its naïveté and complexity, and the way it resisted analysis. But he feared that its bizarreness and its deformations were the main draw for younger admirers who did not perceive the rationality of this approach and perhaps did not recognize the true role that deformation had played in both Bonnard's and Denis's practice. As a method, deformation had functioned to counter the academic norms that Denis had early on sought to pathologize in the "Définition," and had provided an experimental pathway to universal truths deemed difficult to determine, but only so long as its subjective excesses were held in check by a putative objectivity. In the last year of his life, Denis reiterated the rationality of deformation, the systematic logic behind the appearance of no system.

I have not sought to insert Denis's *Décor* into a trajectory of modernist nudes from Manet's *Olympia* to Matisse's *Blue Nude*. Nevertheless, the absence of *Décor* from the story of modernism, whether accidental or intentional, is precisely the point. *Décor* and its "pathological" subjective deformations were irritants that Denis attempted to manage by purifying deformation with objectivity, thus encouraging modernism's discursive split between form and content. *Décor*'s absence from the histories of modernism allegorizes the anxieties and ambiguities, particularly those surrounding the will, that arose as consequences of that split. The debates around deformation, ugly nude women, and the "School of Cézanne" would all soon be rehearsed in the criticism surrounding Matisse and the fauves. The initial reception of *Décor*, Denis's repudiation and positivist circumscription of subjective deformation based on the paradigm of nature's experiments, and perhaps the déjà vu he experienced in 1905 go some way toward explaining the vehemence of his reaction that year.[107] In criticizing Matisse's deformations, he attacked not the deformations per se but Matisse's thought processes in arriving at them. Tracing the fauves' hubris to a belief that everything is knowable and accusing them of willfully ignoring the art of the past, Denis argued that Matisse's deformations were too theoretical, that they amounted to hypotheses without the essential verification of experiment. By ignoring tradition, which Denis argued was the best "safeguard against the vertiginous heights of reasoning," Matisse had taken the reaction against materialism to an extreme and had ignored the humility of second-generation positivists like Taine. "Matisse," Denis pleaded, "let's be objective!"[108]

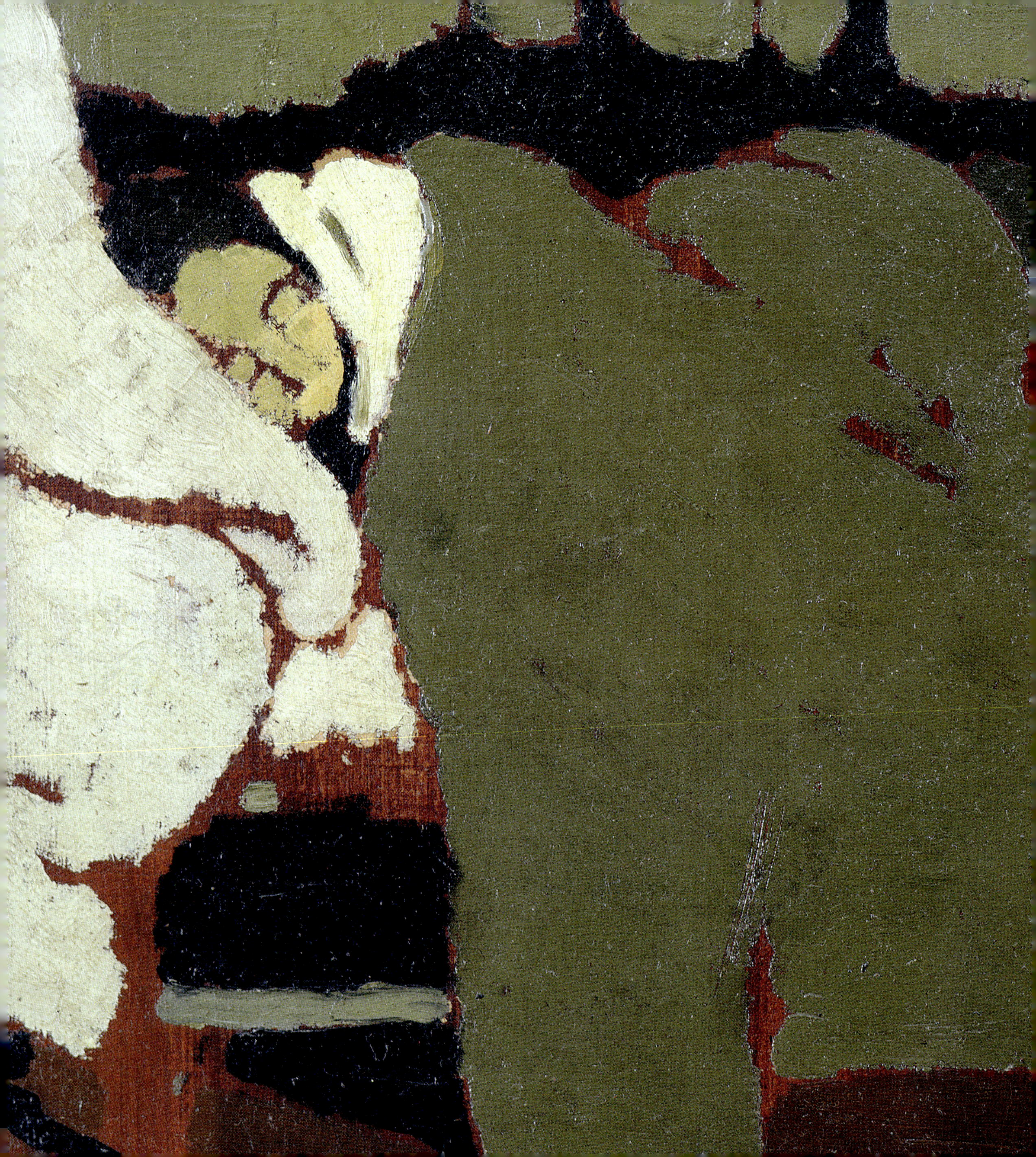

ÉDOUARD VUILLARD'S EXPERIMENTAL ARABESQUES

Among Édouard Vuillard's earliest exhibited works was a painting titled *Woman Sleeping* (*Femme couchée*, 1892, fig. 43). Now catalogued under the less gender-specific title *Sleep* (*Le sommeil*), it is one of a number of works that Vuillard made in the early 1890s depicting a single individual presumed to be a woman sleeping. Sleeping women were hardly unusual in nineteenth-century French art. Ordinarily licensed by a mythological narrative and more often than not depicted nude, they repeatedly offered, as in a number of paintings by Gustave Courbet, opportunities for realist or naturalist voyeurism.[1] Maurice Denis's *Décor*, the centerpiece of his début at the Salon des Indépendants of 1891, seems to fit this category of sleeping or semi-conscious female nudes. In the large study for the painting, Denis had carefully worked through various possibilities for the figure's state of somnolence (fig. 20). In the previous chapter, I considered Denis's *Décor* at length, as well as the gentle pastel parodies that Vuillard made of Denis's ambitious work, in conjunction with late nineteenth-century debates over symbolist practices, and in particular the hyperbolic critiques of deformation. Vuillard's *Woman Sleeping* was a very different proposition, but one nevertheless in dialogue with symbolist theories and determined by the assumptions underlying the experimental methods of the new French psychology.

With *Woman Sleeping*, Vuillard refrained from using the image of a sleeping woman as a pretext for representing an implicitly available female nude. In fact, with the figure tucked up in bed, he gave viewers no discernable body at all, and, the original title notwithstanding, one could hardly be sure that the work depicted a woman at all. That this constituted a kind of violence that Vuillard had enacted upon the figure was visualized in a contemporary caricature that pulled back the sheets to reveal a "Sleeping Woman in

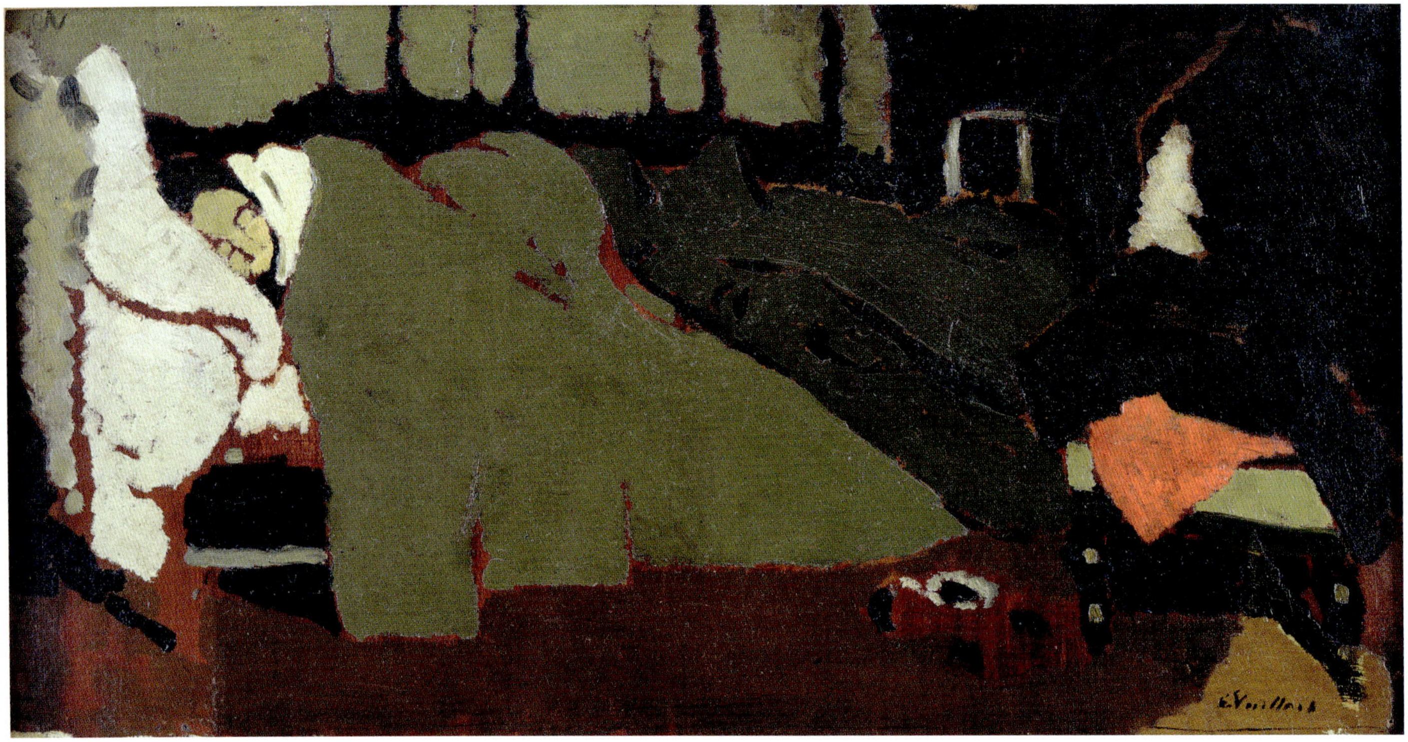

43 Édouard Vuillard, *Woman Sleeping* (otherwise known as *Sleep*), 1892. Oil on canvas, 33 × 64.5 cm. Musée d'Orsay, Paris.

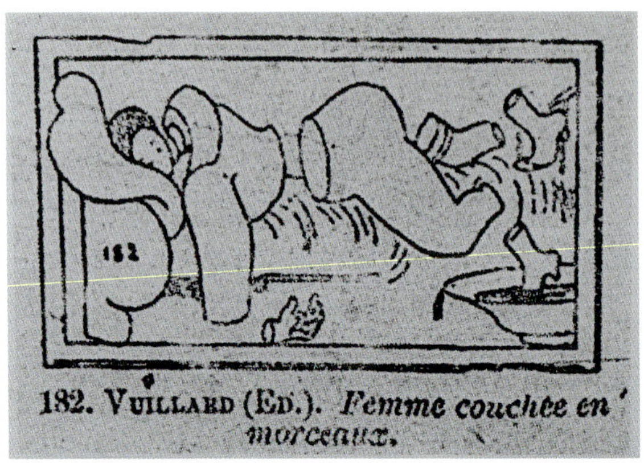

44 Caricature from Mirliton, "Troisième exposition des peintres impressionistes et symbolistes," *Le journal*, no. 9, literary supplement (26 November 1892), 1.

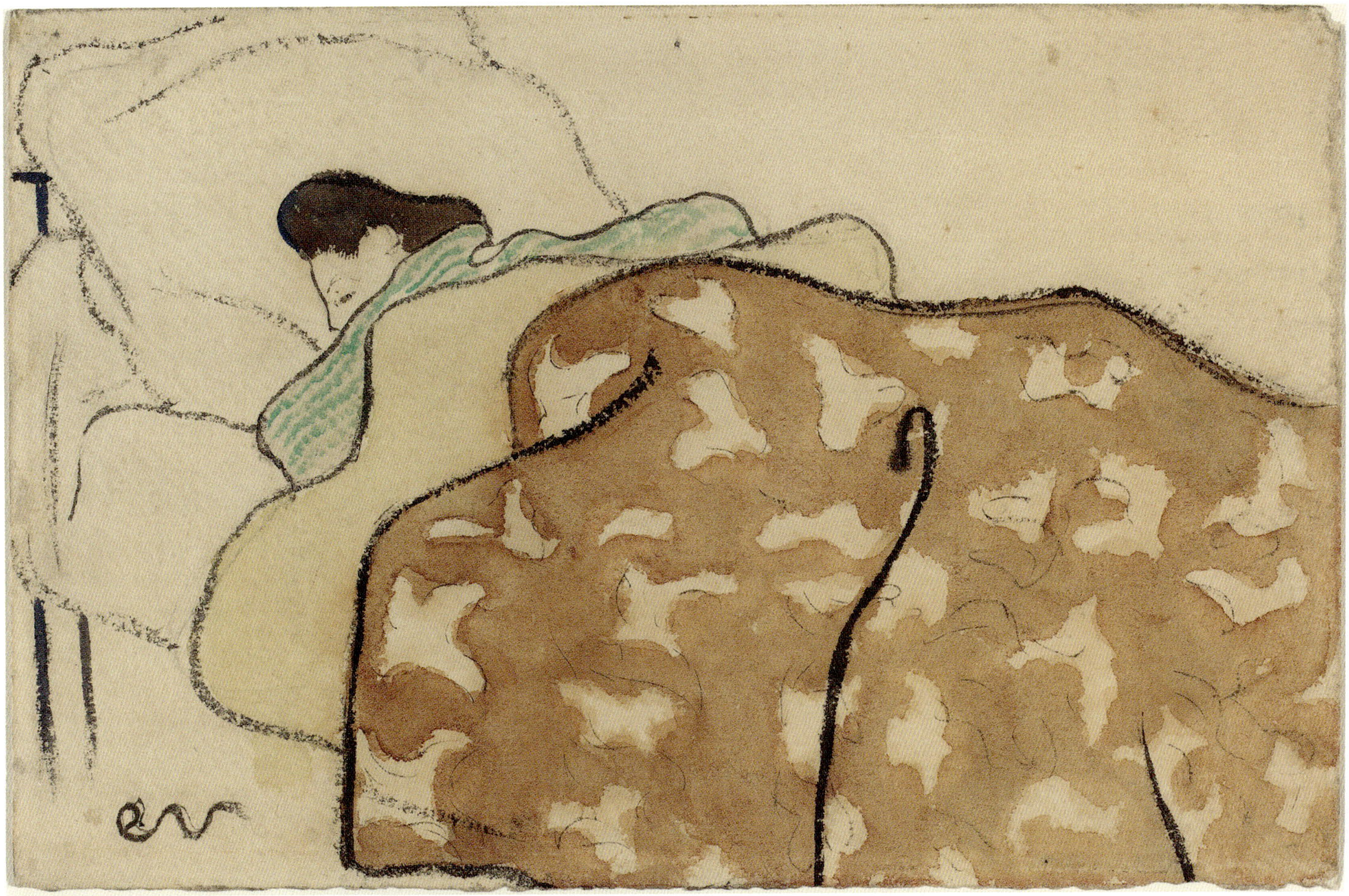

Pieces" ("Femme couchée en morceaux") (fig. 44). A grisly play on contemporary headlines,[2] the caricature highlights one particularly unusual formal element of the painting that constitutes the central problematic of this chapter. In the wake of cloisonnism, synthetism, and Nabi theories extolling the virtues of the arabesque line, Vuillard might have been expected to render the sinuous contours and folds of the bed linens and other elements in *Woman Sleeping* as distinct curving lines emphasizing the planarity of the support, as indeed he did in a related watercolor and charcoal drawing (fig. 45). But upon closer inspection, many of the painting's ostensible lines, especially the curves describing the folds just below the figure's head, are better described as negative linear spaces, areas between larger zones of paint where a thin ground, and even in places the bare canvas, shows through. These formal oddities, where painted areas seem to shrink from one other, opening up gaps in the material surface of the work, appear in many of

45 Édouard Vuillard, *Woman in Bed*, 1891. Watercolor and charcoal on paper, 14.7 × 22.8 cm. National Gallery of Art, Washington, D.C., Ailsa Mellon Bruce Collection, 1970.17.179.

Vuillard's early paintings. But they are especially visible in *Woman Sleeping* and even more manifest in a portrait of Vuillard's and Denis's friend, the actor and soon-to-be theater director Aurélien Lugné-Poë (fig. 4).

In his early journals (also known as *carnets*), which combine text and sketches, Vuillard not only grappled with naturalist, symbolist, and Nabi theories but also interrogated his own psyche with a view to establishing an artistic method.[3] Through this interrogation, he developed—or, more precisely, told himself that he was developing—a method that relied upon his instincts, his habits, and his will: in other words, a procedure that combined both unconscious and conscious mental functioning. This method began with quickly drawn arabesques signifying the direct and immediate observation of nature, "arabesques of observation," to borrow a term from Roger Benjamin.[4] And it proceeded by transforming those arabesques into the curving linear gaps that we see in works such as *Woman Sleeping* and the portrait of Lugné-Poë. Still reading as lines and often taking the place of contour, these gaps in fact resulted from deliberate and often repetitive acts of painting, the very antithesis of the automatic arabesques with which Vuillard began. In this chapter, I argue that this device is best understood as an arabesque of experiment, an element licensed and determined by the late nineteenth century's fetishism of experimentalism and, in particular, experimental psychology's reliance on a pathological method. After a lengthy period of questioning-cum-theorizing revealed in the *carnets*, Vuillard effectively dematerialized the arabesque line, producing a formal element that Denis might have been tempted to argue was pure idea, hypothesis rigorously proved true. But Denis could have no truck with the place that Vuillard had assigned to instinct and the unconscious in his symbolist visual practice. While Denis's experimentalism forged a path toward formalist justifications of modernist deformation, Vuillard's experimentalism reveals the epistemological foundations of the avant-garde's explorations of the unconscious.

Instinct, Habit, and Will

In 1898, well after the heyday of Nabi group activity, Denis received a letter from Vuillard, which Denis later called a "veritable psychological confession,"[5] that disclosed the important roles Vuillard ascribed to instinct and habit over and against the will in his artistic practice. In private, as if he were intimately familiar with Vuillard's *carnets*, Denis had previously critiqued Vuillard's habit of rapidly recording his impressions of the world around him.[6] Denis had come to believe that Vuillard had taken impressionism to a sensationalist extreme, reducing painting to "a kind of journal . . . a kind of stenography of everyday sensations," which had then become a license for any and all symbolist flights of fancy.[7] This "sensualist" obsession with subjectivity, Denis was convinced, derived from an outmoded theory of the role of nature in art, but its "primordial error" was its failure to cultivate a slow, willful, and intellectual creative process and consequently to produce significant works of art.[8] This failure, Denis claimed, was reflected in what he saw as Vuillard's inability to finish anything. By 1898, Denis was increasingly judging art like Vuillard's the unfortunate consequence of the theories that he himself had put forth in 1890. Hoping that his friend had already begun to see the error of his ways, Denis had invited Vuillard to join him in Italy that year to partake in the "lessons in strength" that Denis was then deriving from the Tuscan masters. "What makes a work of art important," Denis insisted to Vuillard, was the "artist's effort, the power of his will."[9]

Vuillard's response to Denis's invitation, his "veritable psychological confession," questioned whether it was really possible to untangle the complex psychological and physiological functioning of the will. In the mode of working that he knew best, Vuillard wrote to Denis, faith in the value of his ideas tended to abandon him, and he often experienced profound doubts and intense lassitude. At those times, he tended to rely more on his working habits, proceeding "in fits and starts" in a process that stripped "formulated ideas" down to "elementary" ones. Often, Vuillard confessed, he hardly knew the will at all and was guided rather by "instinct, pleasure, or . . . a . . . sense of satisfaction." Vuillard surmised that their differences came down to divergent understandings and experiences of the will that he ascribed not only to their distinct personalities and circumstances but also to their different educational backgrounds. Agreeing to disagree in a subsequent letter, Denis conceded that "the problem of the will cannot be posed in the same way for you as for me."[10]

"The problem of the will," as Vuillard intimated when he referred to differences in their education, was a centerpiece of French philosophical training. Although the philosophy curriculum had incorporated positivist methods and the new experimental psychology by the late 1880s, it nevertheless maintained its emphasis, derived from eclecticism, on a knowable self, constructed through the introspective methods still fostered in the classroom.[11] While Sigmund Freud would later give equal if not greater weight to the unconscious in the construction of the self, the lycée philosophy course, as Jan Goldstein has argued, continued to privilege consciousness in subject formation. The will, in opposition to instinct and habit, took its place in the curriculum as a characteristic, even defining, function of the conscious mind. At the same time that the curriculum foregrounded a healthy functioning of the will as constitutive of the self, however, it also included an outsize place for nature's experiments and for discussions of abnormal functioning, including automatism, somnambulism, and other supposed diseases of the will.[12]

Vuillard and Denis had in fact received very similar educations at the Lycée Condorcet in Paris, with one glaring exception. While Denis, along with a number of the Nabis, had taken the philosophy baccalaureate, Vuillard had terminated his formal philosophy training in order to concentrate on mathematics in his final year.[13] So, while Denis

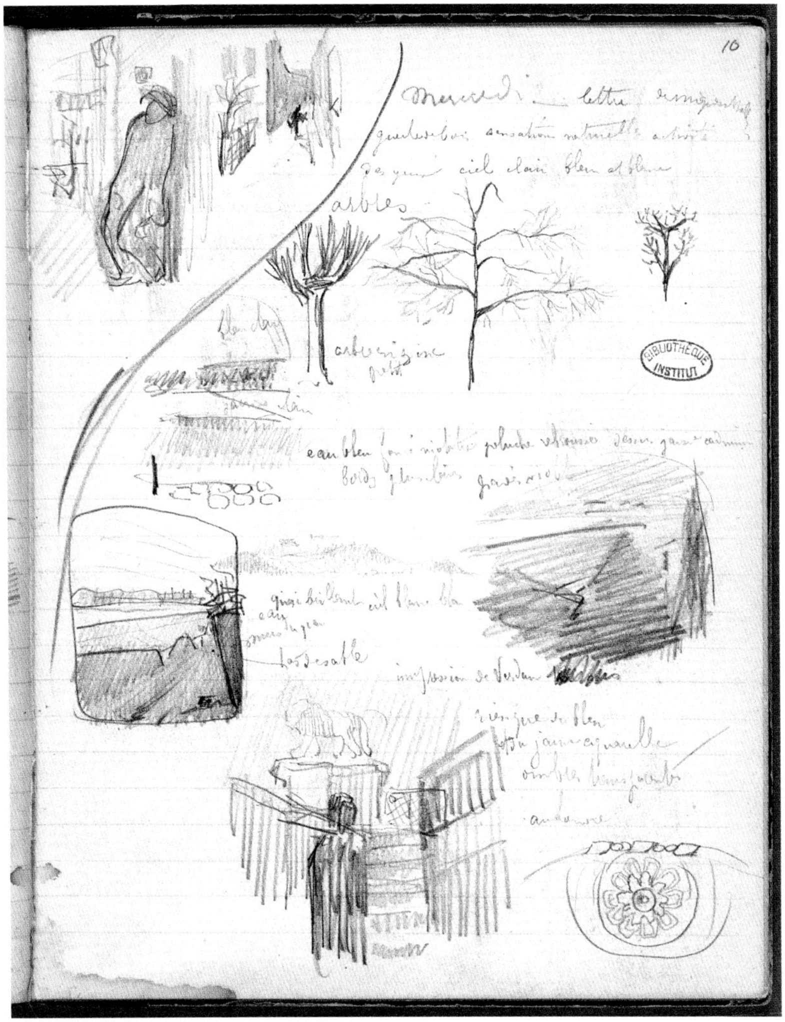

46 Édouard Vuillard, journal, 22 × 17 cm. Institut de France, MS 5396(1): *Carnet* I.1, fol. 10r [21 November 1888].

was learning to construct a robust self through an introspective method that privileged knowledge about the conscious functioning of the will, Vuillard was conscious of constructing a self at one remove from this particularly influential French institution. In 1898, with the benefit of at least a little hindsight, Vuillard had confided to Denis not so much that his will was weak but that he had little experience of the will as Denis had been trained to understand it, and had thus embraced instinctual and habitual work as well as willful creation. Vuillard thereby established himself as deeply invested in, but at a slight remove from, the philosophical foundations of Nabi-symbolist theory as proclaimed by Denis,[14] a remove that encouraged Vuillard to establish an important place in his working method for the unconscious encounter with nature. Vuillard's recognition of his unconscious as a source of inspiration parallels the turn-of-the-century work of experimental psychologists such as Théodore Flournoy and Théodule Ribot, who were soon to identify the unconscious as the seat of creativity.[15]

Vuillard would synthesize unconscious and conscious functions in what remained a highly idiosyncratic naturalism. For despite some art historians' claims about Vuillard's "return" to naturalism after 1900, Vuillard only ever rejected a narrow, almost caricatured view of academic naturalism as the depiction of the externally visible world.[16] Vuillard's Nabi practice sought rather to reorient naturalism toward the self in order to focus on his own perceptions and sensations of nature—not simply to represent them but above all to know and to present them as a form of truth. While, visually, the difference with impressionism was stark, there was but a fine methodological distinction, as Vuillard himself recognized as early as 1895, when he referred to his practice as "interior impressionism."[17]

On 21 November 1888, Vuillard noted in his journal that he had woken up with a "gueule de bois," a French idiom for a hangover (fig. 46).[18] A coincidence, perhaps, but the day before, Denis had written the final exam for his philosophy baccalaureate, and it is tempting to imagine the two friends celebrating Denis's achievement with a night out. Earlier that fall, Vuillard had written to Denis, teasing him about his studying and looking forward to the day when his friend could put philosophy aside and they could go "rambling" outdoors together. In the same intimate letter, Vuillard had complained of his own laziness, nervousness, troubles externalizing his "impressions," and reluctance to exhibit his work. But Vuillard had also confided to Denis that he finally felt that he was becoming a painter, even though he had not yet "given birth" to anything substantial.[19]

Denis's characterization of Vuillard's method as journal-like and "stenographic" well describes the early *carnets*, which contain rapidly done sketches, summary notations, course notes, brief anecdotes, and much longer passages that all reveal a prolonged search for a method to guide his production.[20] Although Vuillard would reject the mantle of public theoretician, his journals shed light on three especially intensive periods of questioning, all beginning before Vuillard exhibited his first paintings in the fall of 1891. During the first period, from late 1888 into 1889, Vuillard interrogated the assumptions and methods of naturalism but also revealed himself to be closely tied to nature as a model. Although he was not among the group of Académie Julian students to whom Sérusier supposedly "revealed" *The Talisman* (fig. 37) in October 1888, at some point later that year or, more probably, during the following year, he had begun thinking about the implications of Gauguin's lesson, which led to his official entry into the Nabi brotherhood, in all likelihood sometime in 1889.[21] During this period, he tried working from memory as an alternative to working directly from nature, but quickly found himself disillusioned with this approach.

The second period coincides with the phase of Nabi theorizing leading up to and immediately following Denis's publication of the "Définition du néo-traditionnisme" in late August 1890. At this point, in the "year of Sérusier," as Vuillard called it, he pledged near-total allegiance to the Nabi-symbolist rejection of academic naturalism and the consequent redefinition of painting as both surface and objective truth. At the same time, however, anxious about originality and his desire to create a body of work (an oeuvre), Vuillard expressed a pressing need for a method that was entirely his own.

The beginning of the third period corresponds with the public exhibition of the first recognizably Nabi works by Denis and Pierre Bonnard at the 1891 Salon des Indépendants, which opened in March of that year. Lengthy entries continued through the summer of 1891, ending just before Vuillard started exhibiting his own paintings that fall.[22] During this period, Vuillard confessed his ongoing dependence on nature and began to synthesize a method that would involve an instinctual, automatic approach to nature, habitual work, and a series of deliberate choices about the formal and material elements of his paintings. *Woman Sleeping* and the portrait of Lugné-Poë, with their experimental arabesques, encapsulate, even allegorize, the method at which he had arrived. When Vuillard began exhibiting his paintings, revealing himself to the public, the intensity and frequency of his confessional journal entries seem to have diminished. He had come to some semblance of a painterly practice that he judged, at the very least, adequate to both Nabi-symbolist theory and his own artistic needs.

Naturalism as Delusion and "Dirty Little Bits of Memory"

From its earliest pages, dating to late 1888, Vuillard's first journal contains small sketches—rapid renderings of outdoor scenes, people, and works of art that attest not only to Vuillard's wanderings through the streets, parks, and museums of Paris but also to his attempts to "fix," as he would say, his "impressions" on paper. Short notes accompanying many of the sketches elaborate on specific scenes: "evening, immense sidewalk, wall on the right, reddish-brown trees, dusty, large woman all black, evening, night effect."[23] These "stenographic" notations often refer to different weather and lighting, suggesting that, like the impressionists, Vuillard took a particular interest in how varying atmospheric conditions affected his impressions of the world around him.

This method of notation also extended to Vuillard's observations of himself. In some of the first entries, he made careful note of instances when his perception had been altered by either psychological or physiological means. On one occasion, he noted working while one pupil was dilated.[24] On another, he tried to describe a certain dusty ocular effect.[25] Elsewhere, he recorded a "bad state of mind" and "considerable irritation."[26] Such notes continued well beyond these early years. In 1894, he noted, "Disorder in my brain," and in 1912, "interior collapse. . . . Nervous."[27] After waking up with a hangover the day after Denis's baccalaureate, Vuillard expressed surprise at the relatively "normal" functioning of his eyes.[28] In the upper left-hand corner of the page, marked off by a curving line suggesting a distinction between the night before and the morning after, he sketched a man slumped against a wall, possibly a self-portrait remembered from the night before (fig. 46).[29] But while Vuillard was observing and recording the world around him, at times relying on his memory to sketch scenes from a previous day, already in late 1888 he had begun to question the validity of both of these approaches. In trying to fix his impressions on the page, Vuillard also recorded a great deal of frustration: "while working lose the impression . . . find it while sketching slowly, didn't fix one single object."[30]

In a lengthier entry from late November 1888, often tantalizingly excerpted, Vuillard expressed serious doubts about the naturalist project under scrutiny. He assumed, in line with Hippolyte Taine's lectures on art, that a subject perceives nature as a set of relationships between forms, colors, sounds, and so forth.[31] It followed, Vuillard argued, that in order to "reproduce" nature, one had to "fix" not only the sensations provoked by individual elements but also one's perceptions of all of the relationships between elements. In recording one set of relations, Vuillard fretted, one inevitably lost the perception of the others, thereby compromising the veracity of the general impression. The problem was inherent in the temporal nature of creating art. In painting, by definition an act of some duration, one could not easily fix the instantaneity of perception unless, Vuillard wrote somewhat facetiously, one could transform the brain into an automatic recording instrument: "If the mental apparatus is not in a state to capture these relationships," he mused, "to hold on to them for a moment and then to record them onto paper or canvas like a somnambulist, it's a waste of time." Far from advocating somnambulism as a creative strategy at this stage, Vuillard's statement was on the whole a lament about the challenges of naturalism. "Doing nature," he had concluded, was therefore only "to upset [and] to delude yourself."[32]

Vuillard's critique resonated with one frequently made by symbolist critics, who accused impressionists of trying to be simple recording instruments, taking naturalism to an absurd extreme of instantaneous observation. Paul Cézanne's famous remark about Claude Monet—"He is only an eye, but good God what an eye!"—underlined the late nineteenth century's perceived limitations of the impressionist project.[33] But while Vuillard sowed the seeds of doubt about naturalism, thereby laying the groundwork for a turn to symbolist theories, he did not reject wholesale all attempts to faithfully reproduce nature; nor, at this point, did he question the desire to paint nature in some form. Rather, Vuillard was already hinting at an ambivalence that would eventually become the root of his continuing adherence to many aspects of naturalism and impressionism, those sensualist approaches to nature that Denis later criticized as outmoded.

Vuillard had already combined working from nature with working from memory—for instance, in his hangover sketch—but henceforth he made a more self-conscious effort to use memory as an antidote to the "delusional" methods of naturalism. For many artists and theorists of art, memory afforded a means of synthesizing sense perceptions before attempting to record them as both an ensemble and an ensemble of relationships. But Vuillard's concerns about naturalism were soon mirrored by an even greater ambivalence about working from memory. In late 1888 or early 1889, Vuillard noted that he was attempting to paint his sister Marie from memory.[34] The accompanying sketches, however, give little indication of this memory work, although a simplified and stylized face, probably a self-portrait,[35] might suggest an attempt to sketch what he remembered of his own countenance (fig. 47). Evidently, Vuillard endeavored to work from his memories of things that were most familiar to him, composite perceptions of the faces he knew best.

Since the middle of the nineteenth century, working from memory had gained an increasingly large foothold in French artistic training, but it had only recently emerged as a key strategy for artists who sought to liberate themselves from the schema of naturalism. Gauguin recommended working from memory as a way to inscribe the self in the work of art, and many in his circle followed his lead in the hope of liberating themselves from the model and orienting their own awareness and the viewer's toward inner states and experiences.[36] But vanguard artists, including those who admired Gauguin, did not universally embrace mnemonic strategies. While working with Gauguin in Arles in the fall of 1888, Vincent van Gogh had tried to apply Gauguin's method, but, holding fast to his belief in the crucial importance of the model, at least for his own work, he struggled to put Gauguin's recommendations into practice.[37]

Just as soon as Vuillard noted that he had begun working from memory, he recorded his disappointment that there had been little or no change in his work.[38] And indeed, the facture, coloration, and overall atmosphere of the many portraits and self-portraits from late 1888 and 1889, including a portrait of his sister (fig. 48), recall the competent and relatively conventional naturalist portraits and still-life paintings that Vuillard had produced throughout much of the previous year. In autobiographical notes written in 1908, he marked 1889 as the year he "started to work from memory," but a few years earlier, probably in 1905, he had recalled that moment in more negative terms, writing that 1889 was the year of "dirty little bits [*petites salissures*] of memory."[39]

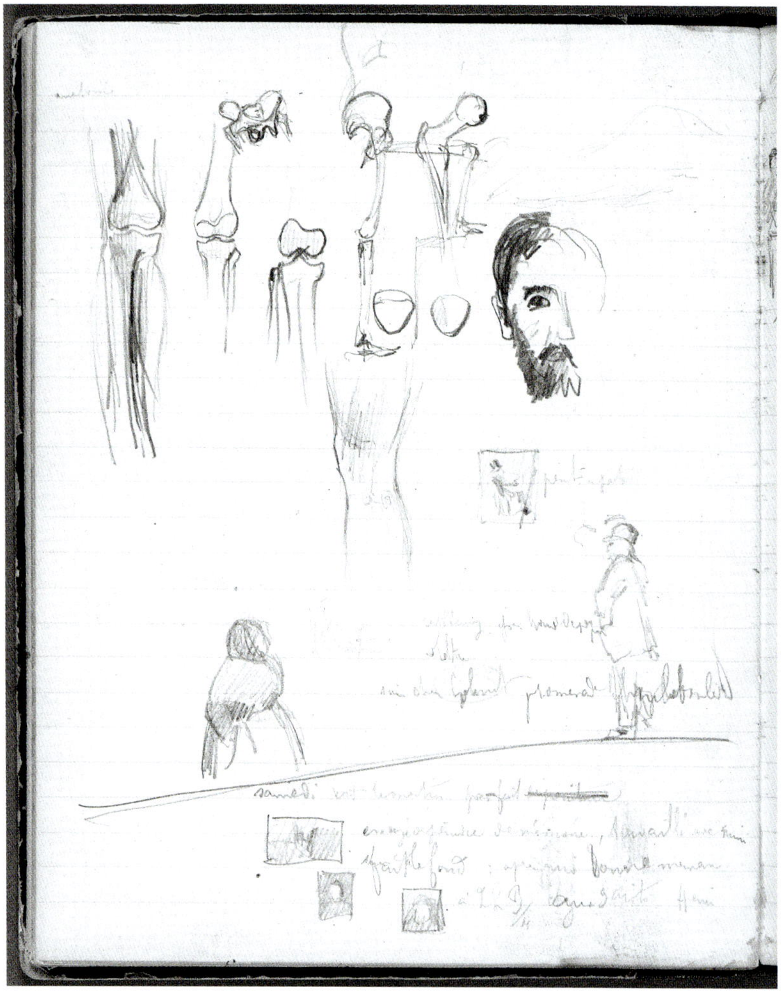

47 Édouard Vuillard, journal, 22 × 17 cm. Institut de France, MS 5396(1): *Carnet* I.1, fol. 32v [December 1888].

Regardless of whether he was aware of Gauguin's ideas in late 1888 or early 1889, Vuillard's interest in and appeal to memory remained firmly in the service of naturalism rather than of symbolism or self-expression. But he found working from memory an inadequate solution to the problem of faithfully representing nature. Despite his qualms about naturalism, Vuillard discovered that, like Van Gogh, he had no real desire to give up working from nature, and he soon admitted that "the idea of working from nature . . . pursues me."[40] Sketches in the journals attest to his ongoing desire to record the world around him, as he instinctively maintained his habits of naturalist notation. By early 1889, still an art student with little or no knowledge of symbolist theory, Vuillard had independently come to a symbolist critique of naturalism, but he had also reaffirmed the central role of nature in his art practice.

"The Year of Sérusier"

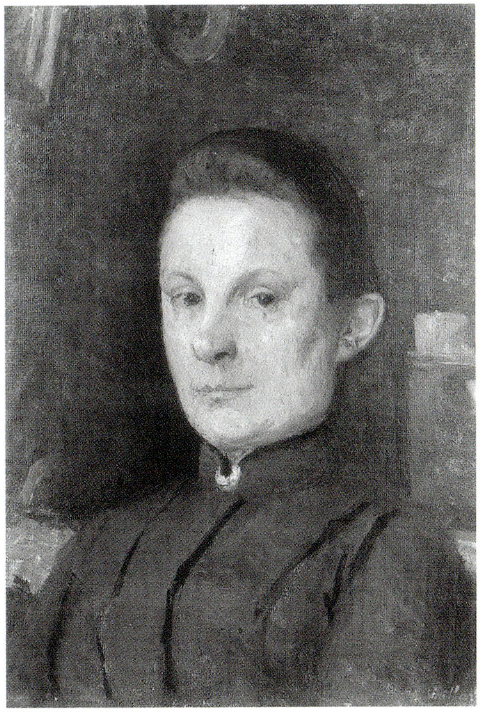

48 Édouard Vuillard, *Portrait of Marie Vuillard*, 1888. Oil on canvas, 32.5 × 24.4 cm. Sold by Christie's Paris, 20 May 2009.

Simply from the number of times Sérusier's name appears in the journals in 1890, it is clear that Vuillard was dealing that year with the full import of Sérusier's ideas, Gauguin's teachings, and symbolist theory more generally, via Denis and in conjunction with his more active participation in the social and artistic activities of the Nabi brotherhood.[41] Vuillard was more implicated in the publishing of Denis's "Définition" than is generally acknowledged. Possibly at the behest of Lugné-Poë, he acted as middleman between Jean Jullien, editor of the journal in which Denis's article appeared, and Denis himself, relaying proofs back and forth between the two men.[42] But more important, as evidenced by the *carnets*, he involved himself in the manifesto's intellectual elaboration prior to its publication in the late summer of 1890.

Vuillard's second concentrated period of reflection in his journals began in the spring or early summer of 1890, with anxieties over originality that were probably sparked by his increasing involvement with the Nabi group and his subsequent concerns over the visual similarity of some of their earliest works.[43] He told himself "not to worry about Sérusier and the gang," because, although they were all working from the same theories and toward similar goals, their works would be different simply because of their individual personalities, and in particular because of the unique ways in which they drew lines: "the same arabesque drawn by 2 different individuals [*caractères*]," he reassured himself, "will provide 2 different expressions."[44]

In July 1890, Vuillard set down an antinaturalist definition of art entirely in keeping with what Denis would soon publish in the "Définition," albeit much less elegantly formulated. His declared desire to break from naturalism now necessitated, Vuillard argued, a new definition of art as something other than a visual representation of the natural, external world: "the work of art starting from a motivating sensation . . . must be cherished by the artist not as a renewal of that sensation," he wrote, but "as decorating a piece of paper [or] canvas."[45] A work of art, in other words, above everything else, is a compendium of its formal and material properties rather than a representation of an

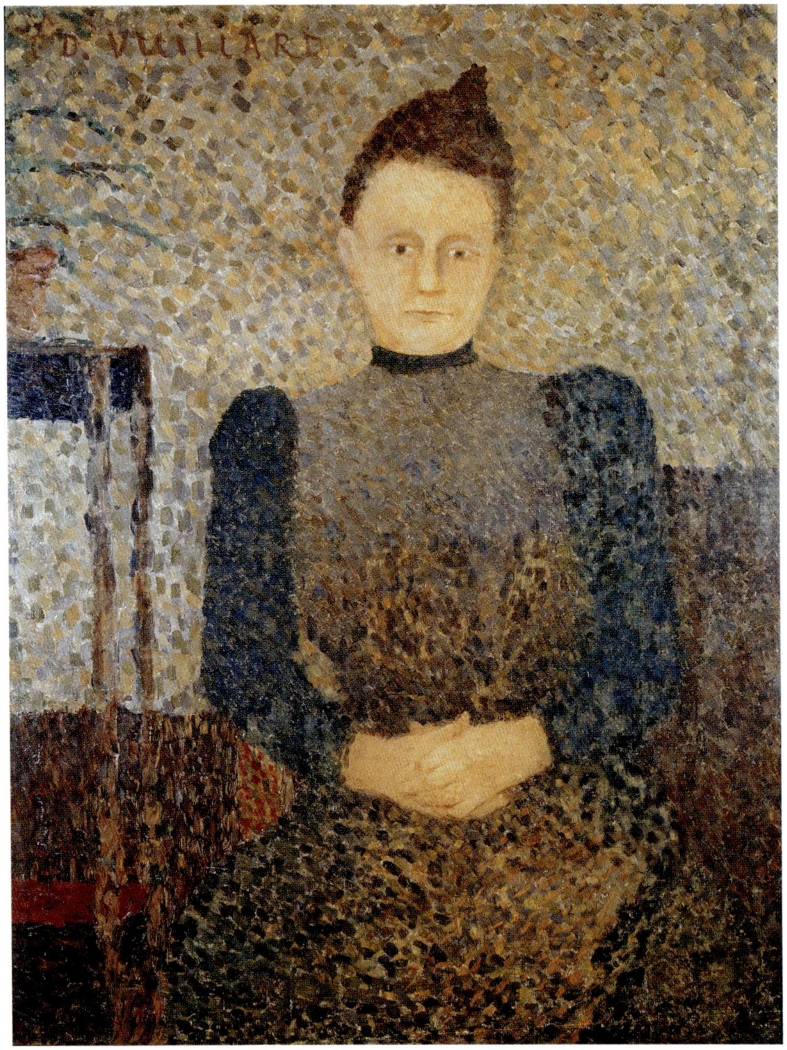

49 Édouard Vuillard, *Schematized Portrait of Marie Vuillard*, 1890. Oil on canvas, 92 × 73.5 cm. Private collection.

object in the world or even an attempt to represent or provoke in the viewer a sensation experienced by the artist before an object. In late August 1890, with Denis's "Définition" newly published, Vuillard refined this statement, providing a much clearer directive, one more explicit in its rejection of naturalism and even more in line with the "Définition": "Let us understand a painting as an arrangement of harmonies, distancing ourselves once and for all from the naturalist idea."[46]

Both Denis's and Vuillard's definitions share much with one offered earlier by Taine: "A painting is a colored surface, in which the various tones and the varying degrees of light and shade are distributed according to choice."[47] While Taine emphasized color in this short passage, in the longer passage from which it is taken, the key formal element is line. As Vuillard would do, Taine declared line to be fundamentally expressive: "lines . . . have a value in and of themselves and they have different effects on us depending on whether they are straight, curved, broken, or irregular . . . just as an arabesque of lines [*sic*] that do not imitate any natural object can be plentiful or meager, elegant or heavy. Our impression varies with their arrangement; their arrangement therefore has an expression."[48] Throughout this section of his *Philosophy of Art*, Taine asserted the seemingly infinite artistic possibilities related to the use of the painting elements of line, form, and color. An artist chooses, Taine insisted, to highlight or to downplay this or that element in order to bring out certain characteristics and expressive capabilities already inherent in the depicted object and its characteristic form. While Denis's statement at the outset of the "Définition" ("a painting . . . is essentially a flat surface covered in colors arranged in a certain order") seemed to minimize the agency of the artist, Taine insisted upon choice as a central element of creation, emphasizing that the painter proceeded by making innumerable decisions about how each element was to be arranged on the surface of a canvas.

At the beginning of September 1890, "still under Denis's influence," he wrote, Vuillard expanded on the ideas of the now published "Définition" to include a similarly hyperconscious awareness of choice, a truly extraordinary degree of volition, in his account of the artistic process. Despite how difficult he was finding it to set aside "the mental habits of naturalist ideas," Vuillard argued that one could overcome the "obstacle" of naturalism by "working with care. . . . Prepare your colors, your palette, and ensure that the forms . . . are precise." "Only what is on the surface of the canvas," he stressed, "affects our eyes . . . but this must be completely conceived . . . (old idea of deliberate [*voulu*]); the application, the execution, therefore, becomes a matter of great patience and care and that is all it is, the work of art." This is the closest that Vuillard would ever come to Nabi-symbolist theory in writing.[49] But, as if to mitigate that dependence, he also insisted emphatically that he had been thinking and working in this way for a while: "Already put into practice 6 months ago (portrait of Mimi in Yellow)."[50]

The most striking aspect of the September 1890 statement, given Vuillard's later skepticism about the will and his claims about the importance of his instincts, is the overwhelming emphasis on conscious activity, on the slow, deliberate, and very careful arrangement of forms and colors on a surface. The so-called *Schematized Portrait of Marie Vuillard* (fig. 49) comes closer than any other extant work of 1890 or earlier to Vuillard's attempts to create something that is at once a portrait of his sister (whom he called Mimi) and a "colored surface."[51] The restricted tonality, deliberate flatness, and reduction of elements to a minimum, however, combine to achieve a stiltedness that was neither particularly interesting nor original, recalling as it did the work of the neoimpressionists. The painting might have put into practice the Nabi doctrine of a "colored surface" and the Tainean corollary of art making as a series of deliberate choices, but Vuillard seemed just as unconvinced of this direction as he had earlier been about fixing his impressions on paper and working from memory. While he expressed his commitment to Nabi theories and the resulting definition of art as an arrangement of formal elements and materials, like many artists seeking to translate symbolist theory into a visual practice that could be recognized and justified as such, he was still struggling with form. While Denis and Bonnard débuted their Nabi works at the 1891 Indépendants, Vuillard held back, waiting until a full year after the publication of the "Définition" to make a tentative début, and nearly six more months passed before he showed a more significant group of paintings.

"I Need a Method of Which I Am Convinced"

His journals suggest that Vuillard underwent a crisis of method in 1890–91, probably stemming from multiple sources: his disillusionment with academic training (he left the École des beaux-arts at the end of 1889 after nearly three years of study); his personal critiques of naturalism and memory; the Nabi critique of naturalism; the very real challenges he was encountering in translating Nabi-symbolist theory into visual practice;

and his ongoing desire to rapidly record the visible world. Overcoming naturalist habits, Vuillard recognized, was difficult, and adhering to a theory of the artwork as primarily a "colored surface" arranged through a series of choices had so far resulted in derivative works of art that he scarcely found convincing. His ongoing concerns over originality—"a petty idea to which I am nevertheless attached, the idea of the *pélichtim* [Nabi code word for the bourgeoisie]!"—fed his need to develop a method derived from, rather than beholden to, Nabi doctrine.[52]

In late October 1890, Vuillard recorded his concerns about his lack of a workable theory and method, terms that he tended to conflate so often that the awkward conjunction "theory-method" is justified, in no small part as a reminder of the unsystematic, even at times contradictory nature of Vuillard's journal reflections.[53] "It is necessary," he told himself, "to have a method for producing where one cannot know the outcome in advance . . . I must imagine, see, the lines, the colors that I place, and do nothing by chance. . . . I must consider all my combinations. But in order for this work to be possible, I need a method of which I am convinced . . . a firmly held theory that would relieve me of the vulgar concern for originality. Only a well-established theory can put my mind at ease and allow me to develop."[54] Evidently, Vuillard anticipated a theory-method as a liberating force for his art, a way to obviate his concerns over originality and final results to a point where he could simply get down to work. Oddly, he situated this theory-method as prior to deliberate practice, almost as a precondition both for working and for his artistic development. He envisioned applying it consistently in every "humble" *ouvrage* such that he could achieve an oeuvre: "The oeuvre has only one method, the sum of all actions. Therefore all *ouvrages* will have the same method." But if this theory-method was to result in original works, Vuillard argued, it must be utterly individual and consequently derive from the ensemble of the artist's mental faculties: "This method depends on the mind [*esprit*], on the harmonious development of the faculties." The rather opaque passage that follows goes on to imply that in order for an artwork to be individual, rational thought ("the mind") must give way to less conscious functioning and become "a slave to the senses."[55]

Evidence suggests that in 1890–91, as he would do periodically throughout his life, Vuillard revisited earlier pages of his journals.[56] In doing so, he came to recognize his continuing dependence on the observation of nature and began to embrace his instinctive desire to record and to note the world around him stenographically. His recognition of the role of nature in his art came, characteristically, in the form of a confession: "It is the idea of working from nature that pursues me," he admitted to himself.[57] In order to justify this instinct, Vuillard turned to the terms of experimental psychology, arguing that because observation was a fundamental psychic fact of existence, an automatic rather than a willed act, it held an inevitable place in any artist's practice. "One observes, one does not will oneself to observe. . . . Pure and simple observation is a simple act of life," he wrote, "and it is this simple and primitive observation that is necessary, that is." Therefore, Vuillard reasoned, the emotions or sensations provoked by observation must constitute "the first condition of a work of art prior to the methodical mind and practical intelligence born from the same soul that is under the influence of that emotion."[58] If observation constituted a primordial, automatic human function, if it derived directly from an individual psyche or "soul," then it could function as a guarantee of originality. Observation, Vuillard concluded, therefore had a fundamental role to play in his emerging method: "Simply pay attention or go walking and *faire ses pieds*—There's a good system."[59]

Not long before Vuillard began to interrogate his own psyche in his journals, Léon Bélugou, the future philosophy correspondent for both *La revue blanche* and the *Mercure de France*, began attending Théodule Ribot's courses in experimental psychology. Faithful student and ardent defender of the pathological method, as I argued in my introduction, Bélugou took copious notes from Ribot's lectures, which support and amplify Vuillard's concerns as expressed both in the journals and in his 1898 letter to Denis.[60] They provide a useful counterpoint to the lycée philosophy curriculum, especially in recording Ribot's views of consciousness and the unconscious as a continuum rather than two separate states, and in Ribot's alliance of inspiration with less conscious states of mind. In the course offered in 1887–88, under the heading "Consciousness and Unconscious States," Ribot assigned the act of seeing to "organic" rather than "conscious" memory, clarifying that vision existed at the threshold of consciousness in normal adults. Observation, as Vuillard would argue a few years later, was in essence for Ribot an act that the body "automatically" "remembered" how to perform.[61]

By 1894, Vuillard was giving a slightly different account of the role of nature in his practice. By this time, he had firmly established observation and the automatic recording of his impressions as the initial steps in his method, but he now clarified the ongoing role of nature by using the rhetoric of experiment and experimental idea(l)s: "I am only confident in ideas and reasoning when they are controlled by direct impressions of nature. I must nourish myself with new observations. . . . I tried this and I was astonished at the prospect for discovery that I discovered. Because an observation that applies to a particular case is applicable to all sorts of facts."[62] Discovery, observation, control, case, facts: this is the late nineteenth-century terminology of experiment. Patently, Vuillard understood the direct observation of nature not only as the starting point of his creative process but also as the ongoing source of necessary controls against which to check his results. Nature, in Vuillard's view, thus supported the imagination by checking its tendency to generalize.[63] Deriding the naturalist "caricatures" of his artistic training, Vuillard echoed Denis's belief that academic naturalism had constrained the positivist search for truth by limiting the observation of nature to external forms. And, like Denis, Vuillard seemed keen to reclaim positivism by way of incorruptible experimentalism in

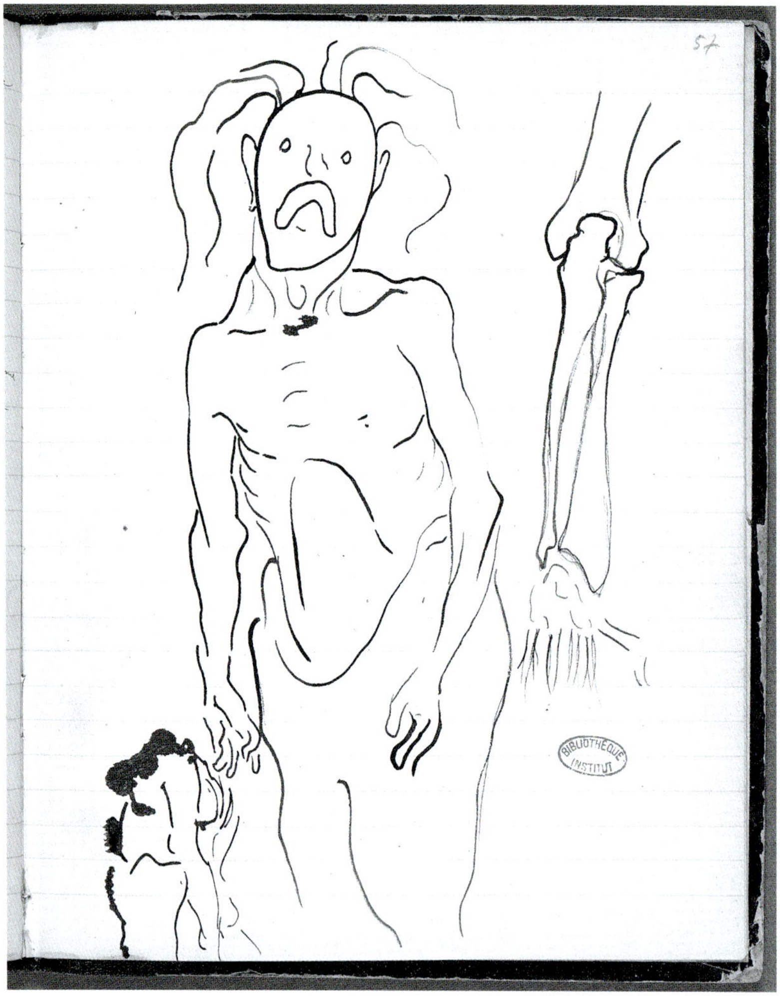

search of a more truthful form of naturalism that was nevertheless in the service of Nabi-symbolist theory. The observation of nature, and of one's own sensations and emotions provoked by nature, were to be the building blocks of Vuillard's "interior impressionism."

As Vuillard would do in his 1898 letter to Denis, Ribot refused to make a clear-cut distinction between conscious and unconscious states, arguing that just as there was only a quantitative difference between the normal and the pathological, there was only a "difference of degree" between consciousness and unconsciousness. In everyday experience, Ribot insisted, it was virtually impossible to distinguish between the two states. Individuals experienced a perpetual imbrication of the conscious and the unconscious that they nonetheless more often than not identified as something like a doubling or a splitting. But in reality, Ribot argued, the notion of two (or more) states was only a scientific abstraction, a structural convenience that gave the lie to what was in fact a continuum. In his 1887–88 course, Ribot previewed the arguments later put forth in his *Essay on the Creative Imagination* (1900), assigning a particularly generative role to the unconscious in inspiration, be it literary, artistic, or even scientific. To prove his point, he quoted from the writings of Goethe and Mozart.[64] He would never have relied on the writings of a young, unknown artist, and certainly had no access to them, but Ribot might as well have cited from Vuillard's journals.

50 Édouard Vuillard, journal, 22 × 17 cm. Institut de France, MS 5396(1): *Carnet* I.1, fol. 57r [September 1890].

From the Automatic to the Experimental Arabesque

At the same time that Vuillard attempted to reason his way to a viable working method, he also sought a way to translate his own emotional and sensorial experiences provoked by the observation of nature, his "impressions," into form. In late 1890, Vuillard reconsidered his somewhat facetious conclusion of two years earlier. Could he in fact turn himself into an automatic recording device, not of the external, visible world, but of the internal experience derived from the observation of nature? If Vuillard were to step through the door he had opened in late 1888, were he to attempt to record his psychological experiences by drawing like a somnambulist, the journals might reveal an accompanying change of form. And indeed, among the final entries in the first journal, those written immediately after the publication of the "Définition," there appear drawings unlike anything Vuillard had produced on earlier pages, evidence of a formal breakthrough, tentative at first, that would be carried over into the second journal (fig. 50).

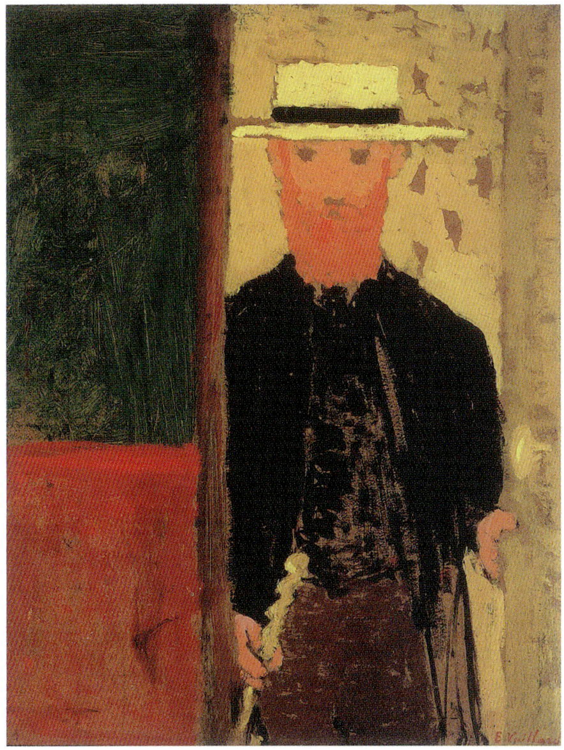

51 Édouard Vuillard, *Self-Portrait with Walking-Stick and Boater*, ca. 1891–92. Oil on cardboard mounted on canvas, 36 × 28.5 cm. Private collection, United States.

Opposite Vuillard's statement from early September 1890, the one advocating a very deliberate, careful approach to the arrangement of forms and colors on a flat plane, appear thoroughly new approaches to drawing, linear signifiers of nondeliberation very different from the tight, naturalistic, and often modeled sketches of earlier pages. They anticipate many of the drawings in the second journal, as well as Vuillard's theater drawings, which suggests that they are intimately bound up with Vuillard's early, radical Nabi style. A strange humanoid figure dominates one page (fig. 50), accompanied by a skeletal arm reminiscent of earlier journal sketches that doubtless relate to the anatomy courses Vuillard attended as a student,[65] and, in the lower left corner, an indistinct shape evoking a hunched, crouching figure. Vuillard initially drew the arm in pencil, like many of the earlier *carnet* sketches, but then retraced it with the pen and ink used for the other two drawings on the page. The shape in the lower left corner is especially free-form, its seemingly accidental inkblots having left a mirror trace on the opposite page.

The large figure, however, proclaims the most significant formal departure. With the most economical of means, with long, thin, rapidly executed strokes of the pen, Vuillard conjured up a highly stylized man-beast. The bare, emaciated torso perched atop thick legs resembles the body of a satyr. A thick, sinewy neck supports a head with small, round eyes, a large outlined moustache or downturned mouth, and the hint of a pointed chin or possibly a beard. The outlines of long ears are visible on either side of a head sprouting a few lines suggesting long, wiry hair. The small, round eye sockets with no eyeballs, the facial hair, the simplified, elongated ear shapes, and the slight upper body all echo Vuillard's early self-portraiture, especially the decidedly schematized *Self-Portrait with Walking-Stick and Boater* (ca. 1891–92) (fig. 51).

In the painted self-portrait, Vuillard played with the viewer's expectations regarding how the material surface of the work might represent its subject naturalistically. The eyes are unexpectedly empty areas of cardboard surrounded by paint, a literal reference to the Nabi claim that a painting is primarily a surface to which paint has been applied, but also a negation of vision, giving material form to Vuillard's rejection of a narrow version of naturalism. The ink sketch declares its separation from the "normal" natural world even more emphatically. The sagging man-beast negates academic naturalism, especially by its placement adjacent to the skeletal arm, a reminder of Vuillard's training in "correctly" representing human anatomy. Like Denis's study for *Décor* on the reverse of a prizewinning drawing of a classical cast, it performs Vuillard's act of freeing himself from old strictures, carrying out a farewell to constraining mental habits. It seems as if Vuillard has figuratively flayed himself, stripped himself of his academic education in the hope of revealing a new interiority.

Were this sketch the single instance of this kind of linear abstraction in the journals, it would be interesting but not especially significant. But because it inaugurates a new, apparently much less deliberated approach to form that dominates the second journal, it signals some of the structures and epistemological conditions of Vuillard's new formal language. It announces that Vuillard will henceforth actively court a somnambulistic approach to nature, not to record it automatically as in a photograph, but to give form to his automatic impressions of nature. In late 1890, he associated this recording of his immediate sensations with the first inklings of an idea, confirming that he conceived of the rapidly drawn trace as the visual signifier of interior observation: "Let's express what I feel (it is a simple designation, a fact pure and simple to designate the thing I have in my head): an expression of tender feelings caused by such and such an object. . . . Then I trace on the paper all at once [*d'un coup*] (in the same way that one speaks poorly when an idea presents itself) the line or the mark imagined and desired."[66] Because it was instinctual, automatic, and therefore seated in the unconscious mind, this mode of expression held out the promise of being utterly individual and could therefore function as the basis of a highly personal method for transforming theory into visual practice.

From here on, Vuillard would figure the starting point of his method as the arabesque. When he began to seriously employ the arabesque in the final pages of the first journal, he often chose to use the more fluid medium of ink rather than pencil, charcoal, or pastel, which had prevailed in the journal up to that point. The second journal begins with pages in which Vuillard explored contour and arabesque using pencil (fig. 52). But he soon returned to ink, now applied with a thick brush. The energy, density, and rapidity with which Vuillard set down these arabesques suggest his enthusiasm for the exercise. Lines intersect with one another; figures emerge out of a tangle of lines, as if the artist turned them into people or animals only after unconsciously moving the brush across the paper. In one of the first arabesques in the second journal, thick ink brushstrokes and shapes resolve into a figure in a long cloak wandering through an indeterminate landscape of curving black forms (fig. 53). Motifs straight out of Japanese woodcuts (fig. 54) suggest that Vuillard may have attended the exhibition of Japanese art held in April 1890 at the École des beaux-arts, but they also connect the arabesque to primitivist, exotic, and decorative forms of mark making.[67] "In the beginning," Denis had written in the "Définition," "pure arabesque, as little *trompe-l'oeil* as possible; a wall was empty: fill it with symmetrical marks of forms, harmonious colors."[68]

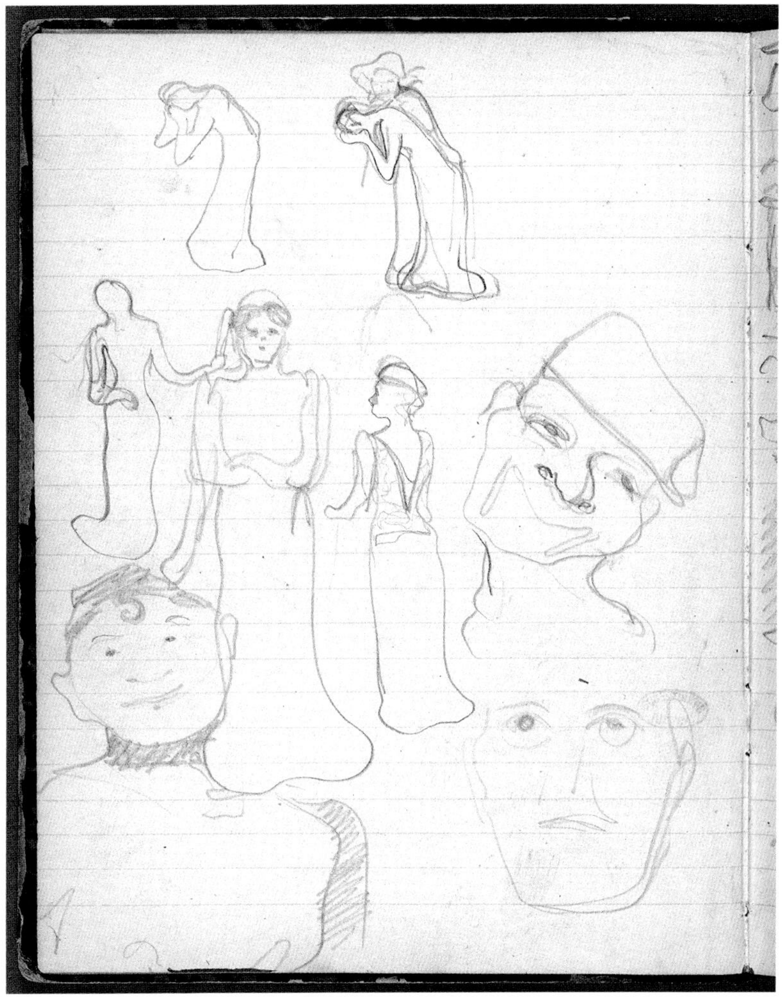

52 Édouard Vuillard, journal, 22 × 17 cm. Institut de France, MS 5396(2): *Carnet* I.2, fol. 3v, ca. 1890.

Like the line itself, the history of the arabesque as a visual device is a sinuous, twisted story. Borrowed from Italian, the French adjective *arabesque,* meaning "of the Arabs" or "in an Arab style," occurs as early as the sixteenth century but was used more often in the eighteenth to describe elements of rococo decoration. The *Encyclopédie,* still recognizing only the adjective form, denoted a European borrowing from Arabic painting and sculpture in which curvilinear vegetal forms proliferated owing to the presumed injunction against depicting humans and animals. The encyclopedia entry emphasized the contemporary use of arabesque forms in decorative works, including royal tapestries, but also pejoratively identified their use as a capricious, less exalted form of decoration, suited only to small, private spaces.[69]

By the early nineteenth century, the term *arabesque* was more frequently being used as a noun, and was associated with the arrival of Islamic architecture in Europe during the medieval period, thus marrying the exoticism of the arabesque with the by then more familiar exoticism of medievalism.[70] German romantic writers and artists such as Friedrich Schlegel and Philipp Otto Runge linked arabesque forms to a quasi-mystical search for the very origins of creativity. The arabesque was, in the words of Schlegel, "the most ancient and primitive form of the human imagination . . . the true mother, the embryo of modern painting."[71] In its supposed echo of the forms and rhythms of nature, it apparently signified a world beyond the visible.[72] Romantics cherished the arabesque above all for its suggestiveness, as a form that could give free rein to the human imagination, and they increasingly ascribed to it the value of musicality. Musical and ballet arabesques are likewise romantic inventions, as the term continued to accrue rich meaning.[73]

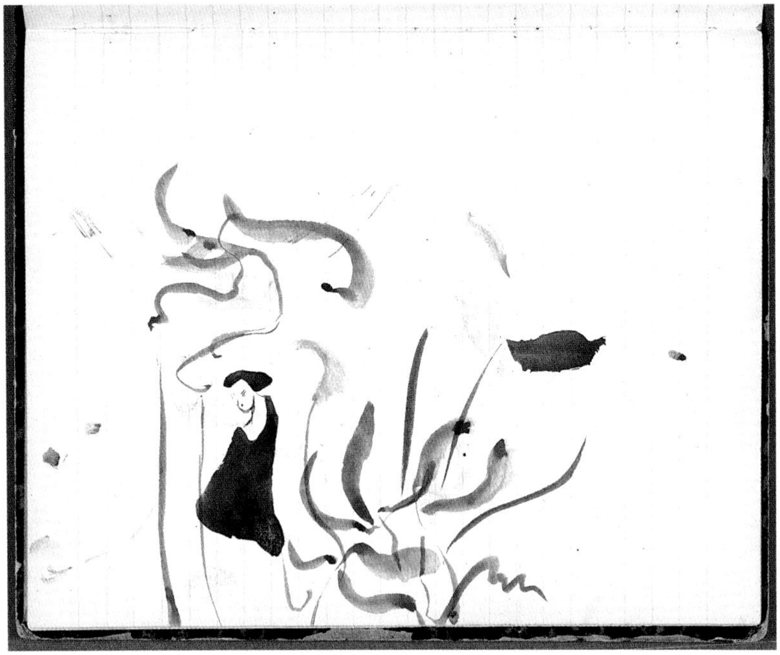

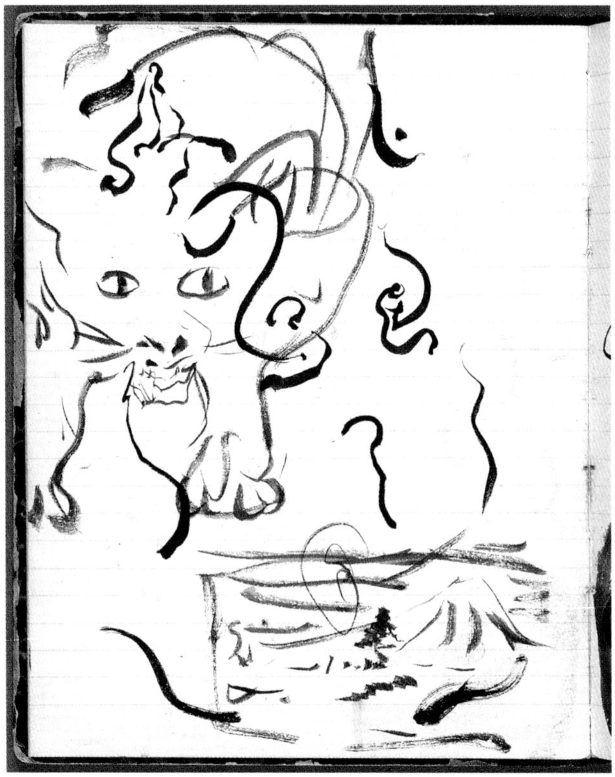

53 Édouard Vuillard, journal, 22 × 17 cm. Institut de France, MS 5396(2): *Carnet* I.2, fol. 5v, ca. 1890.

54 Édouard Vuillard, journal, 22 × 17 cm. Institut de France, MS 5396(2): *Carnet* I.2, fol. 12v, ca. 1890.

Symbolism's widespread promotion of the arabesque line as a multivalent signifier of artificiality, planarity, musicality, primitivism, and the imagination confirms symbolism as, at least in part, a renewal of romantic ideals. Symbolist artists were particularly drawn to what they saw as the ambiguity of the form, the potential, to use Dario Gamboni's term, of a sinuous curving line to enable different, even conflicting, readings as it wound its way across a surface.[74] Artists and critics often spoke of the arabesque as an animate being, as a moving, otherworldly creature with its own mercurial soul. And, as Gurminder Kaur Bhogal has recently demonstrated in great detail, using the Nabis as a key example, the ornamental, curving line enabled links to be forged with the other arts that had courted the allure of the arabesque—poetry, dance, and especially music.[75] Not surprisingly, it held a privileged place among artists such as Gauguin and many of the Nabis who sought to ally painting with musicality under the banner of Wagnerianism.[76]

As for many symbolist and art nouveau artists of the 1890s, the arabesque performed multiple functions for Vuillard. It signified the rejection of modeling and thereby repudiated academic and naturalist iconicity, and it squared with Denis's promotion of almost atavistic form, which held up the arabesque as the formal outcome of a process of devolution that would occur if one could return to the primitive origins of art. It courted a putative non-European exoticism but also signified, as Vuillard made clear, that the primitive, automatic side of himself had been mobilized and captured in visual form. It is the automatic arabesque that has smoothed Vuillard's way into the modernist canon, winding itself like a red thread, from Vuillard's ink arabesques, to his theater drawings, to fauvist arabesques, to surrealist automatism, and finally to the high modernist webs of Jackson Pollock.[77]

Initially, Vuillard developed the ink arabesques of his journals in works on paper that reflected his fascination and involvement with avant-garde theater, showcasing the first steps of his emerging method. These ink drawings, sometimes enhanced with watercolor, captured his actor friends at work, often fixed in awkward poses and contorted facial expressions (figs. 55 and 56). Vuillard's theater arabesques, which repeatedly approach caricature, endeavored to represent the spontaneity and ambiguity of the radical forms of theater being explored at Paul Fort's Théâtre Libre and at Lugné-Poë's Théâtre de l'Oeuvre, the latter of which Vuillard helped found in 1893.[78] But as much as there was important cross-fertilization between Vuillard's theater work and his painting, especially in terms of subject matter, atmosphere, or mood, painting retained its elevated and critical status as a separate realm of ambition for Vuillard, as it did for many late nineteenth-century vanguard artists, including Denis and Munch.[79] When Vuillard wrote of needing a method, he wrote of his need for a method of *painting*. Once he had instituted the arabesque as a signifier of instantaneous observation, how, he wondered (using photography as an analogy), would he move from the "instantaneous" to the painted "Image"?[80]

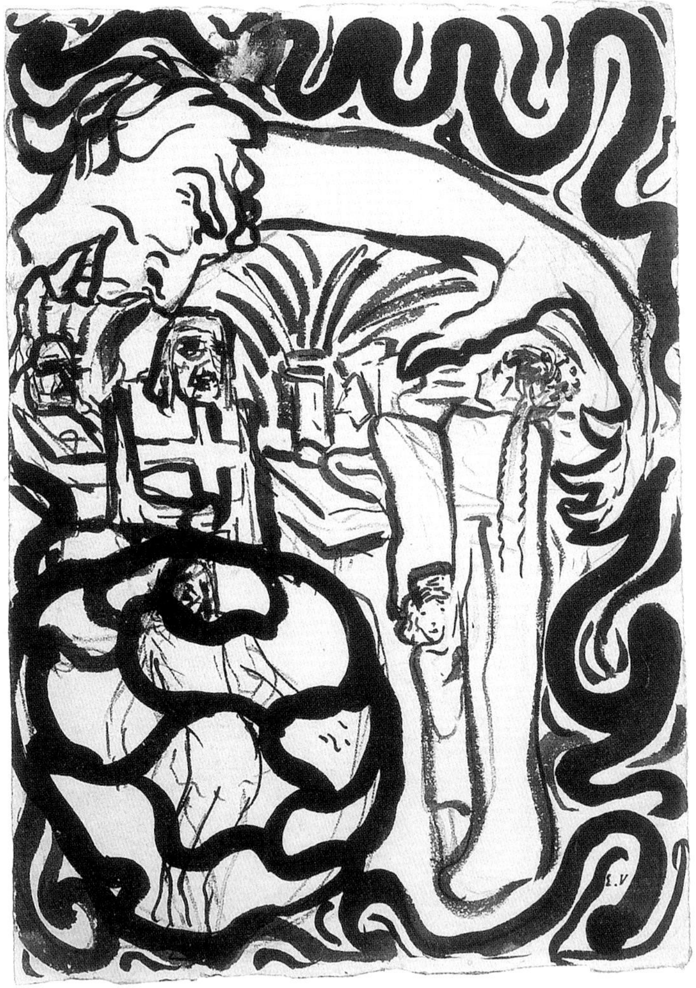

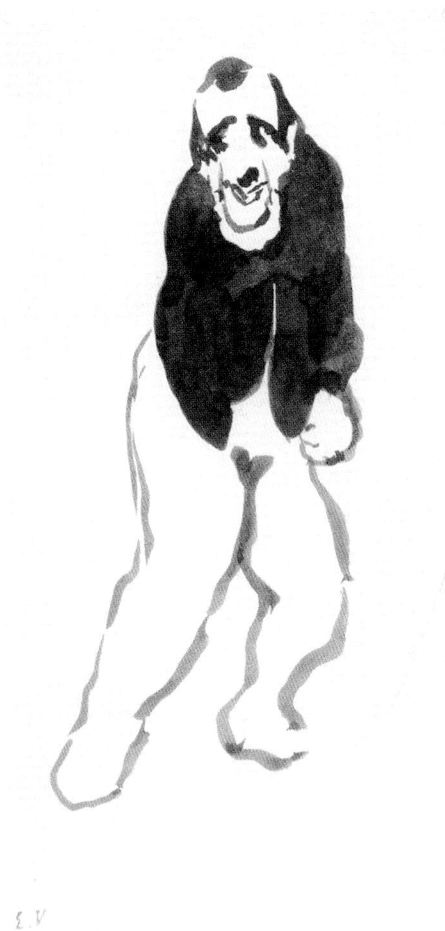

55 Édouard Vuillard, *Grisélidis*, 1891. Brush, pen and ink, watercolor on paper, traces of graphite, 26.5 × 18.5 cm. Private collection.

56 Édouard Vuillard, *Lugné-Poë*, ca. 1891. India ink on paper, 19 × 9 cm. Private collection.

Immediately after he likened the automatic arabesque to the rapid setting down of an idea, Vuillard clarified in his journal that this was only the first step in a creative process that would subsequently demand careful thought and deliberate composition. Painting, in his view, still necessitated making the kinds of slow, deliberate choices dictated by Nabi theory as inflected by Taine. "I develop it, if necessary," Vuillard wrote, "and I compose, a fact quite simple in and of itself to compose, and all my patience (Oh that's easy) is absorbed by a concern to do well and then the possibility of really very prolonged manual work."[81] Painting's ultimate purpose, Vuillard reminded himself, was the production of objective, universal truths and an objective beauty guaranteed by belief in the existence of a god. "If we were simple beings, unspoiled by prejudice and habits, our art would easily be beautiful, because there would be no struggle to distinguish within it the truth from the lie. . . . Would the beautiful then be only subjective? . . . Not if we believe in objective existence, if we believe in God."[82] The automatic arabesque of observation could only ever be a starting point for truth, the basis on which truth could be laboriously founded. Nabi-symbolist theories and Vuillard's own views of painting as *also* involving the will necessitated the transmutation of the automatic arabesque into something signifying the will. Enter the arabesque of experiment, a highly deliberated lack of paint, which transformed the automatic arabesque into a signifier of willed but increasingly repetitive acts of painting.

Woman Sleeping, Man Thinking

Vuillard's *Woman Sleeping* (see fig. 43) and his portrait of Lugné-Poë (see fig. 4) depict two extremes of unconscious and conscious activity while at the same time instantiating radically different gendered subjectivities. The sleeping woman, unconscious, hardly visible, and unidentifiable either as an individual or even as a woman, is the very antithesis of the individualized, albeit caricatured, portrait of Lugné-Poë, which shows the actor hunched over his writing in a state of intense mental concentration. These are far from the only instances of works by Vuillard that include the odd feature that I am calling the experimental arabesque. But not only is the device particularly evident in these two paintings; together they allegorize the two poles of Vuillard's method. In doing so, they mark out a continuum rather than a clear-cut distinction between conscious and unconscious states. Among the earliest works exhibited in the artist's career, they provide the means by which Vuillard's nascent method might be parsed.[83]

Woman Sleeping and its cut-up caricature draw special attention to the painting's form, which is not the same as saying that the subject of sleep is mere pretext for form.[84] But neither is the subject matter a pretext for looking at a female nude, which in and of itself calls attention to the phenomenon of sleep rather than the sleeping *body* as its ostensible subject matter. Sleep was a classic problem for philosophy and, increasingly in the nineteenth century, for experimental psychology. Was sleep a state of mental inactivity and therefore akin to death, or did the mind, as dreams seemed to suggest, remain active during somnolence?[85] If the latter, what accounted for that activity? By the second half of the nineteenth century, the analogy with death was all but relegated to literary metaphor. And sleep's pathological other, somnambulism, had evidenced not only the unconscious but, for psychologists like Ribot, the coexistence of consciousness and unconsciousness in all aspects of psychological life.[86] Ribot had even located personality and some fundamental notion of the self in this coexistence, even though individuals experienced this coexistence more often as alternation and at times, more perniciously, as a doubling of personality.[87]

57 Édouard Vuillard, *In Bed*, 1891. Oil on canvas, 73 × 92.5 cm. Musée d'Orsay, Paris, RF1977–374.

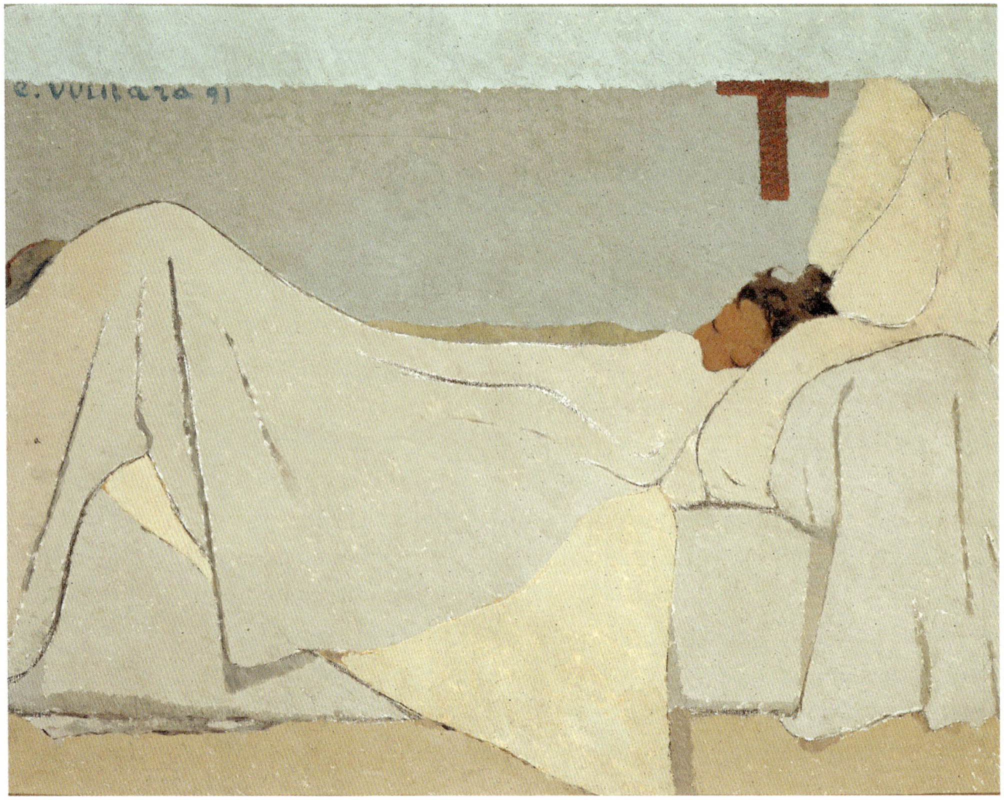

The automatic arabesque would therefore seem to be an ideal form with which to represent the subject of sleep. Indeed, Vuillard composed the related drawing *Woman in Bed* (fig. 45) around seemingly rapidly drawn charcoal arabesques that defined the contours of the figure, the bed linens, and its folds, providing a schematic layout "filled in" with watercolor. These curving lines are maintained in the painting *In Bed* (fig. 57) and to a lesser extent in *Sleeping Woman* (fig. 58). But they have virtually disappeared in *Woman Sleeping*, significantly, I think, the sole painting of this subject that Vuillard chose to exhibit in the 1890s.[88] In *Woman Sleeping*, Vuillard scaled up the automatic arabesque, widened it, and transformed it into negative space by carefully, although not precisely, avoiding painting adjacent areas of paint too close together. At times, Vuillard appears to have filled in these spaces with thin pigment, but he continued to make it clear that there had first been a void, and that the "fill in" was the result of an equally painstaking act of essentially coloring the lines *in* the lines. The effect is not unlike that produced in slit weaving, where a gap between two warp threads opens up when the weaver introduces a new color.[89] Once apprehended, the oddness, originality, and artificiality of these gaps arrest the viewer, but so does the sheer effort involved in producing them.

This effort is especially evident in the portrait of Lugné-Poë, which depicts a state of intense mental concentration by way of Lugné-Poë's hunched shoulders and his tense, misshapen hands pressing against the top and sides of a horizontal surface. In their tensed state, the shoulders and hands signify the concentrated labor of both mental and manual work. Nearly all the linear elements in the portrait, the outlines and folds of Lugné-Poë's smock and the contours of his hands and fingers, are negative spaces through which the cardboard support and at times a thin ground show through. Despite its overt focus on a willful, conscious creativity, the painting retains references to an opposing mental state. Lugné-Poë was known for his stylized, trancelike stage movements, which in 1895 earned him the nickname "the somnambulist clergyman."[90] And some semblance of perspective on the left side of the painting, along with a shape resembling a pillow and the dawn or twilight lighting, makes it likely that Lugné-Poë, pictured in the cramped studio space that he shared in 1890–91 with Bonnard, Vuillard, and sometimes Denis, was leaning not over a desk but over the foot of a bed.

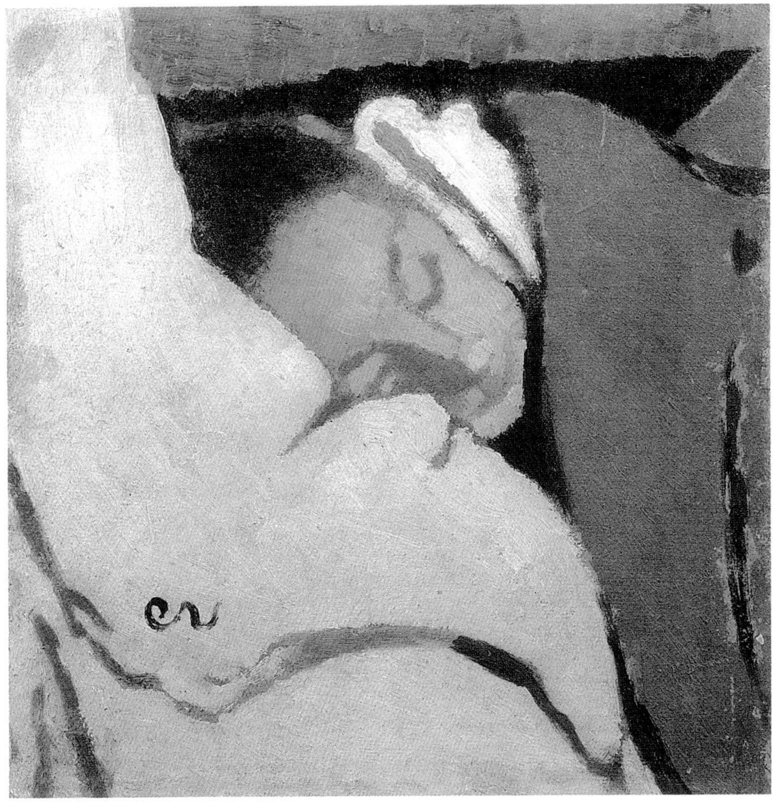

58 Édouard Vuillard, *Sleeping Woman*, 1892. Oil on board mounted on cradled panel, 26.8 × 26.8 cm. Private collection, Washington, D.C.

The negative arabesques are narrower, more even, and more sophisticated than in *Woman Sleeping*, comprising nearly all the contours of the work, in a way similar to the still life painting *Kettle and Pot* from around the same time (fig. 59). Their use is virtuosic, an intentional play on the artist's lively theater arabesques, which include numerous depictions of Lugné-Poë (for example, fig. 56). Vuillard has taken the rapid ink arabesque and slowed it to a crawl, figuring the intense concentration with which Lugné-Poë is shown conducting his own creative work. The most intriguing passage of painting occurs just to the right of the bed's footboard. An attenuated isosceles triangle of light blue, similar in hue to both the background and Lugné-Poë's smock, seems to hover on the surface of the work, confounding the viewer's efforts to distinguish it as part of either the figure or the ground. Further inviting concentrated looking, Vuillard has painted the triangle with parallel horizontal brushstrokes, the material traces of a series of repetitive, initially willful, but eventually habitual acts of painting.

In both *Woman Sleeping* and the portrait of Lugné-Poë, Vuillard offered the viewer an experience of visual play. At a distance, the negative arabesques resolve into contours—positive, curving, seemingly rapidly drawn outlines describing the shapes on the canvas. But especially in the case of the tiny portrait of Lugné-Poë, the viewer must get very close for the "lines" to reveal themselves as labored gaps in the material surface of the work. This viewing experience sets in motion a dynamic oscillation between an imagined positive line and a negative linear space, a perceptual uncertainty that begs questions about both the artist's practice and the visual experience of the work. But the questions provoked are nevertheless resolved into what, for Vuillard, seems to have been a universal truth. Beginning with an automatic, unconscious act, the spark of an idea figured as an arabesque line, Vuillard synthesized that idea and tested it in a willed process that could also be repetitive almost to the point of becoming mechanical. He courted a creative process that relied on both conscious and unconscious functioning, which Vuillard intuited and which Ribot argued could not, in practice, be readily distinguished from each other. This, for Vuillard, was a truth about human experience discovered through a complex and deeply paradoxical process that necessitated the "willed willessness" of the experimenter.[91]

Vuillard's Doubt

His painting was paradoxical: he was pursuing reality without giving up the sensuous surface, with no other guide than the immediate impression of nature, without following the contours, with no outline to enclose the color, with no perspectival or pictorial arrangement.

—MAURICE MERLEAU-PONTY, 1945, TRANSLATED BY HUBERT L. DREYFUS AND PATRICIA ALLEN DREYFUS, 1964

Maurice Merleau-Ponty's description of Cézanne's "paradoxical" practice might equally apply to Vuillard in the early 1890s. But while Merleau-Ponty identified doubt as an inevitable accompaniment to Cézanne's artistic project, Vuillard's doubt (and I would wager Cézanne's as well) might also be read as intrinsic to the artist's experimentalism. It might be sufficient to argue that the creative process that Vuillard developed in the early 1890s courted a formal analogue of the pathological method of experimental psychology insofar as its starting point was an othering of form figured by the automatic arabesque, and that it proceeded by using the observation of nature as a control for the imagination in a series of deliberate and often repetitive acts ultimately aimed at producing objective truths. But still another aspect of Vuillard's method, or, more precisely, his attitude, found a rationale in late nineteenth-century experimentalism. When Vuillard asserted that making art involved a series of carefully considered choices about the arrangement of elements on a surface, he also embarked upon an interrogation of how, in practice, one came to make those choices. Throughout his life, as Gamboni has noted, Vuillard expressed frequent and almost paralyzing doubts about his ability to carry through with his work,[92] but his concerns over his own will to work were especially acute in 1890–91.

As early as the October 1888 letter to Denis, Vuillard had identified the career of a painter as particularly suited to a "hardened idler" like himself. He frequently made statements such as "no strength in the head," and would often use a single word, such as *flemme* or *paresse*, to sum up his state of mind and body.[93] Both *flemme* and *paresse* are usually translated as *laziness*, but the latter also connotes indolence and apathy. Later, he would give more elaborate accounts of his ongoing difficulties in summoning up the strength to think and work: "Annoyed can't concentrate on a point, an idea. . . . Slave to my sensations and a pride condemning [me] to powerlessness. At the studio laziness and pride."[94] Vuillard deprecated himself, often to an extreme degree, over his severe lack of will.

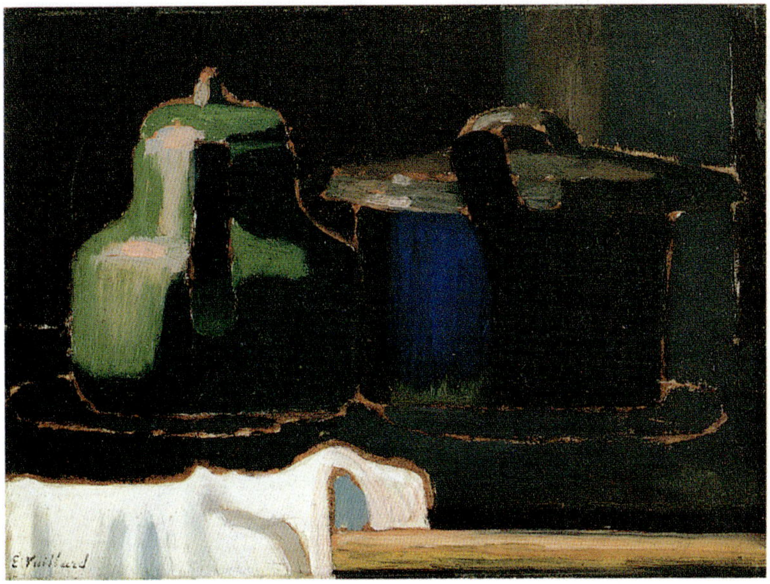

59 Édouard Vuillard, *Kettle and Pot*, ca. 1892. Oil on cardboard, 24 × 32.5 cm. Formerly Museum of Modern Art, New York.

In his journal, Vuillard would come to couch his concern with the mechanisms of his own laziness in terms of the new psychology's understanding of the will as both psychological and physiological, and as one of the most complex and highly evolved human functions. In a series of questions to himself, Vuillard wondered whether his mind "resisted" working because of "physical and moral fatigue." He asked what brought about a change in volition, and he considered whether the amount of will needed to overcome laziness was equal to the amount of laziness itself. "Can one . . . increase this power just as one exercises the biceps?" he asked. He suggested that only life, an "unknown force," could bring about the necessary effort to overcome the overwhelming instinct to resist action. A string of question marks at the bottom of one page (fig. 60) conveys genuine frustration as he probed his own mind: "what about the consciousness of that effort. What is underneath these words efforts, cravings, desire! Confusion of the words and the ideas that they represent!"[95]

These pages fairly bristle with anxiety as Vuillard voiced his torment. But he also implied that he felt it should be otherwise, that he should not only have better control over his own will but should understand his own will better, should have a tighter grasp of the meaning of words such as *effort*, *craving*, and *desire*. Written on 31 August 1890, immediately after Denis's "Définition" appeared, the "should" is of course relative. "Should," he meant, in comparison to Denis and to "Sérusier and the gang," many of whom Vuillard seemed to imagine would be much less confused than he about the functioning of the will. But Vuillard underestimated himself. His skepticism toward an easy comprehension of volition, and his intuition that volition was both physiological and psychological, were more up to date than Denis's professed certainties. For Vuillard, his purported mental laziness was something akin to physical fatigue, an insufficient amount of "moral energy" that undermined his ability to act consciously at any given moment. He understood that a healthy will was continuous with an unhealthy will, each a function of greater or lesser amounts of "moral energy." In *The Diseases of the Will* (1883), Ribot had argued that "in every voluntary act there are two very distinct elements, the 'I want,' which states a situation, which in itself has no effectiveness; and a very complex psychophysiological mechanism, the only place in which the power to act or to resist resides." Like all of Ribot's monographs, his text on the will deployed a pathological method. Among the pathologies of volition that Ribot examined was the "madness of doubt," which he defined as "a constant state of hesitation, for the most vain reasons, with the inability to arrive at a definitive result."[96] In his journals from around 1890–91, Vuillard effectively diagnosed himself as suffering from such a disease.

By 1894, however, Vuillard had come to see his doubt as essential to, even constitutive of, his artistic methods. His lack of will, he had begun to insist to himself, enabled him to develop and sustain the "habit of machine-like work," adding, alongside instinct, another automatic function to his method. This completed a triad of conscious and unconscious mental functions; as Ribot had argued in his 1891–92 Collège de France course, "The Will from a Psychological Point of View," the will was everything besides instinct and habit.[97] The solution to the problem of a lack of will, which made Vuillard's laziness in and of itself useful, was simply to work habitually, to circumscribe the problem of volition by cultivating a different kind of automatic behavior. "In thinking about

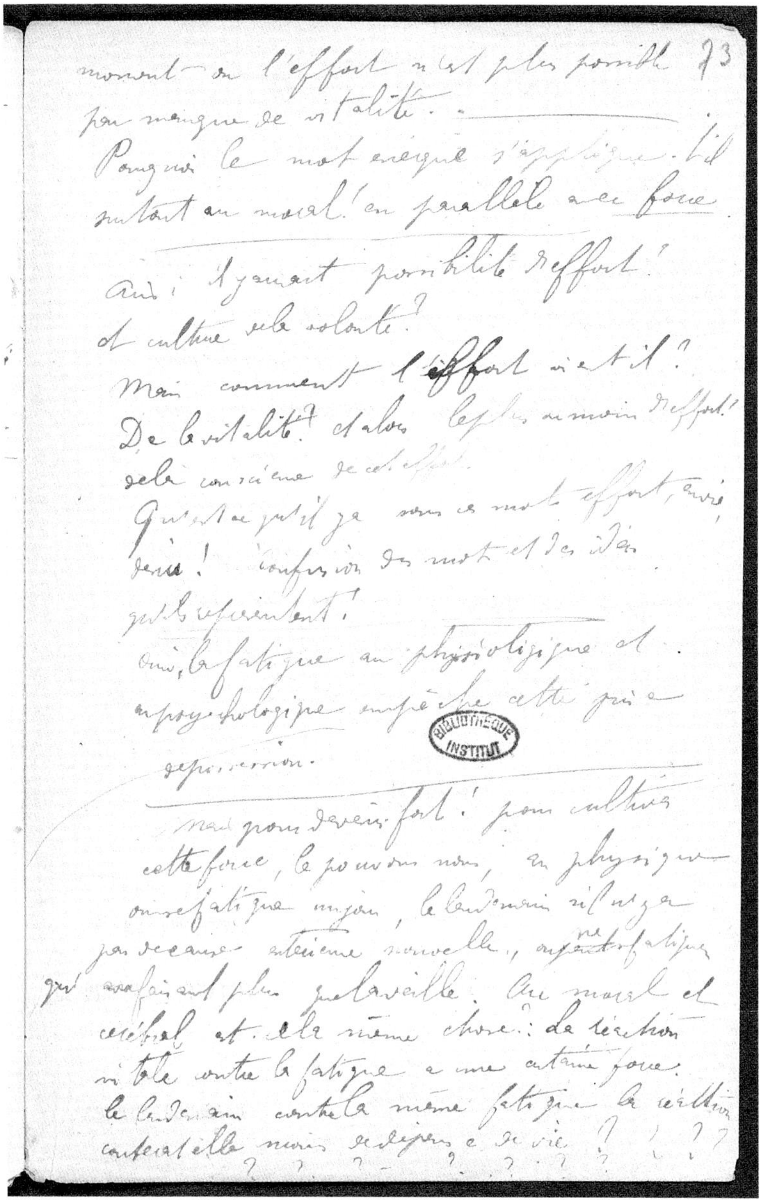

60 Édouard Vuillard, journal, 22 × 17 cm. Institut de France, MS 5396(1): *Carnet* I.1, fol. 73r [31 August 1890].

work," Vuillard wrote in his journal, "I will only be able to put my mind at ease by being delivered . . . by the habit of machine-like work, which my long periods of idleness have not yet been able to destroy. Most of the time, tired from my mind's laziness, my desire for work is only a hope born out of my laziness that renders me easily content. Perhaps habit has its benefits, but the habit of an activity."[98] Vuillard's solution, in 1890–91, to the problem of developing a method at once Nabi-symbolist and wholly individual was a process involving his instincts, habit, and will, and when he found his will to work especially vulnerable to laziness, he exhorted himself to "try to return to the work of 90–91," instead of becoming, "in the face of sensations or new impressions," a "slave to certain dimensions, to certain materials (cardboard, oil)."[99] Vuillard had come to identify thinking coherently, being conscious of his work, as at times detrimental to the new ideas emerging, and he considered habitual work his touchstone. The hard-won, deeply personal method at which Vuillard arrived early in his career remained his lodestar in the 1890s, a source of comfort when he felt doubt and laziness getting the better of him. A lack of will enabled Vuillard to renew his work, to return to the method of 1890–91, and always to see the simple act of working as something quite marvelous.

In his *Introduction to the Study of Experimental Medicine*, Claude Bernard had assigned to doubt a central role: "the unique and fundamental rule of scientific investigation," Bernard wrote, "can be reduced to doubt."[100] Doubt and a professed nescience about final results enabled the experimenter to retain his objectivity, to constantly question the results obtained during an experiment such that rigorous testing could continue and a final truth emerge. The late nineteenth-century image of the experimenter, which Émile Zola had co-opted, was that of both lone genius and humble worker. His flash of inspiration, his hypothesis, was subject to extensive tests and eventually proved true through a repetitive, almost machinelike process during which he willed himself to "willessness." Vuillard cultivated for himself a very similar methodological stance that aped experimentalism, while at the same time situating his practice closer to the unconscious end of the spectrum than any of his contemporaries did. Vuillard seems to have delighted in subverting viewers' expectations about the painted surface in conjunction with subject matter. Where the viewer might anticipate line, she gets negative space or nearly imperceptible boundaries between areas of pattern. Where she expects at least a daub of paint, signifying, for instance, an eye, she gets bare cardboard. Walter Sickert called attention to Vuillard's studied childishness, praising his "purposed negligence" as the crux of his practice.[101] Ultimately, Vuillard's practice, like Cézanne's, was one of paradoxes, but paradoxes central to, among other things, nineteenth-century experimentalism.

In the letter that Denis took as a guideline for his "Définition," Sérusier offered two methods for arriving at truthful symbolist form, or, in his parlance, the "immutable principles." Sérusier argued that an artist could proceed either by a kind of self-deduction, by searching for the immutable principles within the self, or by induction, by observing how other artists had successfully externalized the immutable principles in their works.[102] Denis, more often than not, chose to work from the latter method, canonizing a host of *primitifs* from Fra Angelico to Cézanne. Vuillard, by contrast, strove to deduce a method from an intensive and concentrated exploration of his own interior mental processes. Sérusier had argued that only the artist "unspoiled" by academic training, a modern primitive like Gauguin, could deduce universal truths within the self without the need to check his results against those of past masters. Vuillard's automatic arabesques constituted his attempt to establish the raw data for his artistic practice as he experienced it in the guise of the "unspoiled." While the willful arrangement of colors and lines on a flat surface remained a central feature of his practice, Vuillard, cognizant of having received a philosophical education different from that of his peers, retained a critical place in his method both for his instinctual approach to nature and for habitual work—in other words, for those automatic, unconscious mental functions that late nineteenth-century psychologists opposed to the will.

Vuillard came to fetishize his method as the unique way in which he could work in order to attain truth. In a passage that could have been written by any number of the artists discussed in this study, he made abundantly clear the connection between his practice, his self, and the truth:

> My only desire now is the life of my soul. Work is at once the sign and the product of that life. . . . Truth is life, life is truth, the truth is one or it wouldn't be the truth. Dogmas are only methods for arriving at the truth; any method contemplated by the Spirit offers up a fascinating spectacle[;] it's a harmonious ensemble of more or less pure ideas that makes up the more or less great and vast Beauty of dogma. Dogma is not an end, dogma is a work of art, at least that is how I understand it, it is the result of *expériences* towards a final aspiration that is and can be nothing other than God.[103]

Although he held his own against Denis in their 1898 exchange, Vuillard eventually lost faith in certain aspects of his method, and in particular in automatism and the arabesque of observation as starting points. One might argue that he loved the natural world and everything in it too much not to return to an increased visual correspondence between his painting and the visible world. One might argue that successful decorative and portrait commissions put too much pressure on a method that, at least in part, was originally designed to translate symbolist theories and his own interior experiences into visual form. When Nabi-symbolist theories became less imperative, the method had to change. And it did so in ways that ultimately threatened Vuillard's desire to create a body of work that might without hesitation be called an oeuvre.

The level of intentionality that I have ascribed to Vuillard and to his development of a creative method might be criticized by those who choose to take the artist too much at his word regarding his lack of method. But the work of the 1890s, along with the ruminations of the journals, attests to the careful consideration that Vuillard brought to developing a viable working practice. Examining the method and its underlying assumptions neither negates nor diminishes the success of his results. As theorists of experiment have argued, in the face of discovery, of increased knowledge of truth, method and process are destined to fade into the background—indeed, by design should become entirely transparent to the results.[104] Like Sérusier's *Talisman*, the central subject of such paintings as *Woman Sleeping* and the portrait of Lugné-Poë is the creative process itself, vexing, paradoxical, neither wholly conscious nor wholly unconscious but existing somewhere along the wide spectrum of human psychological functioning.

A final comment about sleeping women and thinking men is in order. An undercurrent of Vuillard criticism has often held that he was not quite man enough. Shy, we are told, harboring secret passions for unattainable women, possibly including his mother, whom he called his "muse," and his sister, Vuillard is nevertheless suspected of having done violence to female bodies by deforming them, a charge that gains him entrée into the modernist canon.[105] My analysis implies that Vuillard's stance on the female body might be part and parcel of a creative method structured by a continuum between automatic and willed states nevertheless perceived as an alternation. As these states were, and often still are, gendered feminine and masculine, as they are in *Woman Sleeping* and the portrait of Lugné-Poë, such an alternation might engender, or be perceived to engender, a process of "androgynization." The uncertain gender of the figure in *Woman Sleeping* thus emerges as a sign of the gender instability inherent in the creative method that Vuillard developed and cultivated for himself, which the caricature of the cut-up *Woman Sleeping* attempted to stabilize by ascribing to the artist masculine violence. Such "critical cross-dressing" was, of course, possible only when performed by a white, bourgeois male member of an avant-garde "brotherhood," but one nevertheless who had less purchase on the robust French republican male self than his colleagues.[106] This is not to say, however, that the performance has exacted no price. In Vuillard's case, one cost has been the stubborn persistence among art historians to address either form or content in analyses of his works, a tendency that renders his symbolist visual practice inscrutable.

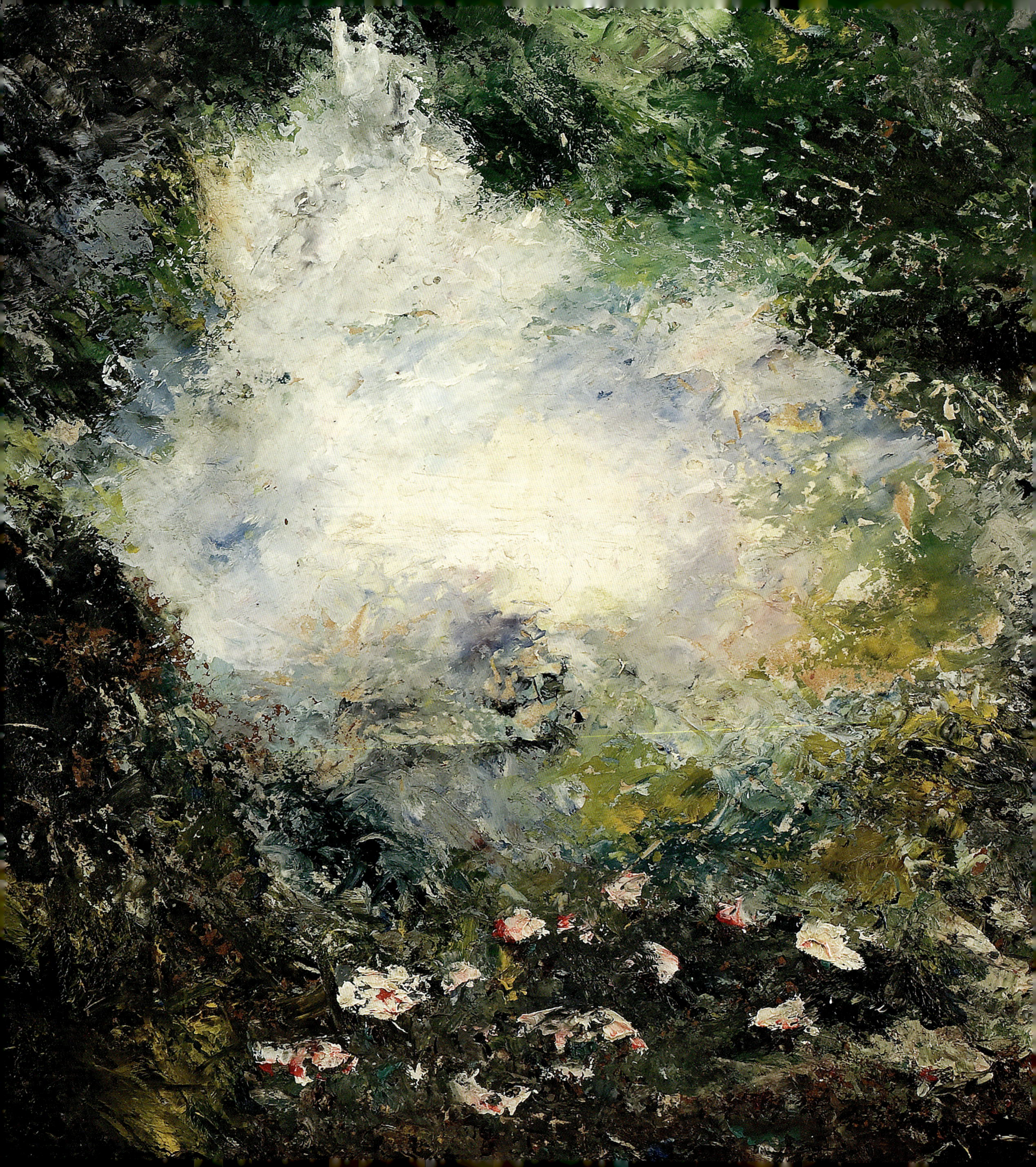

AUGUST STRINDBERG'S NATURALISTIC SYMBOLISM

4

A disgruntled Maria Weyr wrote to her husband sometime in 1893 or 1894 about the habits of her new, unfathomable brother-in-law. He reads nothing but "works of science, . . . Balzac and the French psychologists," she complained. When he writes, he does so as if he is "in a trance," but he actually spends much of his time painting, and "there, too, he is a law unto himself, naturalistic symbolism, as he calls it."[1] Not long before, Weyr's sister, Frida Uhl, had been courted with a painting titled *Night of Jealousy* (fig. 5), a sober, nearly monochromatic work that had been thickly, violently, and seemingly haphazardly executed with a palette knife. On the back of the cardboard support, the artist had dedicated the painting to his new fiancée, explained the motif, and signed his name, "the Symbolist August Strindberg." Maria Weyr soon bade goodbye to the symbolist Strindberg from her parents' house in Austria when he left for Paris in the late summer of 1894. Having spent the previous two years traveling between Sweden, Germany, and Austria, all the while occupied with painting, chemical experiments, and a scientific treatise that he intended to rival Francis Bacon's *Sylva Sylvarum*, Strindberg was preparing to conquer the Parisian avant-garde not only as a playwright but also as a symbolist painter. He was, he told a friend at the time, "the first to paint symbolic landscapes." He had "a whole room full of great symbolist canvases" ready for exhibition, and he even seems to have lined up a dealer.[2] Gearing up for his attack on French symbolist circles, Strindberg was also writing a number of essays in French, follow-ups to his 1887 series of essays titled *Vivisections*. With his paintings, plays, and this new series of *Vivisections*, Strindberg anticipated planting his "seed" in the brains of what he called the especially suggestible Parisians.[3]

Strindberg was one of the few to heed Émile Zola's call to apply experimental methods to his various creative endeavors.[4] Especially in the 1880s, however, Strindberg had maintained a certain ambivalence about Zola's experimentalism, criticizing the French author for relying too much on analysis and ignoring his own self. By the 1890s, however, having immersed himself in French experimental psychology and having taken a particular interest in nature's experiments as a methodological paradigm, Strindberg more fully embraced experimentalism as fundamentally important for, among other things, the style of painting he would call "naturalistic symbolism." In the spring of 1894, not long before Strindberg left Austria for Paris, a French journalist had interviewed a number of foreign writers about Zola's latest bid to join the French Academy. While most of those interviewed expressed a grudging appreciation for Zola but argued that his time had passed, Strindberg conveyed unabashed admiration. He defended Zola against his detractors, argued that the naturalist novelist had already achieved immortality, and demanded that the younger generation of symbolists recognize him as their master. "Is he not a psychologist," Strindberg asked, "who has performed a thousand vivisections on the human Heart with the audacity of a surgeon! . . . Honor him, grateful fatherland! And you, symbolists, honor the master of Symbolism . . . and you, synthetists, honor the greatest of synthetists also superior in analysis."[5] This text might have sunk into obscurity had not a few months later the symbolist journal *La plume* announced the Swede's arrival in Paris by reprinting Strindberg's idiosyncratic plea for Zola to be appreciated as, of all things, the master of symbolism.[6]

Strindberg's experimentation across a wide range of media and disciplines is often identified as the heroic crux of his modernism. And, indeed, in his prodigious output—twenty-two volumes of correspondence alone, compared to Zola's mere eleven—Strindberg became a most energetic entrepreneur of experiment. He imbibed a range of experimental methods, fetishized them, and eventually promoted psychological experimentation as a key feature of his avant-garde modernism. In doing so, Strindberg capitalized on the shifting meanings of *expérience* in the late nineteenth century. Most often, he used the word *experiment* in a scientific sense to refer to the chemical, alchemical, and eventually psychological experiments that preoccupied him at various times in his life. Sometimes he used *experiment* in a more specifically Bernardian way to refer to a doctrinaire method beyond observation that emphasized doubt, objectivity, and synthesis. At times, Strindberg conflated naturalism, experiment, and analysis, even to the point of identifying experiment with a simplistic equation of naturalism and minute photographic detail. At other times, the word *experiment* enabled him to claim distance, in the name of objectivity, from ideas and practices that he knew were distasteful to many of his friends and colleagues. But, most often, Strindberg's use of the term *experiment* shaded from one connotation to another, often combining different notions of experimentation in unexpected and highly creative ways. While the philosopher and psychiatrist Karl Jaspers diagnosed Strindberg's unsystematic approaches to theory, method, and, indeed, experiment as symptomatic of schizophrenic tendencies, I take Strindberg's creative approach to experiment as enabled by, and as a sign of, the not yet fixed discursive relationship between experimentation and modernist vanguardism, which Strindberg registered in the 1880s and came to exploit in the 1890s.[7] More than any other fin-de-siècle figure, Strindberg enables us to see the strangeness of experimental art in the 1890s.

In fact, before the 1890s, Strindberg rarely used the word *experiment* (spelled the same way in Swedish as in English) to refer to his writing or his other artistic practices, more often preferring the Swedish word *försök*. Like the German *Versuch* and the Norwegian *forsøg*, *försök* is a less strictly scientific term that can also be translated as *trial* or *attempt*. In the digitized corpus of Strindberg's work, which includes his published correspondence, the Swedish adjective *experimentell* and the French *expérimental* each appear only once, the former in a late novel and the latter in a letter to Zola, in which Strindberg claimed to have written his play *The Father* "with the experimental formula in view."[8] Instances where the verb form of *försök* appears in close conjunction with the verb *experimentera* confirm that Strindberg understood *försök* and *experiment* as similar but not entirely synonymous terms.[9]

Strindberg's self-imposed shift from naturalism to symbolism gave rise to an intensive period of painting between 1892 and 1894 and to a theory of an "automatic art" based on chance, which was published in his second series of *Vivisections* as "The New Arts! Or Chance in Artistic Production." While his symbolist visual practice was aimed at Parisian audiences, it was cultivated in Berlin, where Strindberg situated himself at the center of At the Black Piglet (Zum schwarzen Ferkel), the avant-garde circle named by Strindberg after the wine cellar the group frequented. At the Black Piglet, which comprised an international group of writers, poets, artists, and scientists, offered Strindberg an ideal environment in which to synthesize his belief in experimentalism, his interest in experimental psychology, and his knowledge of the pathological method in the service of his emerging symbolist practices, which, in the case of his visual practices, centered on visual and material regimes of chance.[10]

While the previous chapters of this study focused on the specific symbolist visual strategies of deformation and the arabesque line, this chapter takes a more wide-angle view on an artist who, while not primarily known for his painting, nevertheless engaged extensively with visual practices through painting, photography, art theory, art criticism, and theater.[11] I thus include here extensive discussion of Strindberg's writings, which might strike the reader as only tangentially connected to Strindberg's visual experiments with chance. But tracing Strindberg's varied and often contradictory deployments of the concept of experiment, as well as those of naturalism and symbolism, is necessary in order to situate Strindberg's painting practice and art theory, not as idiosyncrasies but as underwritten by the increased epistemological value conferred by French experimental psychology on nature's experiments. One of the central arguments of this book is that many of the visual practices produced in the name of symbolism, which were later valued as modernist, were initially envisioned as reorientations rather than as rejections of naturalism. This is very much the case for Strindberg's theory of automatic art and his quasi-abstract paintings, often seen as precursors to twentieth-century automatisms. This chapter reveals that Strindberg's discursive construction of "naturalistic symbolism," the painted form that this "naturalistic symbolism" took, and the essay theorizing this "automatic" painting practice were dependent on an epistemology in which nature's experiments were the most effective and efficient way to construct knowledge about complex human psychological phenomena.

"With the Experimental Formula in View": Naturalism, Experimental Psychology, and Strindberg's *Försökformel*

Zola was undoubtedly the catalyst for Strindberg's earliest attempts to apply his long-standing interest in scientific experimentation to his writing. Already by the early 1880s, having read *Thérèse Raquin* (1867) and *Naturalism in the Theater* (1881), Strindberg had begun pledging his allegiance to the French naturalist author. Strindberg claimed that he and Zola shared a common scientific education that engendered in their work a taste for analysis and a "microscopic gaze which seeks to penetrate to the core of things."[12] In *Naturalism in the Theater*, Zola expressed the hope that, "a man of genius . . . [would] emerge from the crowd to impose . . . a new formula" on drama, and with *The Father*, Strindberg aimed to prove that he was that man.[13] In 1887, he sent Zola his own French translation of the play with a flattering letter, claiming that it had been composed "with the experimental formula in view." As Strindberg summed it up for Zola, the experimental formula would "value the interior action at the expense of theatrical tricks, . . . reduce the decor to a minimum, and . . . maintain a temporal unity as far as possible."[14]

Almost from his earliest encounters with Zola's work, however, Strindberg confessed himself troubled by one aspect of Zola's naturalism. In 1885, he complained that "Young Sweden . . . hides behind an aesthetic word that Zola unfortunately blurted out: analysis. This I would translate as indifference or opportunism on one's own account."[15] Zola had allied experimental method with analysis much more closely in his theater treatise than in his essay "The Experimental Novel." "Each era," Zola wrote in *Naturalism in the Theater*, "has its own formula. . . . We are in an age of method, of experimental science, and we have above all the need for accurate analysis."[16] Strindberg's problem with analysis, or at least with how modern Swedish authors had interpreted Zola's meaning of analysis, was in the injudicious use of boring, minute detail, which rendered, he felt, a dispassionate, two-dimensional reality resulting in "canvases lack[ing] perspective."[17] For Strindberg, analysis that did not take into account the author's own perspective was tantamount to opportunistic falsehood. Echoing Ferdinand Brunetière's criticisms of "The Experimental Novel," Strindberg asked whether it was really possible to experiment on anyone but oneself. The problem, he suggested, was that naturalists tended to have only a "scrappy knowledge of psychology," and thus overreached when they "attempted to describe the all-too-obscure life of the soul." He concluded, "There's only one person's life we really know and that . . . is our own."[18]

Zola's eliding of experimental psychology in favor of a discipline he called "the experimental novel" lay behind Strindberg's initially ambivalent stance toward Zola's experimental method, and indeed, Strindberg was already working to make up for Zola's lapse, intensively studying scientific psychology and laboring to put it in the service of his own writing.[19] In 1883, Strindberg's library contained only eleven books that could be classified under the heading "philosophy and psychology," but by 1892 that number had swelled to seventy-nine.[20] Many of the works acquired in that nine-year period were experimental psychology texts written in or translated into French, including works by Théodule Ribot and Hippolyte Taine. Strindberg told a friend in 1886, with more than a touch of sarcasm, "I now have a whole library on madness, from which it transpires that everyone is crazy apart from the doctors. My delicate nerves and your love of animals are both signs of insanity."[21] This new "library on madness," it seems, had already enabled Strindberg to regard pathology as a kind of construction.

Strindberg's passion for scientific psychology had been ignited during a short stay in France in 1883–84, where he met the Norwegian authors Bjørnstjerne Bjørnson and Jonas Lie, who had been intensively studying French experimental psychology and in particular altered states of consciousness. At the time, Strindberg was working on the poetry collection *Somnambulistic Nights in Broad Daylight*, which he would dedicate jointly to Bjørnson and Lie. A perceptive French critic characterized the poems, in 1884, as "naturalism mixed with the ideal."[22] For Strindberg, the *Somnambulistic Nights* represented a "struggle" with the new psychology, which he hoped would bear fruit after the volume's publication.[23]

Strindberg soon identified those first fruits as a multivolume autobiographical project, envisioned as a corrective to Zola's Rougon-Macquart series in that it turned the scientific gaze onto the self. Strindberg told friends that he was reading so much "psychology, ethics, psychiatry, sociology and economics" that his "head [was] like pulp," but he hoped that all this research would "enable me to analyze myself and discover what makes me tick."[24] In 1886, he sent the manuscript for the first volume, *The Son of a Servant*, to his publisher, claiming that the "psychological material" was "the most important element" and declaring this objective narrative of an individual's psychological development to be an "'evolved' form of the naturalist novel" that he hoped would inspire the "literature of the future."[25] Believing that "literature should emancipate itself from art entirely," he imagined that this new genre would take "The Experimental Novel" one step further. Rather than being merely scientific, literature would itself become a science.[26]

In *The Son of a Servant*, subtitled *The History of the Evolution of a Human Soul*, Strindberg presented his life as a collection of documentary raw material. He used the third person and a journalistic tone to suggest that his descriptions of subjective experiences were objective presentations of fact, and he claimed to have come to this quasi-scientific form as a young man. "Every citizen at a certain age," he recommended, should "deposit his autobiography, written anonymously . . . in the communal archives—Now there's documentation for you!"[27] Strindberg undoubtedly knew that Taine had recommended exactly this kind of documentation as the basis for a new scientific psychology. When Zola published *The Masterpiece* in 1886, the same year that *The Son of a Servant* appeared, Strindberg seized upon the novel as proof that Zola's methods had evolved in the direction of his own, for he saw *The Masterpiece* as above all an exercise in self-analysis that, like *The Son of a Servant*, transformed the novel into a primary document.

In the course of their correspondence in the late 1880s, both Strindberg and Zola recognized that their visions of naturalism were not entirely compatible. While Zola indicated that he had been moved by *The Father* and its "daring philosophical idea," he was nevertheless "bothered" by "shortcuts in analysis" and found the characters too much like "mental abstractions."[28] Nevertheless, Strindberg knew the value of being seen as Zola's follower, especially as he was hoping to have his plays staged in Paris at André Antoine's Théâtre Libre, a venue he would use as a model for his own theater venture.[29] In late 1888, he announced to a friend, "Yesterday . . . I founded a Scandinavian *Försöksteater* (Th. Libre)."[30] Based on Antoine's example of a small theater open only to subscribers, Strindberg conceived his new theater as a closed, tightly controlled system. To Georg Brandes he wrote, "You have here . . . an attempt at *la nouvelle formule* . . . make the pain brief, let the action spend itself in a single movement! . . . I have been confirmed in my belief in my *försökformel*. In every play there is namely *une scène!* . . . Why should I bother with the other flummery."[31] Promoting himself to the influential Brandes, Strindberg claimed ownership of a "new formula," a distilled naturalism based on Zola's experimental method but not exactly synonymous with it. The theater, founded in Copenhagen, lasted less than a year.

Two pieces of theoretical writing linked to the short tenure of the *Försöksteater* sum up Strindberg's attempts to justify his work using rhetoric allied to but not entirely beholden to experiment, and, moreover, reveal the increasing importance of nature's experiments for Strindberg's thinking: the 1888 preface to *Miss Julie* and the 1889 article "On Modern Drama and Modern Theater." The preface, in which Strindberg both paid homage to and sought to supersede Zola, reiterates Zola's claim of an ongoing evolution toward "reflection, investigation, and analysis." Furthermore, Strindberg insisted that the particular innovations of *Miss Julie* in format, lighting, and makeup had all been arrived at "through trials" (*på försök*). And he intended, moreover, to use his audiences for his investigations, presumably at the *Försöksteater*: "In due course I would hope to have an audience so educated that it could sit through a single act lasting an entire evening, but this will first require some further investigations [*undersökningar*]."[32] Strindberg argued that the controlled form of the play would heighten the aesthetic effect, give free rein to the audience's reaction, and allow the individual's imagination to complement and deepen the experience. The manner of controlling scenic conditions in unconventional ways to enhance these effects, which Strindberg called "cropping," was inspired, he insisted, by the "asymmetry and cropped framing of impressionist painting." "I have thereby succeeded," he maintained, "in strengthening the illusion; for not being able to see the whole room or all the furniture leaves us free to conjecture, that is, our imagination is set in motion and we complete the picture ourselves." With the feigned modesty of a scientist placing his faith in the power of his method, he concluded, "I have made an attempt [*ett försök*]! If it fails, there will surely be time to try again [*att göra om försöket*]!"[33]

In "On Modern Drama and Modern Theater," Strindberg criticized naturalist playwrights for relying on the normal and the everyday. Most naturalists were simply realists, he argued, who produced a "kind of mis-conceived naturalism," basing their works on minute everyday detail with no regard for what Strindberg called the "greater naturalism." On this account he attacked Henry Becque's *The Crows* (1882): "There we have the much-longed-for *ordinary* case," he charged, "the *rule*, the universal human norm, which is so banal, so meaningless, so boring that after four painful hours you ask yourself the old question: of what concern is this to me?"[34] Drama, Strindberg contended, should be based instead on real but uncommon circumstances, "an unusual case," he insisted, "but an instructive one, an exception." He proposed studying these exceptions with his *försökformel*, in a tightly controlled manner, with a restricted set of characters, and with only one setting so as better to focus on determining the causal mechanisms of an event. "Our inquiring minds are no longer satisfied with simply seeing something happen," he wrote; "we want to know how it happened. We want to see the strings, look at the machinery, examine the double-bottomed box."[35]

Toward the end of the 1880s, Strindberg became even more disillusioned with Zola, criticizing the 1888 novel *The Dream* as "an opportunistic Academy piece," and even wondering if naturalism was itself "a spent force."[36] But in 1890, when Strindberg read Zola's *Human Beast*, his objections vanished. He was so impressed with how the novel "confirmed" his own "experiences and observations" that he began rereading Zola and pledged to defend the author to a new generation of symbolists.[37] Strindberg's strategy was to position naturalism not as an aesthetic movement but as a scientific discovery, which made any rejection of naturalism and the experimental method mere vanguardist posturing. Recapitulating Zola's insistence on experiment as the scientific confirmation of ideas, Strindberg now idealized both Zola's experimental method and his own *försökformel*, suggesting that they were eternal and could be put in the service of any aesthetic practice, including the new emerging idealisms. By this time, Strindberg was well aware of the latest debates on symbolism in Paris and elsewhere. Entering into this polemic, he would take an unorthodox stance, defending Zola and naturalism as entirely compatible with the new tendencies, as long as Zola, as he would proclaim in 1894, was recognized as a psychologist. In 1892, he implored a correspondent to "re-read Zola! He is the man of today and the future of us as artists, yes even as scientists, for that is what he is!"[38]

Vivisections: Experimenting on Others and Selves

While he was adapting and promoting his *försökformel* for his new theater, Strindberg's enthusiastic reading in experimental psychology resulted more directly in the psychological experiments that formed the basis for the *Vivisections*, which would eventually include his theory of "automatic" art. The first series, published in 1887 mostly in Danish and German periodicals, recounted Strindberg's self-described experiments on others, while the second series, including "The New Arts," was largely conceived as experiments on the self. This shift from the other to the self, or, more accurately, to an other self within, as a source of knowledge, while presaged by *The Son of a Servant*, went hand in hand with Strindberg's self-conscious shift toward symbolism, most fully embraced in the 1896 autobiographical novel *Inferno*. "The only thing that exists is the self (*le culte de moi*)," he wrote to a friend in July 1894; "I know nothing about the world and 'others' except through my self."[39]

Strindberg placed immense intellectual and literary importance on the *Vivisections*, calling them the basis for his "future work" and claiming that he had "discovered" through them "a new, higher, evolved form of literature."[40] Like the autobiographical project, the essays were to be "literature in the modern style," quite literally scientific; he even considered giving the volume the subtitle "A Retired Doctor's Observations (Notes, Dossiers, Stories, Memories)."[41] Describing it elsewhere as a collection of "Psychological Studies," he wrote to one of his publishers, "As the title suggests, my investigation focuses upon living persons. That some of them perish is quite normal with vivisections, when fistular canals are inserted all the way into their intestines."[42] Tellingly, while he encouraged directors to cut his plays as they saw fit, he would insist that the *Vivisections* remain intact.

Strindberg inaugurated the *Vivisections* by investigating his first wife, Siri von Essen, whom he was beginning to suspect of infidelity. There was almost certainly no basis for his suspicions, which were fanned by a growing hatred of the women's movement, but he nevertheless intended to investigate this particular instance of the "woman question" by "researching [*forska*] and experimenting [*experimentera*]."[43] Claiming at the time that his misogyny was "purely theoretical," he intentionally began to inflict "psychic torture" on his wife under the objective banner of experiment.[44] He described the psychological means by which he hoped to extract the name of Siri's lover, how he "read" her mind by fixing her with his gaze, and how he provoked unconscious responses through suggestive conversations, observing her facial expressions as they discussed, for example, massage. "The most complex mind-reading has elicited some vague hints," he noted enthusiastically. "If you are ever moved to conduct an experimental psychological investigation," he told a friend, "it can always be useful for you to have witnessed such an experiment."[45]

Strindberg put these psychological experiments to more productive use in the essays. He had already referred to the autobiographical project as a form of psychological dissection and had conceived of a study of contemporary French peasants under the subtitle "Autopsies and Interviews."[46] But with the *Vivisections*, Strindberg not only co-opted the language of physiology and the long-standing, often caricatured image of naturalist novelists as anatomists (fig. 61); he also echoed Charles Richet and Henri Beaunis, who had used the term "psychological vivisection" to justify the methods of experimental psychology and in particular the use of hypnosis. Richet's and Beaunis's coinage attests to the optimism surrounding experimental psychology at the fin de siècle and the possibility, through studying nature's experiments (such as altered states provoked by hypnosis), of finally rendering knowable human mental and emotional life.[47]

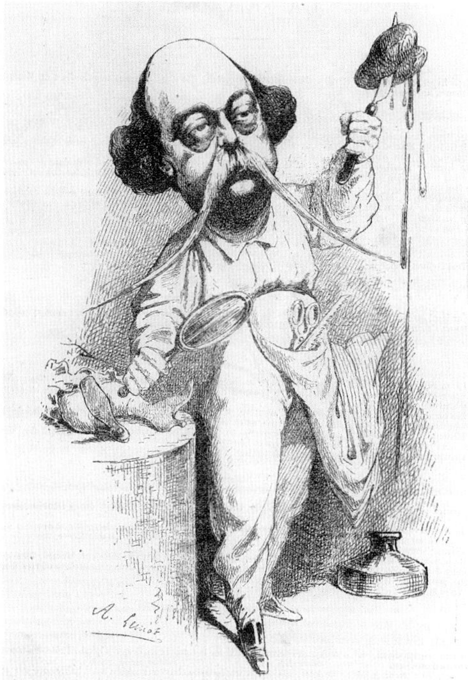

61 Achille Lemot, *Gustave Flaubert Dissecting Madame Bovary*, in *La parodie* (December 1869): 240.

In the *Vivisection* essay "The Battle of the Brains," Strindberg detailed how, after being initially susceptible to the suggestions of a "magnetic" acquaintance, he overcame this hypnotic influence through observation, analysis, and experiment. The text recounts how he eventually planted his own ideas in his colleague's brain through hypnotic suggestion, thus winning a violent psychological battle characteristic of the struggle for power in the modern world.[48] "Mysticism—For Now," an essay that Strindberg referred to as a "strategic lie," took up the tradition of retrospective medicine promoted by Émile Littré and Jean-Martin Charcot in order to give a "physio-psychological account of prayer."[49] Analyzing a recent event in his own family—prayers for a sick child that preceded the child's cure—he concluded that "belief is nothing more than a concentration of wish and desire, heightened to conscious will, and willing is the greatest expression of nervous activity."[50] "Soul Murder (Apropos *Rosmersholm*)" is a wide-ranging discussion of psychological power structures between individuals and society held together by a "web of unconscious deceptions."[51] This particular essay is a veritable hodgepodge of contemporary psychological and psychiatric theory. It includes meditations on the fine line between sanity and insanity, literary examples, such as Hamlet, of what Strindberg called soul murder and soul suicide, and criticism of the asylum as a prison used to lock away troublesome citizens, a critique that echoes the rhetoric of the psychiatric reformers to whom Strindberg and Edvard Munch were close, discussed in the following chapter.

As the tortuous experiments on his first wife implied, Strindberg's claims that he was experimenting enabled him to assert an objectifying distance from what he knew were some of his less palatable views. On numerous occasions, he claimed that his misogyny and his anti-Semitism were merely theoretical, although these claims are at odds with the highly questionable ways in which he treated women and numerous Jewish friends and acquaintances. In the 1890s, as Strindberg more often conducted psychological experiments on himself, this supposedly objective stance of experiment had additional advantages. His so-called experiments with mental illness, hallucination, hypnosis, and the like helped Strindberg to reconcile his own fascination with madness with a frequently expressed fear of being committed to an asylum.[52] He vigorously defended himself against critics who pointed to contradictions in his writing as evidence of insanity by claiming that his experiments were all performed in the name of a greater aesthetic good. Such critiques "stem," he argued, "from my having adopted, or tried out, different points of view in order to see things from many sides! That is rich and humane! Experimenting!"[53] Writing in 1890 to Georg Brandes, who had introduced him to Friedrich Nietzsche's work two years earlier, Strindberg summed up his intellectual development as an ongoing evolution from a metaphysical, moralistic stance to socialism and, finally, to a new experimentalism that he had achieved with "The Battle of the Brains." He would "henceforth . . . experiment," he told Brandes, with Nietzsche's viewpoint, in order "to see where it leads."[54] By 1890, just a year after his breakdown, Nietzsche had come to represent a stereotype of dangerous subversion, a pathogen with the potential to drive readers insane.[55] In claiming to experiment with Nietzsche, Strindberg once again emphasized that he was experimenting with the pathological.

By the end of 1890, Strindberg was making it clear that he had begun to turn Taine's scientific criticism onto himself, and he implied that madness had become central to his method: "What, then, is the method in the supposed madness of my writing? . . . I don't know. I let my mind work freely, . . . one thinks one works consciously, while all along the unknown, subterranean forces of race, heredity, epoch, govern one unconsciously. Popularly speaking, I think one could call me a Seeker, who experiments with points of view." Like the experimenter who embraced doubt as a necessary aspect of his method, Strindberg characterized himself as "a doubter who believes fanatically in the justification of his doubt as a source of wisdom." At the same time, he argued, this seeker was also something like a medium: "A complex of individualities, stemming from many crossbreedings of blood and brains, of many passed stages for which the author only makes himself a 'writing medium' . . . employing all the discoveries of contemporary philosophy, psychology and science."[56] The seeker and the doubter, the objective experimenter and the medium subject, the willed and the unwilled, the conscious and the unconscious—all were bound up in Strindberg's self-construction as a man of the future. And, as he embarked upon a search for symbolist form, which included a quest to produce symbolist paintings, this self-construction would lead him away from the romantic notion of the artist insistently emphasizing his own subjectivity, and toward the late nineteenth-century image of the scientist intentionally courting his less willful mental functioning in order,[57] in Strindberg's case, to objectively investigate the self.

Visual and Material Regimes of Chance

In the early 1890s, Strindberg began a longer period of self-exile from Sweden, where he had had few professional successes. He moved to Berlin in the fall of 1892 and quickly established himself as an influential member of the Black Piglet avant-garde, whose members came together in the wake of Edvard Munch's controversial solo exhibition at the Verein Berliner Künstler, which opened and soon closed in November 1892.[58] In Berlin, Strindberg painted and became deeply involved in his chemical experiments, designed to overhaul the established system of elements. To support his unorthodox scientific experimentation, he sought out a self-identified scientific avant-garde from within the Black Piglet group, but he found less backing than he had initially hoped for.[59] Strindberg's fellow Piglet member, the German poet and writer Richard Dehmel, forwarded early versions of Strindberg's scientific tract *Antibarbarus* to a chemist acquaintance who dismissively recommended that Strindberg first learn proper techniques of experimentation and then catch up on the last ten years of research in the field. Disillusioned by this advice, Dehmel lamented the fact that Strindberg was ignoring his writing and painting for chemistry.[60] Another Piglet member, however, Carl Ludwig Schleich, who himself was attempting to overthrow established ideas in neurology, identified closely with Strindberg's renegade project, believing that "someday the scholar Strindberg will be regarded as highly as the poet."[61]

Members of the Black Piglet circle read and contributed to one of the foremost German publications of experimental psychology of the day, *Sphinx*, founded in Leipzig in 1886 with the unwieldy subtitle *Monthly Journal for the Historical and Experimental Foundation of the Transcendental Worldview on a Monistic Basis* (fig. 62). In its emphasis on experimentalism and its orientation toward French psychopathology, *Sphinx* perfectly encapsulated Strindberg's concerns at the time. And when the journal widened its scope in 1892, changing its subtitle to *Monthly Journal for the Life of the Soul and the Spirit*, this foreshadowed Strindberg's own turn to the occult, especially the scientific brand of occultism that Strindberg would find attractive in Paris in the mid-1890s (fig. 63).[62] *Sphinx* probably

62 *Sphinx: Monatsschrift für die geschichtliche und experimentale Begründung der übersinnlichen Weltanschauung auf monistischer Grundlage* 1, no. 1 (January 1886), front cover.

63 *Sphinx: Monatsschrift für Seelen- und Geistesleben* 7, no. 13 (1892), front cover.

64 "Aus Kerners Kleksographien," *Sphinx* 11, no. 62 (February 1891): 113.

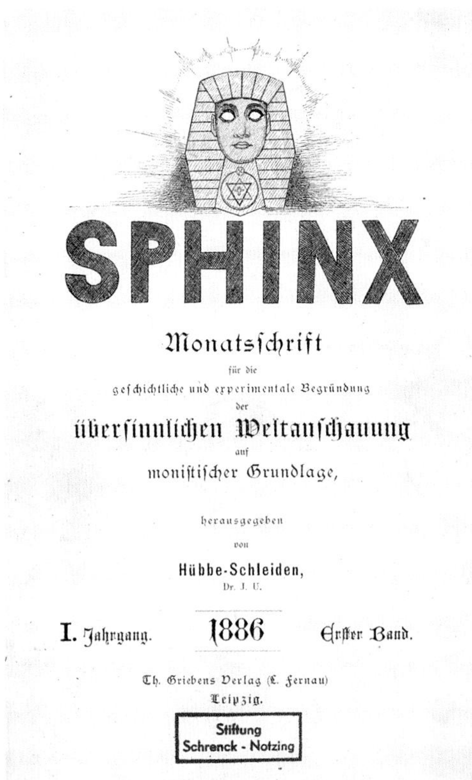

SPHINX

Monatsschrift

für die

geschichtliche und experimentale Begründung

der

übersinnlichen Weltanschauung

auf

monistischer Grundlage,

herausgegeben

von

Hübbe-Schleiden,

Dr. J. U.

I. Jahrgang. 1886 Erster Band.

Th. Griebens Verlag (L. Fernau)

Leipzig.

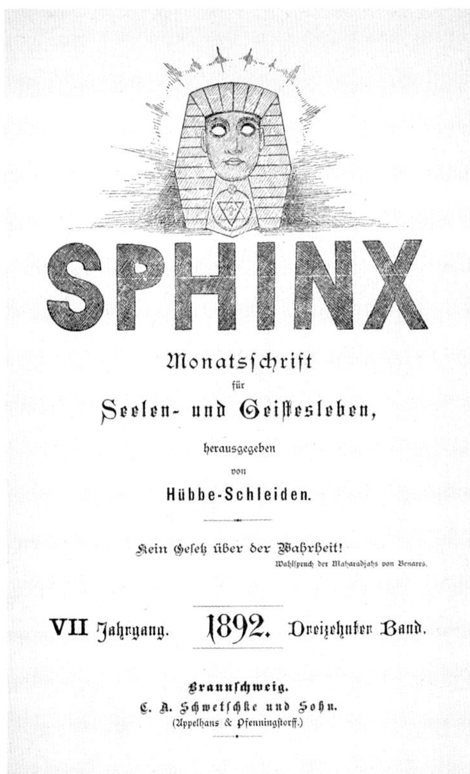

SPHINX

Monatsschrift

für

Seelen- und Geistesleben,

herausgegeben

von

Hübbe-Schleiden.

Kein Gesetz über der Wahrheit!

Wahlspruch der Maharadjahs von Benares.

VII Jahrgang. 1892. Dreizehnter Band.

Braunschweig.

C. A. Schwetschke und Sohn.

(Appelhans & Pfenningstorff.)

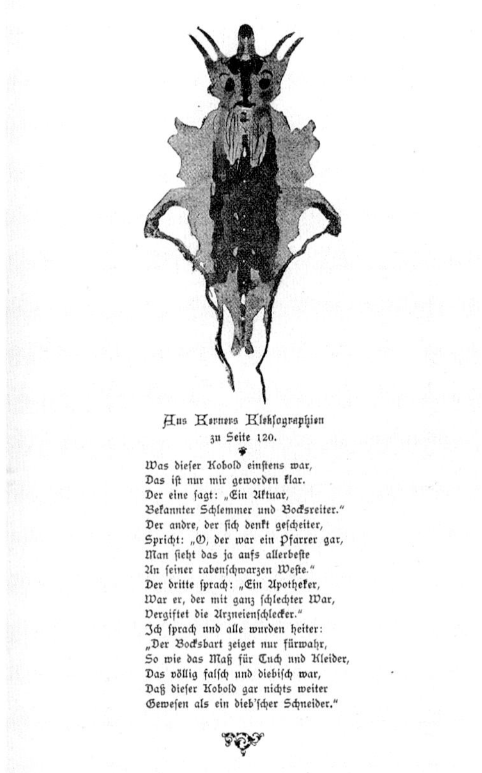

Aus Kerners Kleksographien
zu Seite 120.

Was dieser Kobold einstens war,
Das ist nur mir geworden klar.
Der eine sagt: „Ein Aktuar,
Bekannter Schlemmer und Bocksreiter."
Der andre, der sich denkt gescheiter,
Spricht: „O, der war ein Pfarrer gar,
Man sieht das ja aufs allerbeste
An seiner rabenschwarzen Weste."
Der dritte sprach: „Ein Apotheker,
War er, der mit ganz schlechter War,
Vergiftet die Arzneienschlecker."
Ich sprach und alle wurden heiter:
„Der Bocksbart zeiget nur fürwahr,
So wie das Maß für Tuch und Kleider,
Das völlig falsch und diebisch war,
Daß dieser Kobold gar nichts weiter
Gewesen als ein dieb'scher Schneider."

introduced Strindberg to Justinus Kerner's "Klecksographs," which may have been Strindberg's earliest encounters with graphic representations of chance. In 1886, *Sphinx* had celebrated the hundredth anniversary of Kerner's birth with an article by Carl du Prel, a frequent contributor to *Sphinx* whom Strindberg greatly admired. Five years later, when the Klecksographs were published in a book, the journal reproduced them alongside Kerner's poetry (fig. 64).[63] Kerner had ascribed his discovery of these accidental inkblots to his increasing blindness in the 1850s. Folding the paper, rubbing the wet ink together, and then unfolding the paper produced symmetrical images in which Kerner, a devotee of Franz Anton Mesmer, saw the work of spirit forces. With additional touches of ink, he subsequently enhanced the blots to bring out the specific features of the fantastical creatures he perceived (fig. 65).

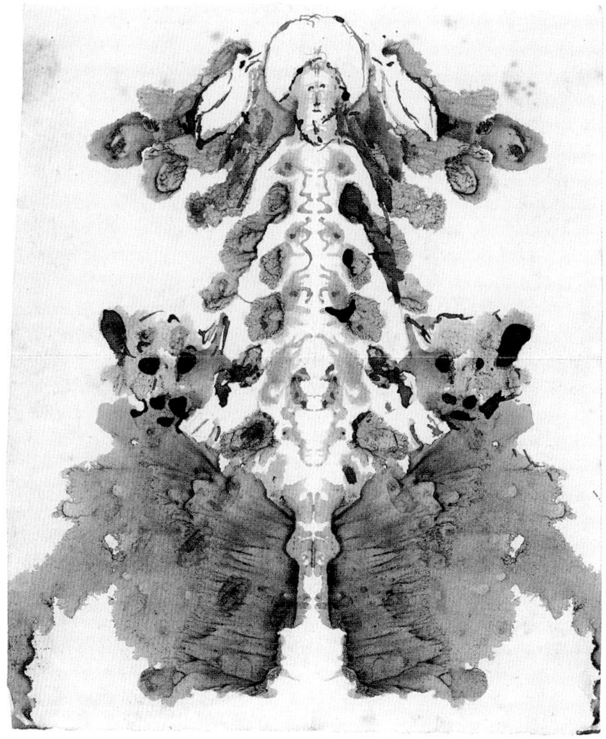

65 Justinus Kerner, *Klecksograph* (Inkblot), 1850s. Ink on paper. Städtische Galerie im Lenbachhaus und Kunstbau, Munich, G18402.

Strindberg himself dated his fascination with the visual and material products of chance to a mid-1880s visit to a colony of Swedish artists in Grez-sur-Loing in the forest of Fontainebleau, where he had been intrigued by what his painter friends did with the chance remnants of paint left over on their palettes. This interest in chance deepened during his Berlin period, spurred on by ongoing engagements with photography that led Strindberg to explore photographic practices that did not involve a camera. One result of this interest would be his own "Klecksographs," included in his *Occult Diaries* (fig. 66).[64] Like Kerner, Strindberg wrote that he was beginning to suspect that accident and coincidence were signs of a spiritual force that could be investigated scientifically through phenomena such as inkblots, cloud formations, "celestographs," and pieces of burnt coal. Such chance resemblances and coincidences were given profound and complex meaning in *Inferno*. In clouds, Strindberg saw the head of Napoleon. In a photograph of the poet Paul Verlaine on his deathbed, he saw a demon. In burnt coal, he saw fantastic creatures that he captured in small charcoal drawings (fig. 67). And on *Sphinx atropos*, death's head moths, he saw not only skulls but also Chinese characters.[65] Strindberg increasingly conceived of chance as tightly bound up with the natural world, and he came to see cultivating chance as akin to emulating nature's own processes. Experimenting with nature's chance occurrences would thereby become the basis for Strindberg's "naturalistic symbolism."

Strindberg often wrote of painting and science as twin antitheses to his creative writing, with the latter remaining through much of his life his main preoccupation and most reliable source of income. He frequently associated both painting and science with

66 August Strindberg, *Klecksograph*, from *The Occult Diaries*, 23 February 1898. Ink on paper. National Library of Sweden, Stockholm, MS SgNM 72, 1898, 54.

67 August Strindberg, charcoal drawing representing partially burnt coal, 1896. Charcoal on paper, 8.9 × 12 cm. National Library of Sweden, Stockholm, MS SgNM 15:6,5.

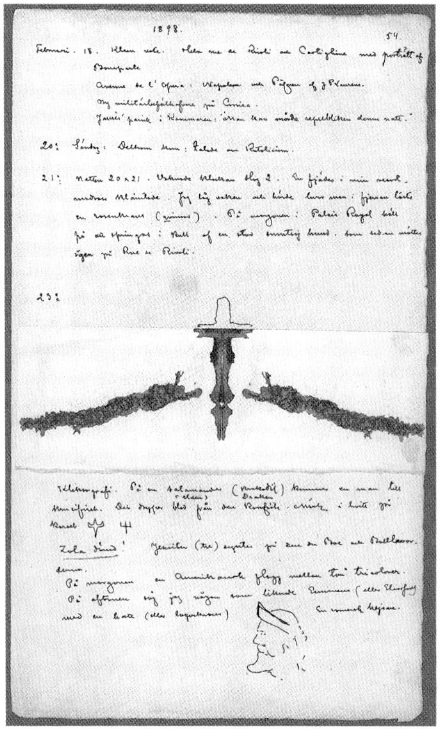

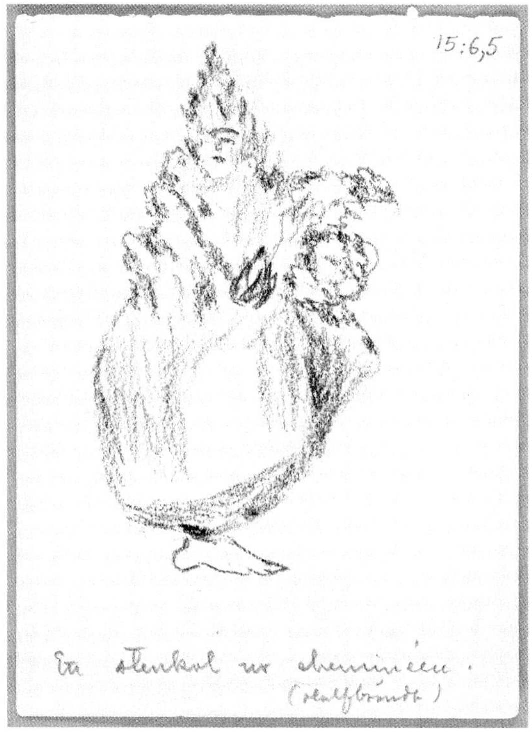

periods of psychological crises. "I can't deny," he wrote in 1892, after embarking on a second intensive period of painting, nearly twenty years after the first, "that this resurgence of my first, youthful love, natural science, seems to me an autumn flower which must bloom before winter comes, just as my old flame, painting, had to reappear before I died. What battles haven't I fought to suppress this passion for research [*forskningslåga*], which was ruining my writing and bringing my family to *misère*."[66] Strindberg had gone to Dalarö, on the eastern coast of Sweden, in the spring of 1892, where he often painted solitary flowers on desolate beaches (fig. 68).[67] In Berlin, no doubt affected by Munch's efforts to establish a symbolist visual practice that Black Piglet member Stanisław Przybyszewski would soon term "psychic naturalism," Strindberg continued the painting campaign begun in Sweden, but now explicitly under the rubric of symbolism.

Night of Jealousy (fig. 5), the painting given to Frida Uhl as an engagement gift in 1893 and the first that Strindberg identified as symbolist, carried a much more evocative title than Strindberg's earlier paintings. The title suggests that the work be read as allegory, but it differed in other ways from works done earlier at Dalarö such as *Flower on the Shore* (fig. 68). While both use the thicker paint application that Strindberg favored during his second painting campaign, applied with a palette knife rather than a brush, the Dalarö

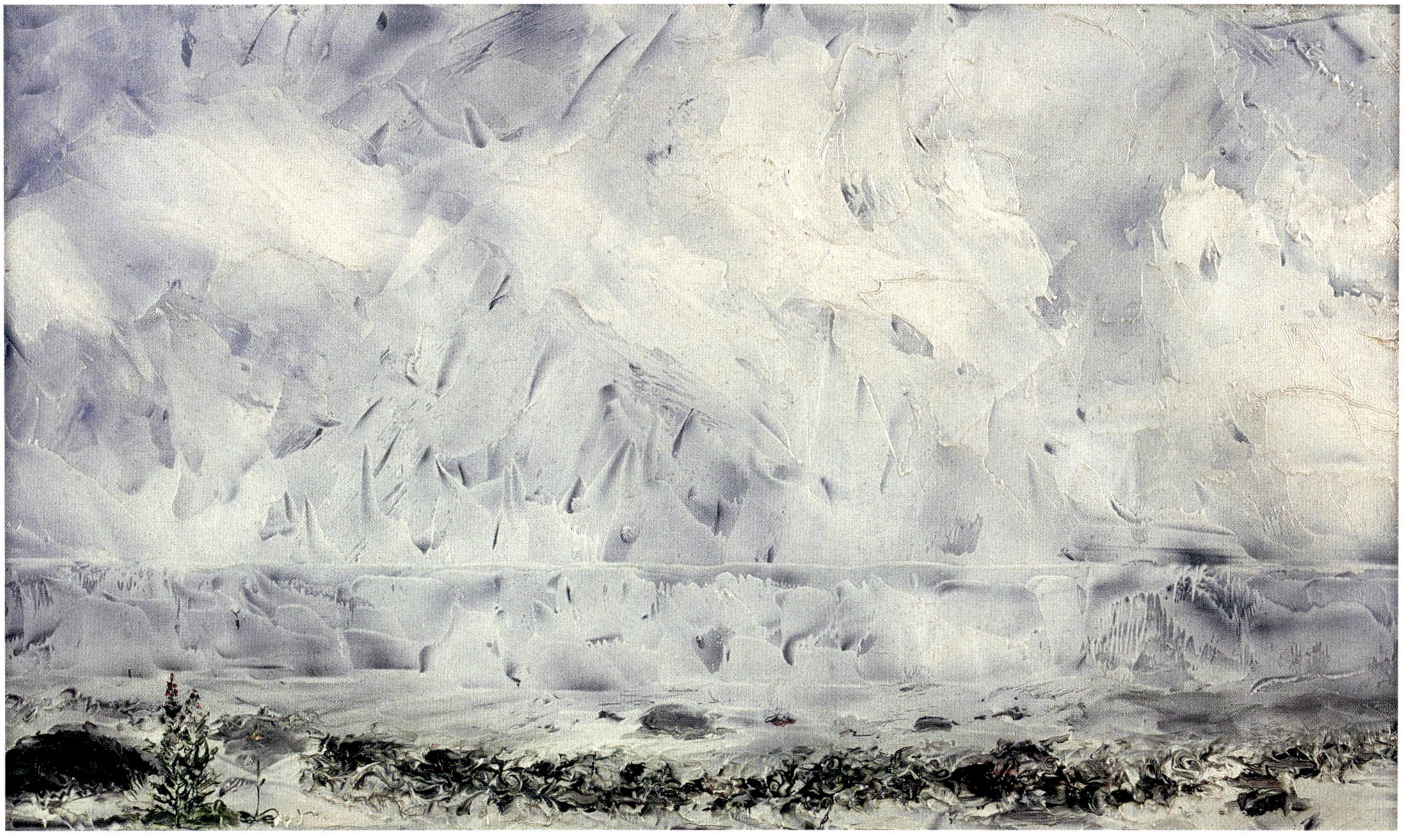

picture maintained conventional proportions of sea and sky, and used a lighter, more pastel palette. *Night of Jealousy*, by contrast, with its violent paint application, unusual composition, and somber colors, rejected landscape conventions almost to the point of abstraction. In 1894, Strindberg specified that he was the first to make "symbolist" or "symbolic landscapes," not "symbolist paintings." This was a crucial distinction, since the most familiar symbolist visual practice for Strindberg would have been that of Munch, who at that moment in Berlin was concentrating on themes of love and death as he worked intensively on his *Frieze of Life* motifs. While Munch's "psychic naturalism" usually involved a radical thinning of paint, such that motifs fused with the material support of the work,[68] Strindberg's symbolist paintings emphasized the thickness of paint rather than the support, which in the early 1890s was rarely canvas and more often cardboard or zinc.[69] The support, in other words, rarely became a critical component of the work, although it did in the case of *Palette with Solitary Flower on the Shore* (fig. 69), with which Strindberg inserted himself into a long tradition of virtuoso paintings on palettes. The palette painting also signified the method theorized in "The New Arts"; the essay began

68 August Strindberg, *Flower on the Shore, Dalarö*, 1892. Oil on zinc, 24.7 × 43.5 cm. Malmö Art Museum, Malmö, Sweden.

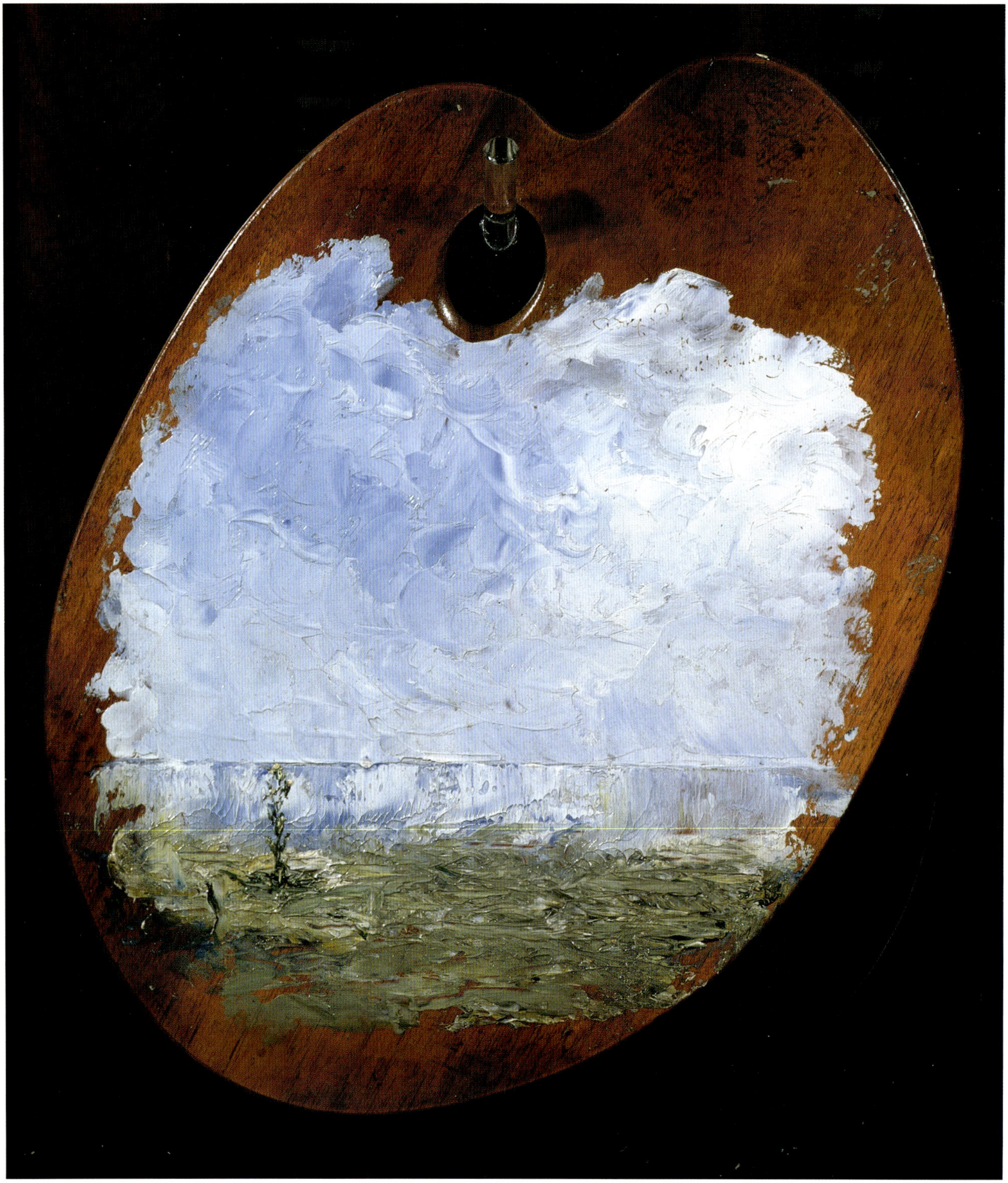

with Strindberg describing his fascination with palette scrapings and also emphasized his use of the palette knife rather than the brush. Significantly, Strindberg signed this work by scratching his name into the paint in the upper right-hand corner, fusing himself with this defining tool of the serious artist.

Strindberg highlighted the naturalism of "naturalistic symbolism" most obviously in his choice to focus exclusively on land- and seascapes, which no doubt hearkens back to his earliest encounters with vanguard painting in the form of impressionism. At the same time, he also conserved the equation, stretching back to Gustave Courbet, between realism or naturalism and raw materiality in the form of roughly applied paint, often given the plasterlike quality of having been applied and then smoothed over with a palette knife.[70] The symbolist components of Strindberg's practice, in other words, inhered less in the motif, and even less in the surface quality of the works, than in how Strindberg described his method, which started as an encounter with the material products of chance and proceeded to cultivate chance, bringing a number of the paintings from the early 1890s, including *Night of Jealousy* and *High Seas* (fig. 70), very close to abstraction. Perhaps the most radical of all of Strindberg's paintings, and possibly the last painting he did in Paris in 1894, *High Seas* used a drier, more impasto paint application, and incorporated an additional technique of chance. Holding a flame to the work, perhaps from a Bunsen burner used for his chemical experiments, Strindberg burned portions of the paint surface, creating a sooty effect in some areas that could not have been entirely consciously controlled. Like Édouard Vuillard, Strindberg associated the acts of painting and viewing with both conscious and unconscious mental practices, with both the willed and the unwilled. Strindberg may have had his theory of automatic art in mind before he traveled to Berlin. But within the Black Piglet circle, he found the means with which to produce his first symbolist paintings and to begin theorizing "naturalistic symbolism" as reliant on chance and automatism.

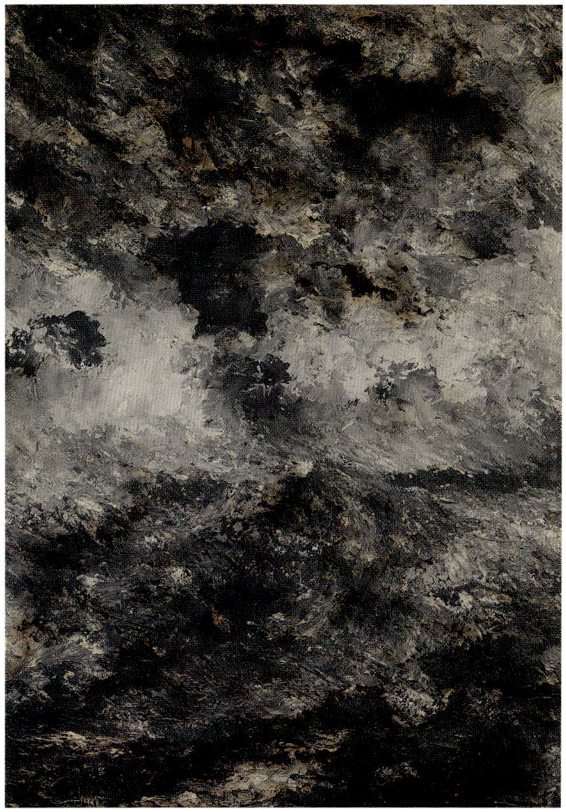

69 (*opposite*) August Strindberg, *Palette with Solitary Flower on the Shore*, 1893. Oil on wood, 38 × 34 cm. Private collection.

70 August Strindberg, *High Seas*, 1894. Oil on cardboard, 96 × 68 cm. Sven-Harrys Konstmuseum, Stockholm.

"Deranged Sensations" and "The New Arts!"

Before Strindberg left Austria for Paris in the late summer of 1894, he submitted two *Vivisections* essays to *La revue des revues*, including "The New Arts! Or Chance in Artistic Production," as advance guards for his assault on the capital. A third essay, "Deranged Sensations," began appearing in *Le Figaro*'s literary supplement in November 1894, the same month that "The New Arts" was published. "Deranged Sensations" dramatically recounts the psychophysiological effects wrought by a train journey to Paris, and reveals Strindberg's continuing dependence on a pathological method. Strindberg explained his intentions in the essay to the Swedish painter Richard Bergh. Curious about its effect on Bergh's "delicate nerves," Strindberg characterized "Deranged Sensations" as the product of a hard-won "favorable mood" in which he had tried to "anticipate the capacities of a future, more highly developed mental life." At his regular Parisian haunt, he claimed, it astounded his audience as "new, *extraordinaire* but mad." "Since my talent wasn't in question," he boasted to Bergh, "nor its originality, my madness only served as seasoning, and I'm now addressed as 'Cher Maître.'"[71]

Strindberg introduced the essay by conjuring up a range of associative mental images that he said he had experienced during the train trip to Paris, a city he described as the "fair and factory of battling brains." The train carriage was, he wrote, a "torture cell" filled with passengers "snatched from their environment" and packed in all together. Arriving in Versailles, he fell asleep, but soon woke up in an abnormal state, his faculties of perception and memory profoundly disturbed by the physiological effects of the long journey. "The jolts of the train had stirred up the pulp of my brain," he wrote, "so thoroughly that I lost the ability to follow my thoughts. It was as if my connecting fibers had been broken, because I felt as if my head was empty; in each moment, I wanted to remember something that I could not recall." A walk outdoors confirmed that his perception, and especially his vision, had been altered as a result of the voyage. He described the odd feeling of walking toward Versailles while the château appeared to remain the same size. The next day, retracing his steps, he manifested the same perceptual problem, but he now explained it as a psychophysiological disturbance, affecting his ability to perceive perspective in normal and conventional ways.

Strindberg—or, more accurately, the narrator of the story—then apparently experienced what might be termed a pathology of memory, which he diagnosed by identifying incongruities between what he saw before his eyes and the mental image he retained from his first trip to Versailles in the 1870s. He explained the gap as his memory's being affected by his knowledge of the "grandeur" of Louis XIV's reign and by his familiarity with modern architectural practices strikingly similar to those of Versailles. Suffering from what he determined was an irrational fear of the space between himself and the château, he found himself unable to approach the edifice any more closely, repelled by an unknown force. An act of will, however, enabled him to overcome this fear by "returning to [his] philosophical ideas," recalling that other, similar abnormal perceptual phenomena had so far resisted explanation. This rationalization allowed him finally to enter the grounds of Versailles.

Once inside, Strindberg wrote that he experienced auditory hallucinations but wondered, hopefully, if they might not be explained by the evolution of his body and mind toward the modern. Are "my nerves . . . evolving towards more refinement," he asked, "and my senses becoming more subtle? Am I going to have a new skin? Am I on the verge of becoming modern?" Returning to his lodgings, he slept for days, as a flood of sensations, memories, and emotions washed over him. Puzzling over cloud formations, and concerned that his house was moving for no reason, he continued to try to make sense of the odd phenomena he was experiencing. Finally, he decided to emerge again into the world: "stripped of the clothing of a civilized man . . . I see with the clearsightedness of a savage. I listen and I sense like a Red Indian!" He entered a woods, contemplated a pine tree, and when he reemerged saw only what at first appeared to be a vast desert, a scene that soon resolved itself into "the City, the great city, the greatest city in the world . . . the little dirty houses of buyers and sellers. Paris!! . . . It is really Paris!! Hail!"[72]

At once a meditation on altered perception and an homage to the dizzying sensory effects of Paris and its environs, the essay deployed the pathological method as a way of creatively approaching the world and of sloughing off civilization in favor of more primitive forms of perception. Strindberg thus positioned himself as both subject and object of the text, the first-person narrative of his own psychological experiences often colliding with the more objective-sounding assessments of those experiences. The text is constructed as a journalistic account of self-experimentation, a testing out of experience under different pathological conditions in order to achieve the pure perception of the noble savage, which in the end results in the essay itself, a product of the author's modern primitivist imagination. "Deranged Sensations" declares not only that Strindberg had learned much from studying the pathological states of mind he either experienced or was capable of inducing, but that this method of investigation could result in a creative product. This particular stance toward experimentation, dependent on nature's experiments, synthesized Strindberg's psychological research of the 1880s with his self-conscious attempts to fuse naturalism and symbolism.

While "Deranged Sensations," in its objectifying descriptions of Strindberg's apparently willful attempts to make rational sense out of unwilled phenomena, implicitly combined the conscious and the unconscious to bring forth a creative product, "The New Arts" gave a more explicitly theorized account of how Strindberg envisioned the conscious and the unconscious combining to produce art, and painting in particular. In a letter to a friend, which included a description of a number of the paintings that he had sent ahead from Austria to Paris, he claimed to have "invented a new (that's to say, old) kind of art," which he called "l'art fortuite."

> It is the most subjective of all art forms, so that in the first place only the painter himself can enjoy (= suffer) the work because he knows what he meant by it, as do the chosen few who know the painter's inner (= outer) a little (= a lot). Each picture is, so to speak, double-bottomed, with an exoteric aspect that everyone can make out, with a little

> effort, and an esoteric one for the painter and the chosen few. It should be pointed out that the pictures were painted in a half-dark room, and cannot on any account stand a full light; they appear best in strong fire-light or a half-dark room. All the pictures are painted using only a knife and unmixed colors, whose combination has been half left to chance, like the motif as a whole.[73]

The letter contextualized Strindberg's ambitions for his painting and its accompanying theory among the symbolist visual practices that had been developed by vanguard groups such as the Nabis and At the Black Piglet. The idea of a privileged coterie whose members were to be initiated into an understanding of difficult, esoteric works echoed Maurice Denis's intentions for his painting *Décor*, which Denis imagined would be understood on three levels, from the simplest to the most philosophical, the last being accessible only to the elect.[74] This is one of the central paradoxes of symbolism, namely, that it ostensibly courted universal modes of communication, but that its practitioners often acknowledged that they anticipated communicating with only a select few, those who were already men of the future. Strindberg's Neoplatonic vocabulary—including the terms *exoteric* and *esoteric*—was commonplace among Black Piglet members, as well as among the more occult-curious symbolist groups in Paris and elsewhere. A review in *Sphinx* of a recent book by contributor Raphael von Koeber on Arthur Schopenhauer's philosophy emphasized the distinctions Koeber had made between exoteric and esoteric natures, while Édouard Schuré's *The Great Initiates* reminded readers that "all the great religions have an exterior history and an interior history; one obvious, the other hidden."[75]

At the outset of "The New Arts," Strindberg declared that so-called primitive modes of making art, including craft, were due more to random acts of nature than to the willed acts of human beings. He linked these supposedly primitive modes with his own encounter with a set of panels he had seen while staying in Grez-sur-Loing in the 1880s. Using that journalistic tone so familiar from both his novels and his essays, he enumerated the banal subjects of the panels: "portraits of ladies; a) young; b) old, etc. Three ravens on a branch. . . . You know immediately what they are." Immediately recognizing a work's subject matter, however, offered little aesthetic pleasure. Pleasure in looking derives not from mimesis, Strindberg maintained, not from asking, What does it represent?—but from posing an altogether different and much more fundamental question: "What is it?"

Strindberg was soon told by his painter friends at Grez-sur-Loing that what he was actually looking at were "palette scrapings," the remnants of whatever paint colors were left on the artist's palette that could themselves serve as the basis for further inspiration if the artist was "in the mood" to "make some kind of sketch." By mimicking nature's own randomness, these palette scrapings constituted, Strindberg wrote, the basis for a more "natural" mode of creation. Harmony was already guaranteed, since the colors on the palette were left over from a previous, more conventional work. Liberated by the chance arrangement on the palette from having to make conscious choices about things such as color harmonies at the outset, the human soul could thereby express itself more freely: "the painter's soul is able, with all its creative force, to look for the contours. . . . The hand moves the palette knife randomly." The result, Strindberg argued, was an ideal and "charming hodgepodge of unconscious and consciousness," which allowed the artist to reject mimetic representation in favor of an art produced in the same way as "capricious nature." The artist remained appropriately unconcerned with, even unaware of, the final results, and the painting could therefore well and truly be called "natural art."

Similarly, Strindberg insisted, the pleasure in viewing derived not from instantly recognizing the motif and acknowledging the artist's skill in accurately representing it, but rather from the process of looking while letting the imagination wander through the different possible answers to the question "What is it?" To further clarify this kind of aesthetic experience, Strindberg described the pleasure he derived from seeing an object in the landscape that he could not immediately identify, giving a detailed account of how he understood his consciousness and unconsciousness to be functioning throughout the psychological experience of trying to figure out what he was looking at.

> At first, it was a cow, then two peasants embracing, then a tree trunk. . . . This fluctuation of impressions was pleasing to me . . . an act of will and I no longer want to know. . . . I feel the curtain of consciousness rising. . . . No, I don't want it to. . . . Again. . . . Now, it is a lunch outdoors, they are sitting down to eat. . . . But the figures are motionless as in a panopticon. . . . Ah! darn! the charm vanishes . . . finished. . . . It is an abandoned plough on which the laborer has thrown his outfit and hung his bag! Everything has vanished. There is nothing more to see. The source of joy has gone!

For Strindberg, it was obvious how this pleasure, which stemmed from the experience of perceiving an ambiguous object, related to the aesthetic experience promised by modern painting.

> Does that not offer a striking analogy with modernist paintings, so incomprehensible for the "philistines"? At first one only sees a jumble of color; and then it starts to look like something. It looks like. . . . But no, it does not look like anything. Suddenly, however, a point is fixed like the nucleus of a cell. It grows, the colors cluster around it and accumulate. Rays form and spread into branches, into twigs just like ice crystals on the window . . . and the image presents itself to the viewer, who witnesses the act of the picture's procreation. In this way, the painting is always new, changing with the light, it never tires, it is endlessly rejuvenated, endowed with the gift of life.

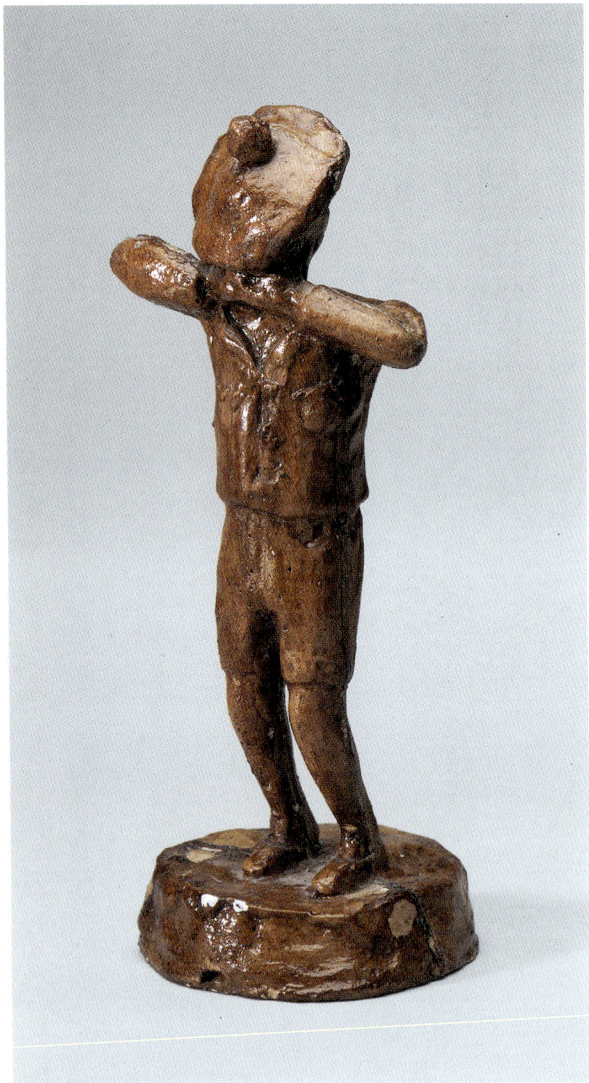

71 August Strindberg, *The Weeping Boy*, 1891. Gilded plaster, 20 cm high. Private collection.

Aesthetic value, in Strindberg's view, thus inhered in the work's polysemous quality, from the extent to which the beholder could see different things depending on his particular "mental state."

Although he claimed to have discovered aesthetic pleasure in palette scrapings a number of years earlier, Strindberg noted that he only "improvised" the "theory of 'automatic art'" while molding a small clay figure (fig. 71).[76] What he initially intended to be a young supplicant was transformed into a weeping boy when, in frustration, Strindberg clapped his hand down on the figure's head, metamorphosing the head into a tam-o'-shanter. After a few final adjustments, Strindberg declared the work "perfect," and in a way it must have been, since he re-created the figure at least once more. Strindberg's interest in modeling, however, appears to have been short-lived. Instead, he translated his deliberate attentiveness to chance into a method of painting by logically linking his experience in making the statuette to the pleasure of perceiving nature as infinitely suggestible, resulting in a theory that attempted to synthesize the conscious and unconscious functions involved in creation.

Using the painting *Wonderland* (fig. 72) as his key example, he claimed to start from a motif not in front of him but already synthesized by memory and by knowledge of the conventions of painting. Strindberg's motifs were always, it should be said, very conventional—landscapes, seascapes, and forest interiors such as *Wonderland*. Appreciating the pleasures of ambiguity and the role of the unconscious, Strindberg elaborated a painting process in which he paradoxically unwilled himself. He reworked initial marks, remained attentive to randomness, stayed open to suggestibility, and cultivated a chance application of paint. His goal was to mystify the motif, make it less fixed, and through the entire process bring forth a living, breathing, and ultimately "natural" painting. The theory of automatic art encouraged the artist to grope for form by paying attention to the automatic processes of perception and then to elaborate on those processes through the conscious functioning of the will. This was not unlike the multistep method of painting that Vuillard had developed a few years earlier, explored in the previous chapter. Strindberg's automatic art sought to bring forth visual form through psychological experimentation on what he assumed were his own more primitive mental faculties in order to achieve a more pure, more true kind of art, a "naturalistic symbolism" not so much representing nature as emulating nature itself.

Strindberg conceived of his symbolist visual practice as an entirely new kind of painting, an advanced art for the future based not on mimetic representation but on the processes of nature investigated scientifically. He had no illusions, however, about the staying power of this new style and readily acknowledged that all vanguard art was destined to be superseded. Strindberg's theory of painting instituted a willful attentiveness to automatism, to chance, to the Tainean *petits faits* of psychological experience that since the 1870s had been crucial to experimental psychology's methodological imperative of nature's experiments. He explicitly connected such phenomena to his own primitive mental functioning and to supposedly primitive, less evolved forms of art making, revaluing both as wellsprings of an entirely new kind of art. In 1894, Black Piglet member Franz Servaes declared that Munch was a more radical modern primitive than Paul Gauguin, in that Munch did not need to travel to Tahiti in order to experience the primitive; he was able to plumb the depths of his very own soul.[77] Although Strindberg would publicly defer to Gauguin as much more savage than himself, he nevertheless aspired to be recognized among the era's modern primitives, courting, in the essays "Deranged Sensations" and "The New Arts" and in his painting practice, the kind of praise reserved by Servaes for modern primitivism.[78]

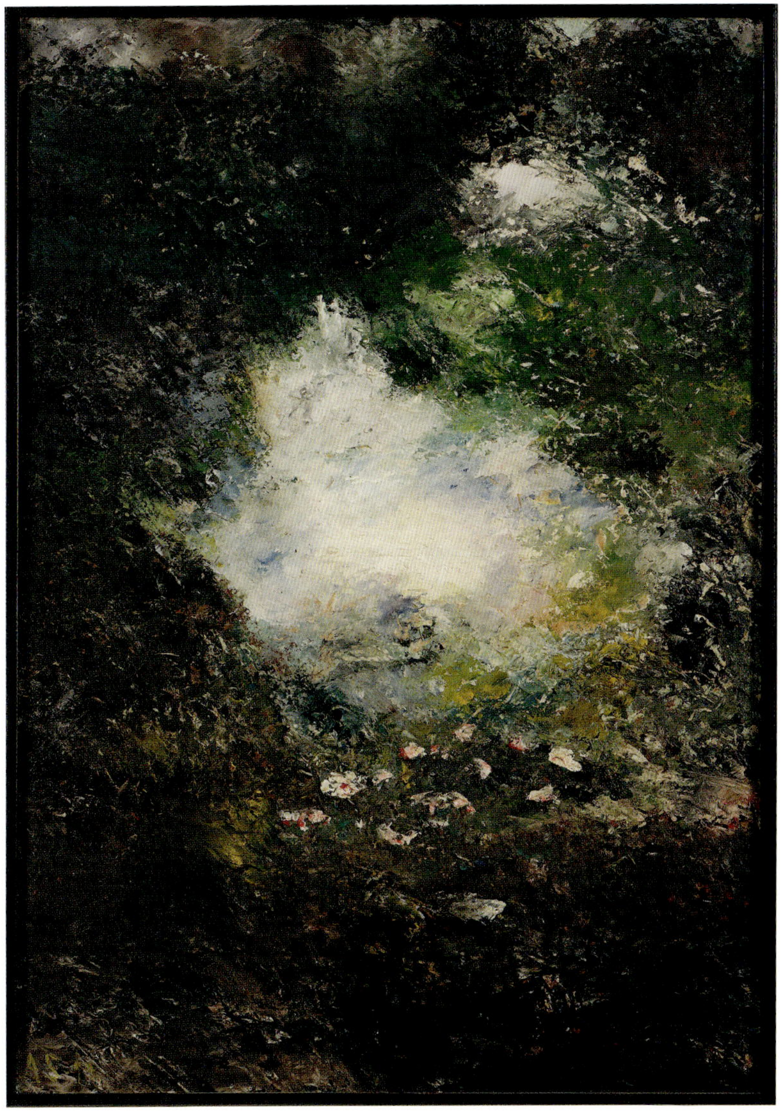

72 August Strindberg, *Wonderland*, 1894. Oil on cardboard, 72.5 × 52 cm. Nationalmuseum, Stockholm.

Strindberg, Entrepreneur of Experiment

Strindberg repeatedly identified himself and his extraordinary faculties with the future and with an avant-garde revolution to come.[79] In this way, he strategically adopted a rhetorical reversal similar to that used by Denis in his own manifesto for a new art, the "Définition du néo-traditionnisme." The normal present, meaning normal philistine tastes, normal bourgeois values, and normal naturalism, were now judged abnormal.

And all that was abnormal—the primitive, the insane, the automatic, and the unconscious—was newly valued, not simply as decadent appeals to antiestablishment values but as pathways to truth. This reappraisal of the pathological as useful, as a functional point of access, was made possible by positivist epistemologies that saw the pathological as a starting point for the scientific production of knowledge.[80]

The most obvious example of Strindberg's method of experimentation on his own pathological states remains the autobiographical novel *Inferno*, in which Strindberg recounted how he encouraged his "other self" to come to the surface so that he could investigate his own alterity.[81] The method was entirely in line with the French experimental psychology that he had been reading so enthusiastically since the 1880s. The experimental method privileged an objective exploration of pathologies as the basis for studying human physiology, sociology, and psychology. Only the explicit emphasis on the self, encouraged by discursive constructions that situated symbolism as "objectifying the subjective," was ostensibly new. By 1896, Strindberg was proclaiming himself the "Zola of the Occult," a "naturalist-occultist" who desired to "see first with my outer eyes and then with my inner eye."[82] He still understood himself fundamentally to be a man of scientific method, but now more than ever he pictured himself on the cutting edge of science, poised to take the work of Jean-Martin Charcot, Claude Bernard, and others much further than ever imagined, and to align himself with the period's most radical investigators of the psyche. In 1894, he already felt that he had "scented this imminent major fin-de-siècle current in science."[83] "I am not a spiritualist," he told one of his translators, "but it has been observed that in my research I have gone beyond routine science and established the transition to scientific occultism."[84] Situating himself in the lineage of the best-known nineteenth-century experimentalists, he emphasized that the domain might have changed but that the scientific methods of experiment remained unassailable.

Strindberg recognized that psychological self-experimentation could effect a synthesis of naturalism and symbolism, and he elevated this "naturalistic symbolism" to an art form that had as its aim new knowledge of universal experience. In the autobiographical novel *Gothic Rooms*, written in 1903, Strindberg summarized his experimental attitude in the guise of his main character, Arvid Falk: "He experimented with points of view, and like a conscientious worker in a laboratory set up control experiments, adopting tentatively an opposed position." The stance of the experimentalist, Strindberg valiantly claimed, might have its perils. Conflating the heroic scientist with the tragic romantic genius, Strindberg brought experiment and vanguardism together: Falk "was a vivisectionist who experimented with his own soul, always suffered from open wounds, until he gave his life for science."[85]

Ultimately, "naturalistic symbolism" was a kind of alchemical product. Strindberg attempted to reorient the natural sciences and scientific methods toward a synthetic art that would insert the self more completely into the natural order of things, and would in turn bring the spiritual and art back into the fold of the natural sciences. The formal innovations at which Strindberg arrived, including the near abstraction of paintings such as *Night of Jealousy* and *High Seas*, were like precipitates of this alchemical process. In other words, as in Paul Cézanne's work, the deliberateness of the formal innovations in Strindberg's writing and painting is uncertain. For someone who desired above all to know the truths of the universe, beginning with the self, experiment offered a way to that ultimate destination. The means of reaching it could be either scientific or artistic, but they were not in and of themselves constructed as the goal.

After *Inferno*, Strindberg relinquished many of his scientific aspirations, for the most part ceasing the chemical experiments that were "bringing [his] family to *misère*" and returning more seriously to writing plays and novels. Scholars have often interpreted this shift as the result of a newfound psychological stability after he returned to Sweden for good in 1898. But another explanation presents itself. By the early years of the twentieth century, having built up a robust rhetoric of artistic experimentation, having established himself as fundamentally engaged in his writing and his painting not just in *försök* but also in experiment, Strindberg could safely sever scientific experimentation from his experimentation in other domains.

One of Strindberg's major contributions to modernism, therefore, was his entrepreneurial attitude toward experiment, his diffusion of experiment, largely by analogy, to all his practices, whether in the realm of science, literature, the theater, photography, painting, or sculpture. In 1904, in *Black Banners*, a biting critique of contemporary literature, Strindberg reminded his readers that he had built his entire oeuvre on his own experiences, for "experience is capital."[86] Strindberg's thinking remained very much governed by the French language, in which *expérience* carries the double meaning of experience and experiment. Given that he so successfully consolidated the rhetoric and practice of experiment with the rhetoric and practice of vanguardism, he might as well have written that experiment was capital. The extraordinary products of Strindberg's experimentation, like Cézanne's, have taken on the status of almost accidental artistic discoveries—the surprisingly modern language of Strindberg's plays and novels, the existentialist intimations of a number of the essays, and the protosurrealist emphasis on chance and the automatic in the paintings and art theory.[87] But such supposedly unintentional "discoveries" are just as readily understood as constructed by and constitutive of nineteenth-century experimental cultures.

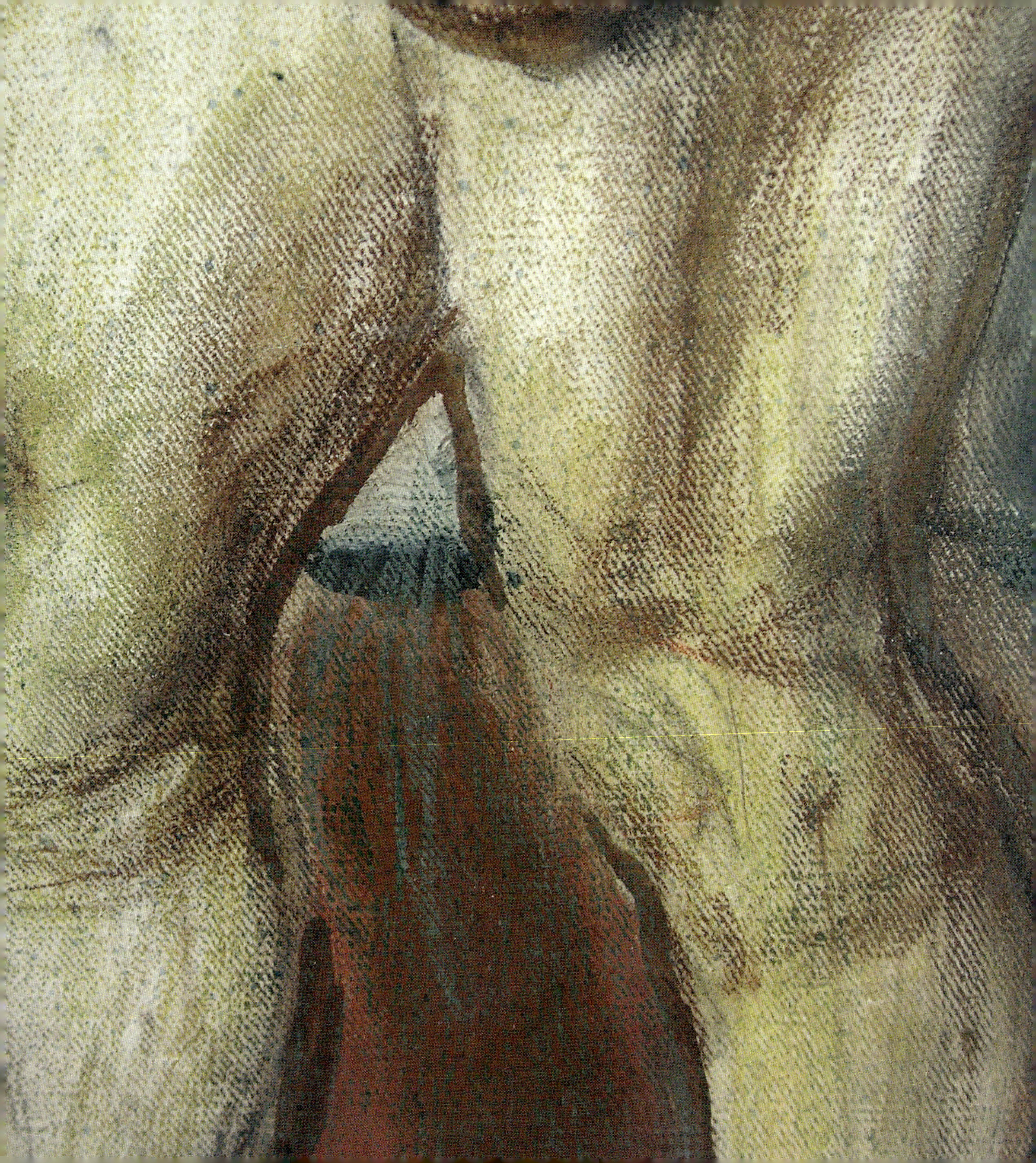

5

MADNESS AS METHOD

The Pathological Experiments of Edvard Munch

Sometime in 1893, Edvard Munch revisited a self-portrait that he had begun before traveling to Berlin in the fall of the previous year (fig. 73).[1] Around an essentially naturalist depiction of his own impassive countenance, Munch added a red background overlaid with wide black arabesques and, hovering above his head, a menacing and grotesque female face. On his own painted visage, he enacted a startling and seemingly irrational violence, scraping away large areas of paint on the cheeks and forehead to expose and gouge into the wood panel beneath. Munch never exhibited the portrait in his lifetime, but it nevertheless acted as a declaration, to himself and probably to a small group of supporters, that he had finally sloughed off naturalism in favor of symbolism and a new artistic identity.

This extraordinary self-portrait capped nearly four years of striving in fits and starts toward symbolist form. In 1889, having achieved notoriety at home in Norway and garnered state support for his virtuosic naturalism, Munch had set off for points south, where he had quickly assimilated the latest French modes: impressionism, pointillism, and eventually the synthetist flat planes of the school of Pont-Aven.[2] By 1891, the Norwegian painter and critic Christian Krohg was already identifying Munch's work with "Symbolism, the latest tendency in French art," and by 1893 Munch's transition from naturalism to symbolism appeared more or less complete.[3]

Rather than a frightful mask or a depiction of Medusa,[4] the grotesque face at the top of the self-portrait is more convincingly read as an exaggerated *masqueron*, a decorative element typically found on gilt frames common to the late nineteenth-century neo-baroque interiors of Charles Garnier.[5] Convalescing on the French Riviera throughout much of 1891 and 1892, Munch spent considerable time in one of Garnier's most sumptuous

73 Edvard Munch, *Self-Portrait*, 1893. Tempera on unprimed wooden panel, 70 × 44.3 cm. Munch Museum, Oslo.

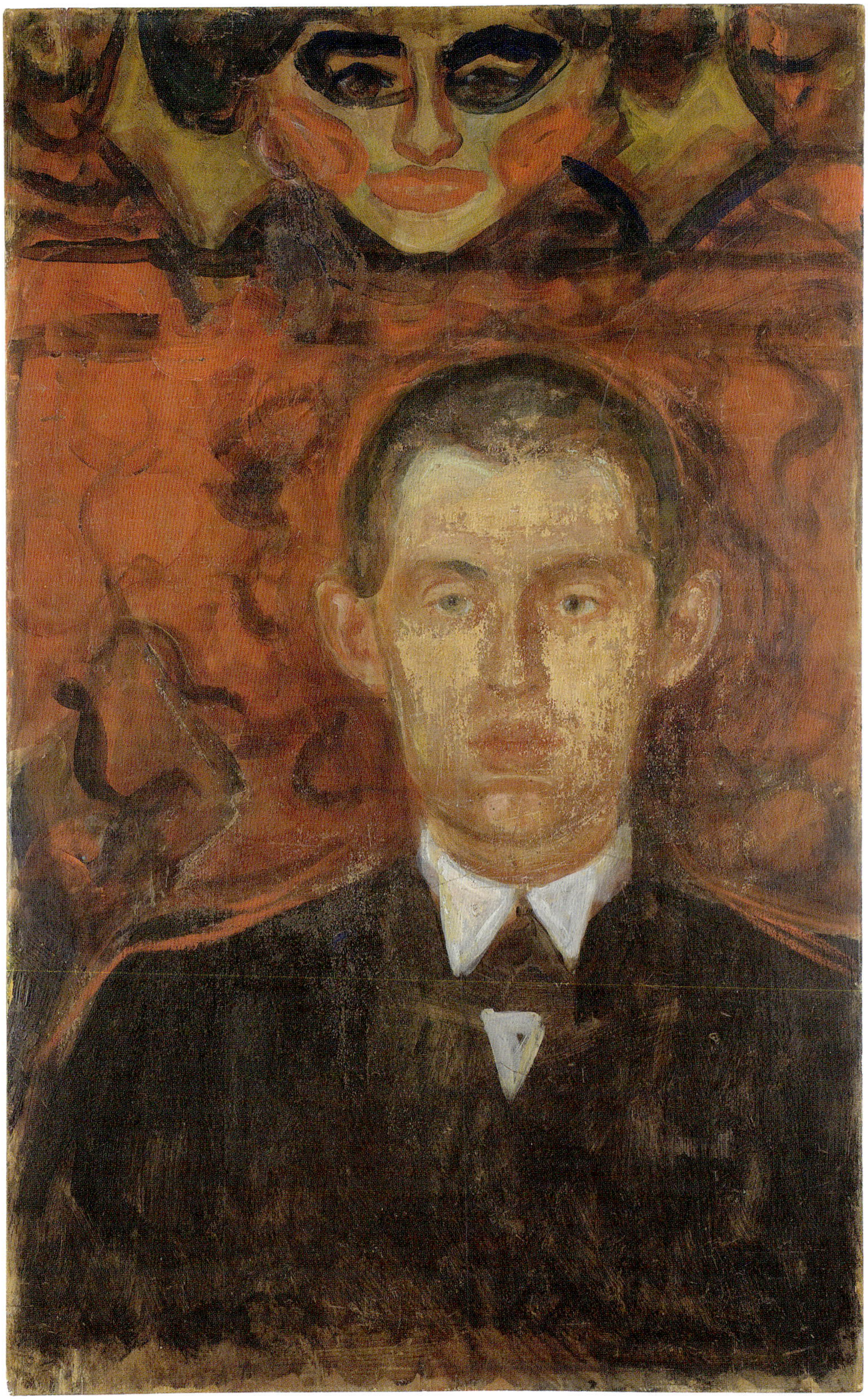

architectural creations, the casino at Monte Carlo, where he observed and possibly experienced what was known in the nineteenth century as gambling mania. Depicting himself in evening attire under a *masqueron* and against a stylized red damask wall covering—situating himself, in other words, at the casino—Munch used this self-portrait to identify Monte Carlo as a key locus in his search for symbolist form. The seemingly irrational violence visited upon his face implied that the irrational had played a role. And indeed it had, for at a crucial juncture, Munch had attempted to paint the irrational madness of gambling using mad, irrational form.

Munch's work has often been seen as a signifier of the artist's madness, as in and of itself insane, or as having the potential to provoke insanity in others. As early as 1890, a Norwegian art critic claimed that Munch's proclivity for cool blue tones was symptomatic of the artist's neurasthenia.[6] Five years later, a Norwegian medical student argued publicly, probably with Munch in the audience, that his compatriot's art stemmed from mental illness and was therefore, like the artist himself, abnormal.[7] Much more recently, in 2005, nearly a year after the theft of two paintings from the Munch Museum in Oslo, a consultant for a risk-management firm advocated unprecedented security measures because, she said, Munch's works "talk to you with such an intensity of expression that you have to worry about unstable people being provoked to vandalism."[8] Most art historians, however, no longer assume transparency between Munch's work and his biography. Reinhold Heller and Patricia Berman, for example, read the artist's putative madness as at once mythic and strategic, as instantiating a privileged position of truth telling and as helping to promote a romanticized vision of the mad genius.[9] This now orthodox view of Munch's madness as tactical, however, does not fully address how madness might have been useful for Munch's artistic practice—how, in the artist's words, his "invalid's mentality had a very favorable effect" not just on his career but also "on [his] work."[10] This chapter argues that Munch's madness comprised a nexus of strategies designed not only to construct a persona or justify a new style but, more fundamentally, to make art in the first place and to make it legible and of consequence to Munch's first audiences of the 1890s.

Munch in all likelihood completed his self-portrait under a *masqueron* toward the beginning of an exceptionally productive two years in Berlin, during which, as a member of the At the Black Piglet (Zum schwarzen Ferkel) group, he painted what would become his most recognized works. A community of intellectuals enthralled by Nietzschean philosophy, monism, French experimental psychology, and psychiatry lauded Munch for having attained symbolist form above all with *The Scream*, the painting that, in words inscribed amid the red paint streaks of the sky, "could only have been painted by a madman" (fig. 6). Having achieved symbolist form in Berlin, Munch's methodological dependence on madness persisted when he returned to Paris in 1896. In *Women in Hospital*, a painting representing the institutionalized insane, Munch turned pathological form to political ends, using his symbolist practice to paint something akin to propaganda for an emerging psychiatric avant-garde with very close ties to symbolism.

Munch's symbolist art of the 1890s constitutes the most sustained example of an artist working to unify form and content in order to communicate what he and his supporters promoted as universal truths, and doing so by deploying a method determined by the epistemological assumptions of nature's experiments. As Munch explored the other within the self, and how the self could become other through madness constructed as mental illness, he developed a visual pathological vocabulary by imagining what psychological alterity might look like. Munch's pathological method was certainly a useful tool for his artistic experimentation, but it was one thoroughly rooted in nineteenth-century experimentalism. This chapter explores Munch's primary relationship with madness—madness as method—and analyzes how this method developed within the larger intellectual, artistic, and social contexts in which Munch actively participated. From his Monte Carlo paintings, to *The Scream*, to images of institutionalized mental illness, Munch strove for truthful and universal portrayals of human beings by situating himself as a scientific observer of "experiments prepared by nature."

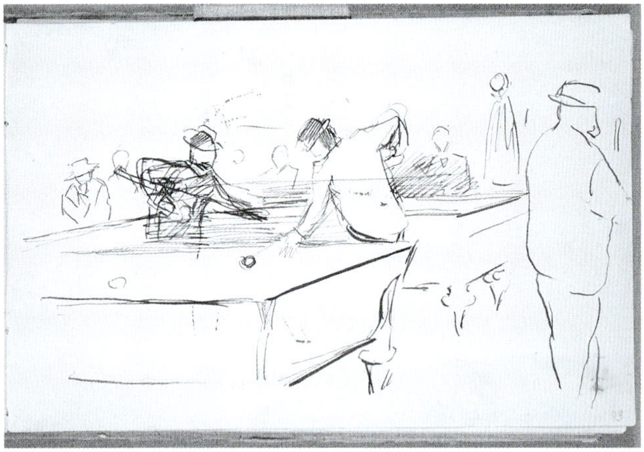

74 Edvard Munch, *Billiard Players*, ca. 1890. Ink on paper. Munch Museum, Oslo, T129, p. 43.

Pathological Symbolism and the Lure of the Roulette Table

When G.-Albert Aurier published one of the earliest critical appraisals of Vincent van Gogh in the *Mercure de France* in January 1890, identifying the Dutch artist's supposed mental pathologies with his sincerity, his individuality, and indeed with his symbolist painting, Munch was living on the outskirts of Paris and familiarizing himself with the writings of French symbolism.[11] At the time, he was sketching and painting typically naturalist motifs, including street scenes, café interiors, and cabarets. But he was also already exploring how to invest these observed scenes with heightened psychological content. Especially attracted, as he wrote, to the "intense green of a billiard table," Munch fixed on the motif of the pool hall and worked to imbue it with different moods.[12] One sketch (fig. 74) conveys a lively social atmosphere similar to Jean Béraud's popular belle époque paintings of billiard halls (fig. 75), but other sketches displace human presence from the center of the room to present a less animated scene. One shows a forlorn broom propped up against an unused table (fig. 76), and in another, a man strikes a pose associated with melancholy (fig. 77). Together, these sketches recall the mood of Van Gogh's 1888 painting of a bar with a billiard table, *The Night Café*, which Van Gogh called one of his "ugliest" works (fig. 38).[13] In *The*

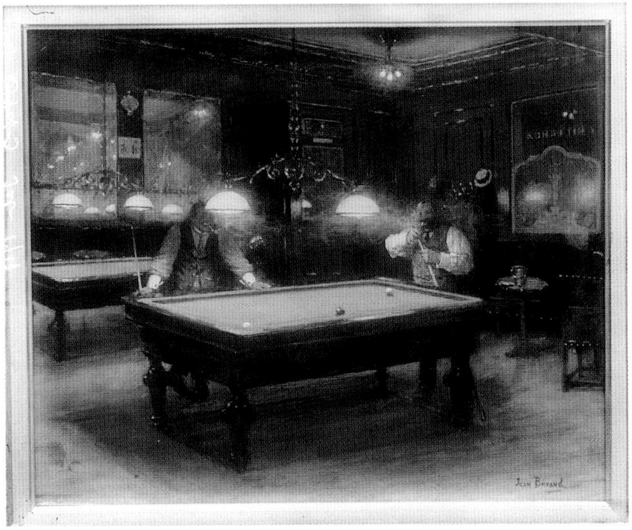

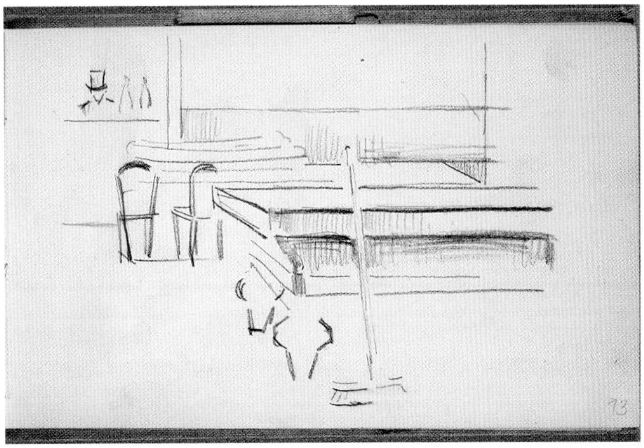

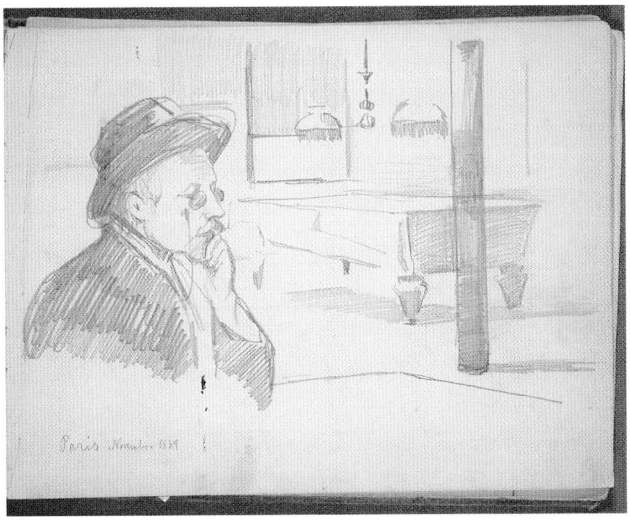

75 Jean Béraud, *The Billiard Players,* before 1919. Oil on board, 35 × 44 cm. Whereabouts unknown. Sold by Christie's New York, 27 October 2004.

76 Edvard Munch, *Billiard Hall,* ca. 1890. Pencil on paper, 11.7 × 19.5 cm. Munch Museum, Oslo, T127, p. 13.

77 Edvard Munch, *Man in a Billiard Hall,* ca. 1889. Pencil on paper, 23.1 × 30.8 cm. Munch Museum, Oslo, T126, p. 3.

Night Café, Van Gogh used form and color to represent psychopathology, "terrible human passions," in his words, "the power of the dark corners of an *assommoir* . . . where you can ruin yourself, go mad, commit crimes."[14] The emptiness at the center of the composition, the exaggerated perspective, the clashing colors, and the harsh artificial lighting, completely at odds with the warm glow of Béraud's convivial scenes, all combine, as Van Gogh intended, to recall, and even perhaps to induce, extreme psychological responses.

Munch's notes on the billiard motif concentrated on how the expansive green baize of the billiard table offered a unique opportunity to experience the physiological phenomenon of the retinal afterimage, and how this instance of abnormal perception could be exploited to capture both the "impression of a moment" and a "mood."[15] Unwittingly, Munch's notes echoed the recommendations of Paul Gauguin, as related by Paul Sérusier to Maurice Denis and other Nabi artists: paint what you see, even if you see colors you believe to be unnatural.[16] But this *Talisman* symbolism was to be enriched by a pathological approach to symbolism in keeping with Aurier's reading of Van Gogh. In the early 1890s, Munch was already trawling for his own psychologically rich motif, one in line with Van Gogh's *Night Café* as a place where one could ruin oneself or go mad. The reds and greens of the billiard table soon gave way to the reds, greens, and blacks of the roulette tables of Monte Carlo.[17]

Early in 1891, Munch began traveling regularly to Monte Carlo, not only for entertainment but in order to observe, in the words of his frequent traveling companion, the "many painterly situations from the gaming room."[18] As was becoming typical of Munch's working method, he approached the motif not only through visual means but also via written ones, elaborating upon and then attempting to distill an experience to its essentials. Munch's many texts on Monte Carlo vacillate between different narrative modes, from journalistic to confessional, from dispassionate observation to interior monologue. The author, narrator, and subject are often at odds; in one instance, an affected stream of consciousness reveals the speaker as delusional. Nonetheless, the notes help trace a shift in Munch's concerns from a preoccupation with class and social alienation in the casino to an interest in the subjective experiences of gambling mania, a shift accompanied by a transition from more naturalist to more symbolist form.[19]

Nineteenth-century critics of gambling considered the mixing of classes and genders to be among the chief moral hazards of the gaming hall. "All types and variety of types meet at Monte Carlo," wrote one such critic, "the millionaire Russian noble throwing down his notes beside the smug *bourgeois* who carefully ventures his little economies; the immaculate county squires elbowing the not too-stainless *prima donna* . . . aristocrats and adventurers; countesses and *cocottes*. . . . Where king Roulette holds sway social and moral distinctions cease to exist."[20] In two ink sketches and the first two of his three roulette paintings, Munch concerned himself with the potentially dangerous consequences of this social leveling. Béraud had done something similar in the painting *Monte Carlo (Rien*

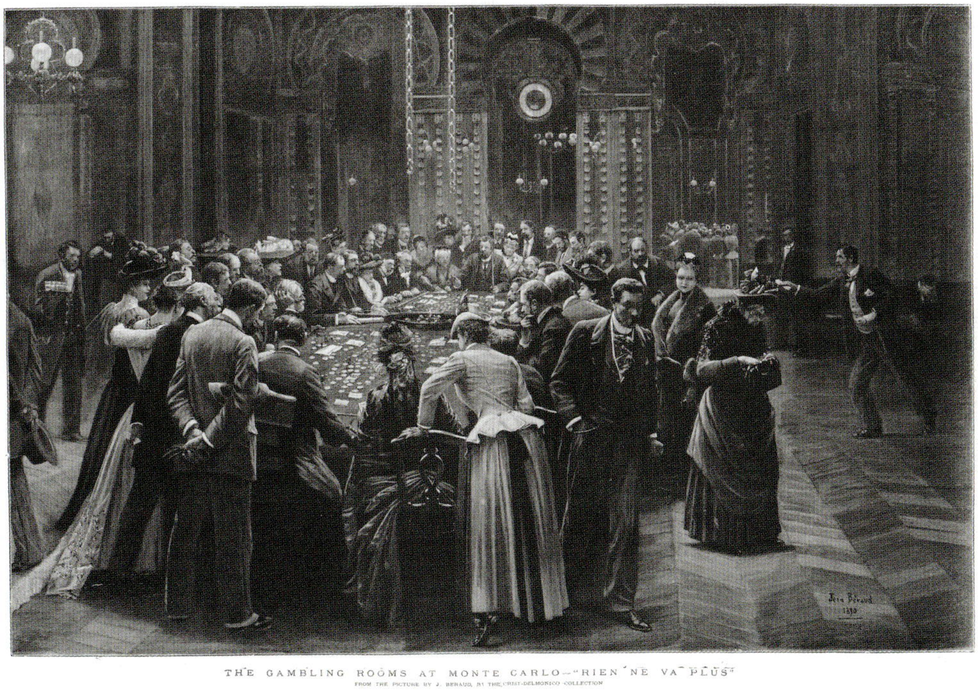

78 Jean Béraud, *The Gambling Rooms at Monte Carlo—"Rien ne va plus."* Engraving after *Monte Carlo (Rien ne va plus!)*, 1890, oil on canvas, 103 × 131 cm. Private collection.

ne va plus!), a popular success at the Salon du Champ-de-Mars of 1890, where Aurier noted—and condemned—its extreme photographic naturalism.[21] Béraud's dynamic composition, disseminated by a double-page engraving in the illustrated weekly the *Graphic* (fig. 78), highlighted the grand interior of the casino and the bustling crowd swelling and contracting around the roulette table. The work is populated by the types noted by the commentator above, many of whom seem to be experiencing various deleterious effects of gambling; mouths gape, heads rest wearily in hands, a fur-draped *cocotte* stares directly out of the scene, and a man in shadows on the far right cradles his head in a gesture of despair.

Munch's two ink drawings distill class divisions into schematic signifiers such as evening wear, pomaded hair, indistinguishable clothing, and swollen hands. Despite their close physical proximity, the figures gaze in different directions, a sign of their disconnection (figs. 79 and 80). The first of Munch's roulette paintings likewise offers up an image of social alienation. It removes the glittering surroundings of Béraud's painting and uses a muted naturalist palette of dark grays, brown-grays, green, and ruddy oranges to portray a crowd of faceless and expressionless drones (fig. 81). Arms reach simultaneously out on the left of the canvas to place bets, lined up in parallel as if they were levers on a machine. In looking at the composition, it is worth recalling that Walter

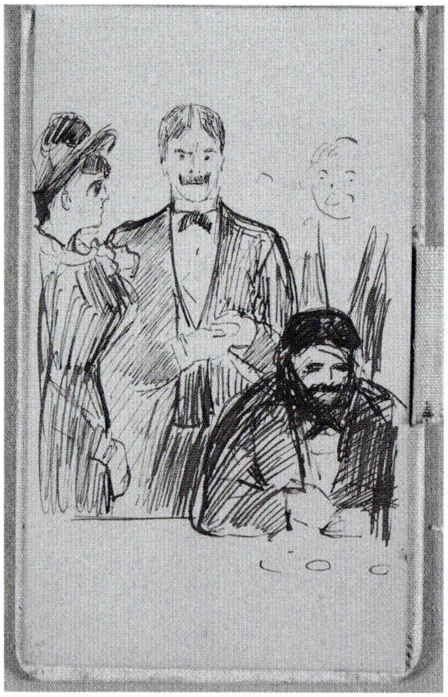

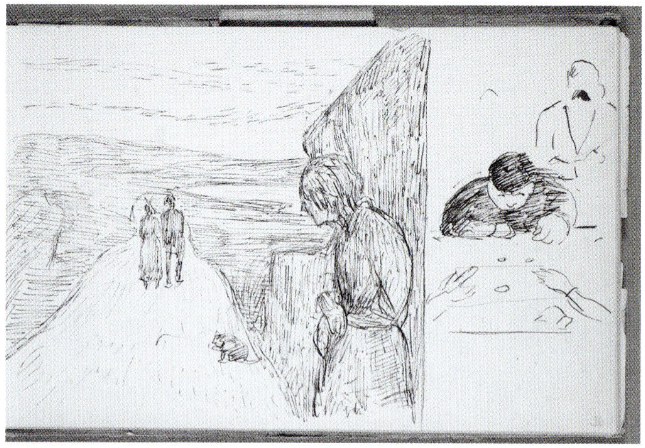

79 Edvard Munch, *At the Roulette Table,* ca. 1891. Pen and ink on paper, 9.9 × 16.8 cm. Munch Museum, Oslo, T128, p. 32.

80 Edvard Munch, *Along a Road (The Lonely Ones)* and *At the Roulette Table*, ca. 1891. Pen and ink on paper, 17 × 27 cm. Munch Museum, Oslo, T129, p. 36.

Benjamin figured the gambler as the mirror image of the industrial laborer, in his mechanized actions and eternally delayed wish fulfillment.[22]

Munch's second roulette painting pushes the crowd into the background in order to distribute class and gender difference across the three foreground figures (fig. 82). On the left, a ruddy-haired man in thinly and sketchily painted gray clothing sits in stark class difference to the elegantly attired man standing on the right, who sports a thickly painted white cravat and waistcoat. Munch portrays him quite literally as more materially substantial than the man on the left, who is now the only figure reaching across the table to place a bet. The enigmatic woman seen from behind with the gravity-defying hat retains her ambiguous position at the center of the composition. Blocking our view of the table, it is not clear whether she is gambling or merely observing. These first two roulette paintings seem to reflect Munch's initial impressions of the gaming rooms—the crowds, the class and gender mixing, the social alienation, and the slightly sordid atmosphere—but their formal strategies, still closely tied to naturalism, largely fail to represent the gambling mania that had begun to fuel Munch's imagination on his repeated journeys to Monte Carlo.

Gambling Mania and Pathological Form

In his reworked notes, Munch heightened the psychological drama of his experiences in Monte Carlo by describing in detail the symptoms of madness displayed by those

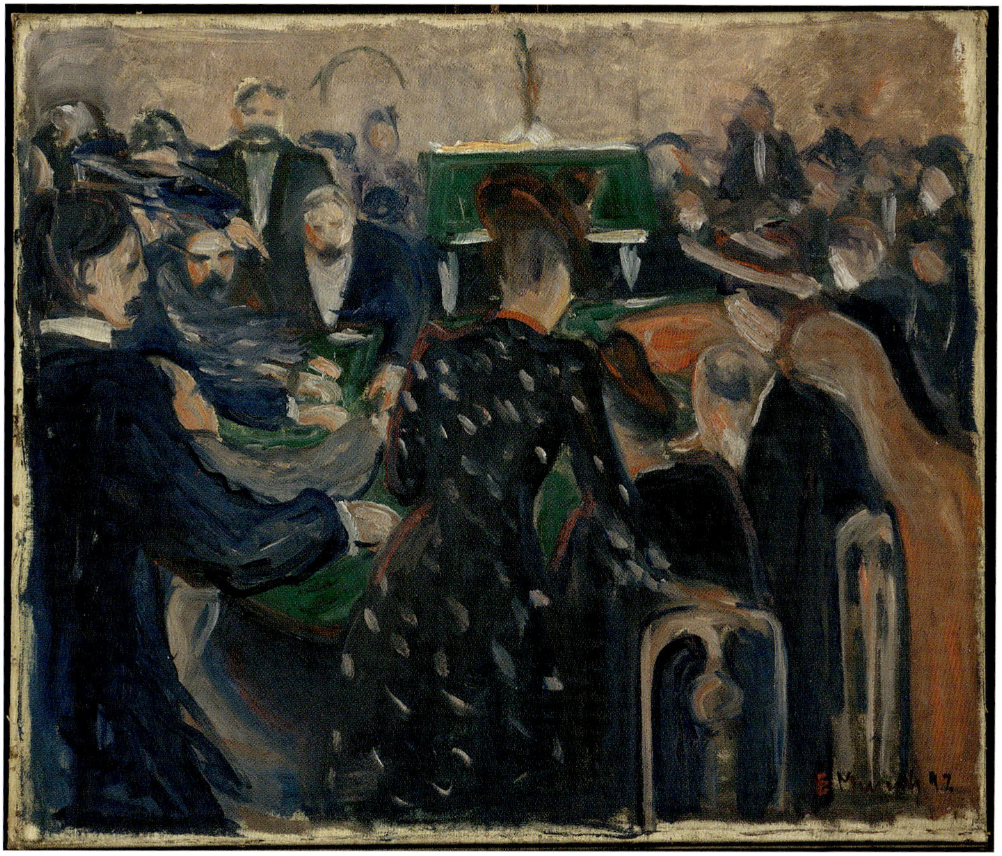

81 Edvard Munch, *Roulette I*, 1891–92. Oil on canvas, 54 × 65 cm. Munch Museum, Oslo.

caught up in playing roulette, including his own apparently intensifying obsession with gambling. He wrote of his sleeplessness, dizziness, inability to work, altered sensory perceptions, tendency to fix his gaze, and overwhelming compulsion to return again and again to the gaming halls in the hope that his luck would turn. Aping Fyodor Dostoevsky, who had written the novella *The Gambler* in order to settle his own roulette debts,[23] Munch identified the uncontrollable urge to gamble as both a physiological and a psychological illness that threatened the integrity of the self. "Is it contagious," Munch wrote; "are there bacteria in the rooms of Monte Carlo"? Gambling, he observed, produced obsessive hallucinations—"always I see before me the emerald green table"—and provoked thoughts of suicide: "how tempting it is despite everything—to just forget everything—everything that tortured me—everything that gnawed away at me recently—if I lose everything what will I do?" At one point, he wrote of himself as temporarily "cured," content simply to walk in the gardens of the casino, ruminating on his art and saying a friendly hello to the guards. But in contemporary parlance, Munch, or rather the text's narrator, was on the verge of a relapse.

82 Edvard Munch, *Gamblers in Monte Carlo*, 1892. Oil on canvas, 72 × 92 cm. Private collection.

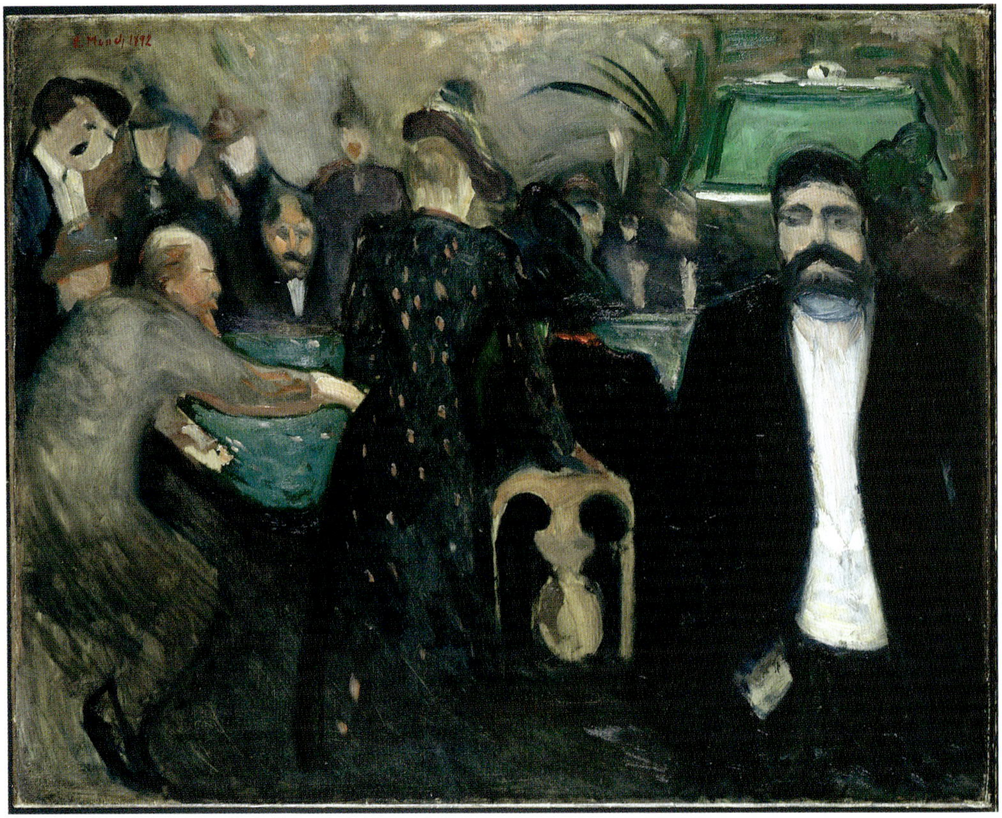

> I went back into the room to meet my friend. He was standing at the table—overheated and nervous. I'm mad—completely mad—I can't stop myself—then he throws down money on the table and loses and loses. I borrowed some francs from him—I placed my bet—and I won again. Now I felt that destiny was my friend—the money was pushed towards me—my pile got bigger and bigger. My trouser pockets—waistcoat pockets—every possible pocket was bursting with gold and silver. Is it my imagination or reality?—everyone seems to be talking about *rouge* and *noir*—about Monte Carlo. . . . In the railway carriage—in the train corridor I hear the words constantly repeated—*rouge—noir—gagner—perdre*—roulette. I dream about it at night—if I hear a conversation in the corridor it is about Monte Carlo—*rouge—noir.* My cheeks are burning—I have an eye infection. Nevertheless—it is an intoxication—it's good to forget—all your worries. I awake in the morning in a pool of sweat—and before a sliver of light at the window heralds the break of day, I hear talk of money in Monte Carlo.[24]

The air in Monte Carlo, in the words of one commentator, was "thick with systems," especially in 1891 and 1892, when Charles "Monte Carlo" Wells broke the bank a record number of eighteen times during an extraordinary run of luck.[25] Numerous books and pamphlets detailed supposedly scientific methods of predicting where the ball would drop, and the casino encouraged belief in such methods by publishing the numbers in a daily bulletin.[26] Munch described a means of breaking the bank, but he clearly signaled the method as delusional, just another symptom of gambling madness:

> It is so obvious—why had no one thought of it before me? Simply wait—just be patient—when one of the colors comes up a number of times, then bet on the other color. If you lose—just wait for the next turn. . . . I will become rich—it is so obvious—I shall be able to devote myself to my work—pure art. . . . Yes, I will wait. I waited until the next suite came up—bet—and lost. This was strange. I decided not to play for a while—just to take notes and study the game. . . . Black came up 6 times in a row. In all probability, the red had to come up now. I put 20 on red. A lot of people had put down large sums of money on the same color . . . *Noir*.[27]

Munch's "method" was, of course, ridiculous. Indeed, this is precisely what is so "obvious," that the gambler's desire for control, his belief in the possibilities of a method, and his apparent certainty that he could beat the system were all invoked as symptomatic of the gambling mania to which Munch represented himself as fully in thrall.

Munch exhibited the first two naturalist roulette paintings in his scandalous Verein Berliner Künstler exhibition in late 1892, and it is likely that only after this critical moment in his career did he attempt to give visual form to these psychologically more intense notes. *At the Roulette Table in Monte Carlo*, probably completed sometime in 1893, more fully dispenses with naturalist form, deploying the roulette motif to paint madness and in the process coming much closer to the visual idiom that Munch would practice at the height of his symbolist phase (fig. 83).[28] The first-person subject of Munch's roulette writings imagined his gambling method not only as a means of supporting his family but as a way to devote himself to his art, not unlike Dostoevsky, who viewed gambling as at once pathological and artistically productive, even transcendent.[29] The method that Munch described was nothing more than a patient attentiveness to the binary of red and black and a carefully timed bet on a probable reversal of that binary pairing. Linking this strategy with "pure art," Munch allegorized a properly symbolist method of painting as a reversal of the naturalist modes with which he was fully conversant and, perhaps more important, with which he was *known* to be fully conversant. Munch thereby positioned himself to invert "normal" formal idioms and to stake a claim on their reversal. In approaching Munch's symbolist practice, it is useful for the art historian to imagine naturalism's other.

At the Roulette Table in Monte Carlo shifts the woman at the center of the first two compositions to the lower right corner of the picture and tips the table up to provide a more unobstructed view of its surface, which is now placed on a dramatic plunging diagonal.

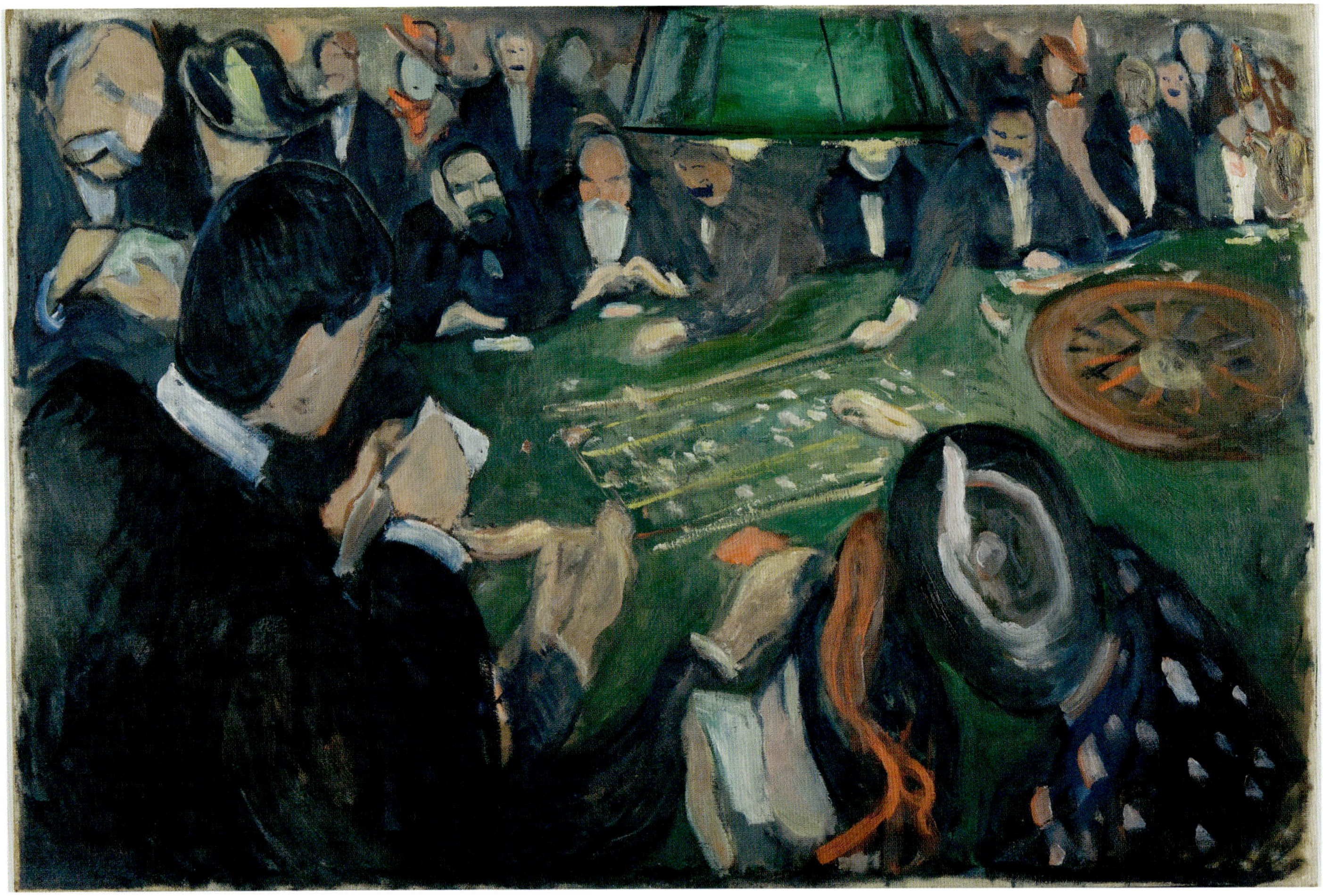

83 Edvard Munch, *At the Roulette Table in Monte Carlo*, 1893. Oil on canvas, 74.6 × 116 cm. Munch Museum, Oslo.

The playing board and the bright red roulette wheel, echoed by the oval form of the woman's hat, have become the main actors in the scene, the wheel presumably inhabited by the "mystical creature" that Munch imagined lived inside.[30] The keyed-up color rhythms of green and red, alternating with black and white, enliven the muted naturalist palette of the earlier canvases and parallel the way in which Munch's notes used a staccato voice to present his observations from the gambling rooms. Saturated red punctuates the scene, circling the table like blood splatters from the wheel, describing the flowing ribbon of the woman's hat in the foreground, a hat and scarf in the left background, and bright neckties or boutonnieres in the upper right corner. Not only does this make the entire composition seem to spin; it also suggests that the red paint strokes of the wheel, its "bacteria," are being transmitted with the violence of a centrifuge, beginning to infect the players who will eventually contribute to the blood-soaked reputation of the casino.

Significantly, Munch depicted no discernable coins on the table. Patches of color might resolve themselves into coins in the viewer's mind, but they might also, like the wealth to be gained in Monte Carlo, be only figments of the viewer's imagination.

Tipping the roulette table toward the picture plane enabled Munch to enact more fully a complex play of deformed hands that he had first explored in one of his ink drawings (fig. 80).[31] Instruments of easy money or ruin, hands concentrate the action of gambling. They place bets, rake in winnings for the bank, take notes, or are held tightly to the body. The exaggerated perspective of the table allows many of the hands in play to appear etiolated, nothing more than pathetic, withered appendages. This is especially the case with the female figure in the spotted dress and oval hat, the only figure present in all three paintings, whose tiny hand protrudes from her hat like a useless decorative feather. A later, engraved version of the composition using a more exaggerated perspective makes this pathological play of hands even more explicit (fig. 84). In the print, more hands appear on the table, and Munch more literally deformed them by using a variety of very deeply and very shallowly etched lines. In fact, most of the sites of action in the print, including the circular contour of the roulette wheel, many of the hands, and the croupier's stick, were engraved very lightly into the plate, creating an actual void at the center of the composition. By barely engraving the hands, Munch made them both deformed and unformed. The most extreme case is once again the feather-hand, barely visible as it protrudes from the woman's hat to come into contact with the faint line of the croupier's stick. Almost impossible to discern, the misshapen instrument of the gambler's downfall, the hand that places the bet is *other*, both in its form and in the technique of its representation.

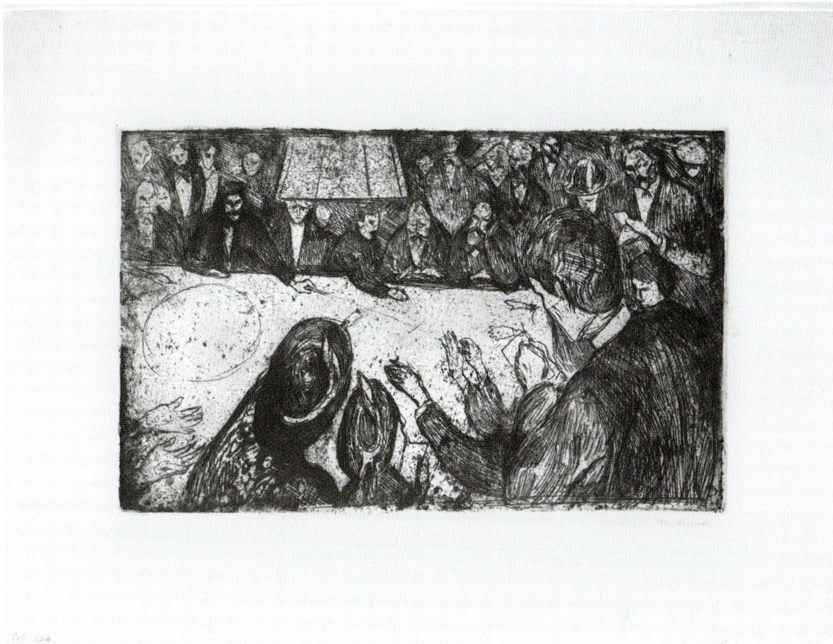

84 Edvard Munch, *The Roulette Table*, 1903. Etching, 26–26.8 × 43.3–44 cm. Munch Museum, Oslo.

The 1893 roulette painting, however, stops short of the pathological form of the later etching. The painted surface remains for the most part even and consistent, with Munch pathologizing the work through spatial distortion, deformation, the use of color, and caricature. Indeed, it is among Munch's earliest uses of caricature to distill formal essences for the purposes of generalizing, a strategy linked to physiognomic illustration, an echo of which appears in the female profile in the lower center of the painting that might be compared with a phrenological illustration of an "idiot" (fig. 85).[32] *At the Roulette Table in Monte Carlo* also includes one of Munch's earliest painted avatars of himself, the large figure on the left of the painting in *profil perdu* with an angular jaw and distinctive hairline, a figure also seen in *Death in the Sickroom* of the same year (fig. 86).[33] Viewed from the back, Munch's avatar takes notes or possibly sketches, indicating his roles as both the gambler with a system and the artist who observes. But the gambler-artist is himself under observation, not only by the painting's viewers but by an older man in the upper left corner who is also, it seems, taking notes. This final version of the roulette motif implicates the viewer in the psychological drama of the game, suggesting that she is already gambling, peeping over the figure's shoulders in order to get a good look at the table before placing her bet, perhaps connecting with one of the disembodied arms reaching from the lower edge of the canvas.

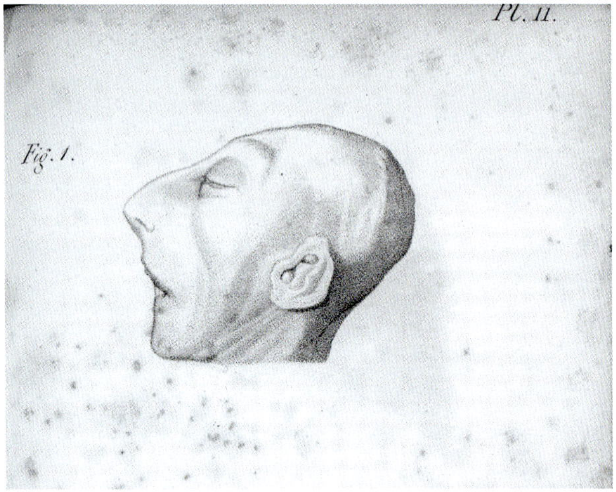

85 *Idiot, 25 Years Old,* plate II in J. G. Spurzheim, *Phrenology, or The Doctrine of the Mind; and of the Relations Between Its Manifestations and the Body,* 3rd ed. (London: Treuttel, Wurtz, and Richter, 1825). The Thomas Fisher Rare Book Library, University of Toronto.

In comparison to the diluted oil paint and tempera, as well as the mixed and intermedia approaches that would become hallmarks of Munch's painting practice in and around 1893, the surface of *At the Roulette Table in Monte Carlo* remains fairly conventional, eschewing the particular dialectic of form and content that would become hallmarks of Munch's symbolist form.[34] But it reveals the emergence of another aspect of symbolist dialectics, Munch's desire to use the inextricable link between the normal and the pathological, posited by experimental psychology, to reject the norms of naturalism in order to represent the psychologically fragmented modern self. The self-portrait introduced at the beginning of this chapter, which I linked with Munch's time in Monte Carlo, not only points to the importance of Munch's experiences in Monte Carlo for his emerging pathological symbolist practice; it also performs for the viewer that modern fragmented self in its (un) natural habitat of the casino, revealing the singular experience of staking one's last coin as a risky but aesthetically productive encounter with madness.

The Scream as Symbolist Experiment

Although the event on which Munch based his *Scream* motif took place on the outskirts of Kristiania (present-day Oslo), Munch wrote the first extant text based on his memory of that event while in the south of France, traveling almost daily to the casino in Monte Carlo and reflecting on various forms of despair, including the despair provoked by gambling that in extremis could lead to suicide: "I was walking along the road with two friends—the sun was setting—I felt a wave of sadness—The sky suddenly turned

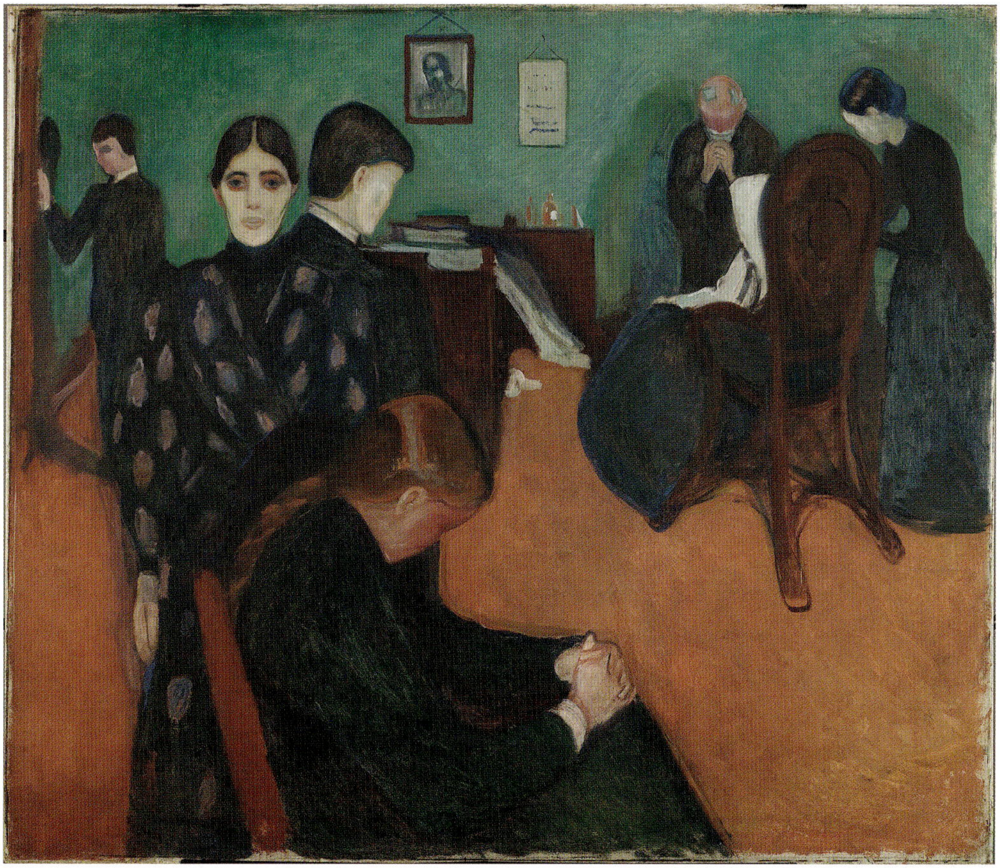

86 Edvard Munch, *Death in the Sickroom*, 1893. Oil on canvas, 134.5 × 160 cm. Munch Museum, Oslo.

blood-red. I stopped, leaned against the railing dead tired—looked out at the flaming clouds that hung like blood and swords—over the blue-black fjord and city—My friends walked on—I stood there, trembling with fright—and I felt a loud, unending scream through nature."[35] Munch took this memory, and probably the text itself, with him to Berlin, where, in the crucible of the Black Piglet group, he was recognized and celebrated for having finally achieved symbolist form. Black Piglet members, among them Stanisław Przybyszewski, August Strindberg, Richard Dehmel, Carl Ludwig Schleich, and Max Asch, further exposed Munch to a host of new and often radical scientific, philosophical, and social ideas, including French experimental psychology, all of which doubtless drove Munch to develop a more thoroughly experimental and pathological approach to his symbolist visual practice, an approach exemplified by the best-known version of *The Scream* (1893) (fig. 6).[36]

Sympathetic to Munch's efforts visually to communicate universal truths about modern human experience, including love—which, Przybyszewski claimed, in an appropriate metaphor, "rolls around . . . the perception of happiness like a roulette ball"—the Polish writer supplied a model for viewing Munch's symbolist works and reflected on their genesis in his 1894 article "Psychic Naturalism" ("Psychischer Naturalismus"). Identifying Munch's work with French symbolist aims of "synthesizing nature," Przybyszewski judged Munch's paintings to be "revelations of a naked individuality . . . creations of a somnambulant, transcendental consciousness . . . vulgarly called the unconscious." More crucially, Przybyszewski likened Munch to an experimentalist, a scientist working in the laboratory, making "[chemical] preparations of the soul" with which the artist and the viewer could study the modern self's most intimate psychological functions. But Munch's works, Przybyszewski contended, were not just any "preparations," but were "preparations of the animalistic, irrational soul" in which "every conceptual process has ceased to operate."[37] In Przybyszewski's view, in other words, Munch painted psychological alterity with a view to experimenting, offering his paintings up as objects of investigation from which both artist and viewer could derive knowledge about the inner lives of human beings.

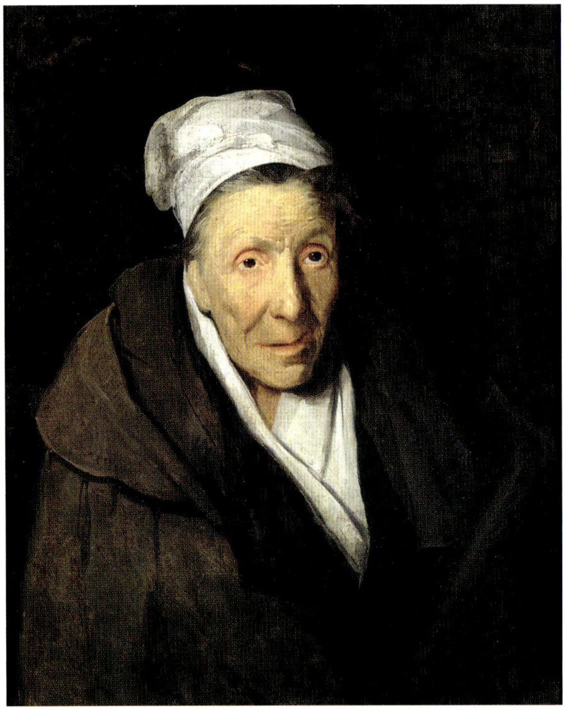

87 Théodore Géricault, *The Woman with Gambling Mania*, 1819–24. Oil on canvas, 71 × 65 cm. Musée du Louvre, Paris.

Przybyszewski was well positioned to make such claims, having formally studied medicine, neurology, and experimental psychology in the 1880s and 1890s.[38] In his memoirs, he maintained that reading Théodule Ribot, Hippolyte Taine, and Herbert Spencer had in fact prepared him for Friedrich Nietzsche, whose philosophy he viewed not as a paradigm shift but as a natural extension of the new scientific psychology.[39] Nietzsche's philosophy had only recently become a topic of intense conversation among the Berlin-based avant-garde, introduced to the group by the Swedish writer Ola Hansson through Georg Brandes's lectures.[40] Not long before Przybyszewski publicly identified Munch's paintings as chemical preparations, the art critic Willy Pastor echoed Nietzsche's metaphor of pathology as a magnifying glass on the normal in a lengthy and very positive review of Munch's solo exhibition. Pastor insisted that the artist painted by training a microscope on his own psyche, directing a scientific instrument to examine, "deeper and deeper," the "gaping abyss" of his soul.[41] Under Przybyszewski's editorial direction, his own article and Pastor's were republished in July 1894 in the volume *The Work of Edvard Munch*, which also included a text by Black Piglet member Franz Servaes that amplified his colleagues' visions of Munch's works as both creations of and the basis for studying the less highly developed mental functioning of the unconscious. Servaes praised Munch for having "the certainty of a sleepwalker" who is able "to view and to experience the primitive aspects of human nature," by which he meant the less highly evolved unconscious mind.[42]

Przybyszewski's claim that Munch's paintings were scientific "preparations," along with Pastor's comparison of Munch's practice to using a microscope on the inner self, and Servaes's insistence on Munch's ability to access the primitive parts of his own soul, represented Munch's works as a form of psychological experimentation very much in line with the new experimental psychology.[43] As nature's experiments dictated, the nature that Munch's paintings synthesized was an abnormal one, an "animalistic, irrational soul," a primitive form of humanity summoned to investigate universal psychological processes. Evidently, Pastor's, Przybyszewski's, and Servaes's articles squared with Munch's understanding of his own practice. He took pains to ensure that friends and family both at home in Norway and in Berlin read Pastor's text. And according to Przybyszewski's much later and possibly fanciful recollections, Munch had been so thrilled with "Psychic Naturalism" that he presented *The Scream* to the Polish writer the day after reading the article.[44] But if it was to be understood as a form of psychological experiment, how, visually and materially, did *The Scream* signify as a symbolist "experiment prepared by nature," as a scientific "preparation" of an "irrational soul"?

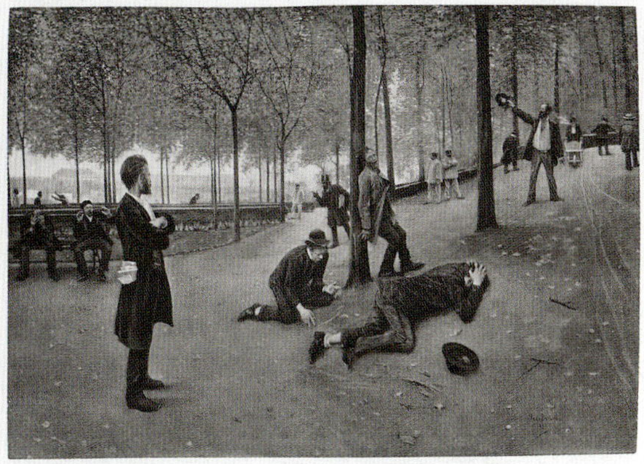

88 Jean Béraud, *The Insane*, 1885. Oil on canvas, 112 × 142 cm. Whereabouts unknown. Engraving in Henry Havard, *Salon de 1885* (Paris: Librairie d'Art, 1885).

According to Servaes, *The Scream* revealed the "world as a madhouse . . . in insane colors rudely screaming together."[45] But far from being dashed off in a mad frenzy, Munch had deliberated on the painting for nearly three years, reworking the original text, producing a number of drawings, and aborting at least one painting before arriving at the 1893 composition, which nevertheless "could only have been painted by a madman," as the tiny pencil inscription in the top left corner of the work declares.[46] Insanity was hardly a new subject for painting. Théodore Géricault had painted portraits thought to be of monomaniacs, including a gambling monomaniac, in the early 1820s (fig. 87), and Béraud had more recently rendered the stereotypical gestures and physiognomies of the insane with his characteristic precision (fig. 88). But Munch overtly rejected what Przybyszewski called the "shallow" aesthetic of naturalism in order to produce a work that looked as if it had been made not by an artist named Edvard Munch but by an unknown madman. In order to do so, he must have had some notion of what the art of the mentally ill looked like, a set of visual schemata shared with his ideal audience members, most obviously members of the Black Piglet.[47] Apart from Przybyszewski and Strindberg, who both continued to read voraciously in medicine and psychology during the Berlin period, a number of Black Piglet habitués could have helped Munch gain access to asylums in and around Berlin, where he could easily have seen works created by patients, which were then beginning to be much more widely collected in asylums, although still largely for the purposes of diagnosis.[48]

Like Strindberg, Munch may have received an invitation to visit an asylum from one of his Berlin patrons, Max Asch, according to some a gynecologist and to others an expert in mental illness.[49]

Black Piglet members would also have been familiar with reproductions of drawings by the mentally ill and by mediums thought to be mentally ill in the periodical *Sphinx*, the eclectic fin-de-siècle journal introduced in the previous chapter, which sought to investigate the relationships between "Man, Nature, and the Universe."[50] *Sphinx*'s original subtitle, *Monthly Journal for the Historical and Experimental Foundation of the Transcendental Worldview on a Monistic Basis*, announced, among other things, its lofty scientific ambitions. The journal was connected to both the Munich-based Psychologische Gesellschaft and the Berliner Gesellschaft für Experimental-Psychologie, organizations that regularly corresponded with the Society for Psychical Research in London and the Société de psychologie physiologique in Paris. These societies tended to distance themselves from more spiritist organizations, including theosophy groups, staking their reputations on rigorous scientific experiments that aimed to establish new truths about human psychology. Many Black Piglet members, including Dehmel, contributed to *Sphinx*. Two months after it appeared, Przybyszewski's "Psychic Naturalism" was favorably reviewed for the journal by Franz Evers, a regular writer for *Sphinx* and erstwhile Black Piglet member who owned some of Munch's earliest etchings.[51]

In August 1888, in an article titled "Somnambulic Drawing," Gustav Gessmann discussed two very different works of mad creativity by a young male patient under the care of Viennese doctors who claimed at times to channel a dead painter named Seleny.[52] While apparently acting as a medium for Seleny, the patient labored over highly composed drawings such as *Symbolic Representation of the Soul's Uplifting* (fig. 89), whose technically accomplished detail, symmetry, and almost hyperrealism recall scientific illustrations. But when not under this influence, the patient experienced convulsions, between which he sketched rapidly and with little to no deliberation, creating drawings such as *A Spirit Head* in less than an hour (fig. 90). Gessmann described how each drawing reflected different symptoms of the patient's psychopathology. The first displayed a mania for geometric detail and an obsessiveness that, in conjunction with the claims of mediumism, identified the patient as delusional. The other presented expressive peculiarities and an overall irrational abstraction that revealed the patient's extreme emotional instability. Gessmann took special interest in the patient's more frenzied work, and gave an account of how he had produced drawings such as *A Spirit Head* by hatching the surface to build up form. Out of what had previously seemed to be a dark, indistinct chaos, a head or a landscape would suddenly emerge as if it had emanated directly from the patient's soul. These more expressive drawings, Gessmann contended, rejected any traditional and willed process of art making, in which, for example, one might draw the outline of a head and then fill in the features. The insane mind, Gessmann implicitly argued, worked in ways foreign

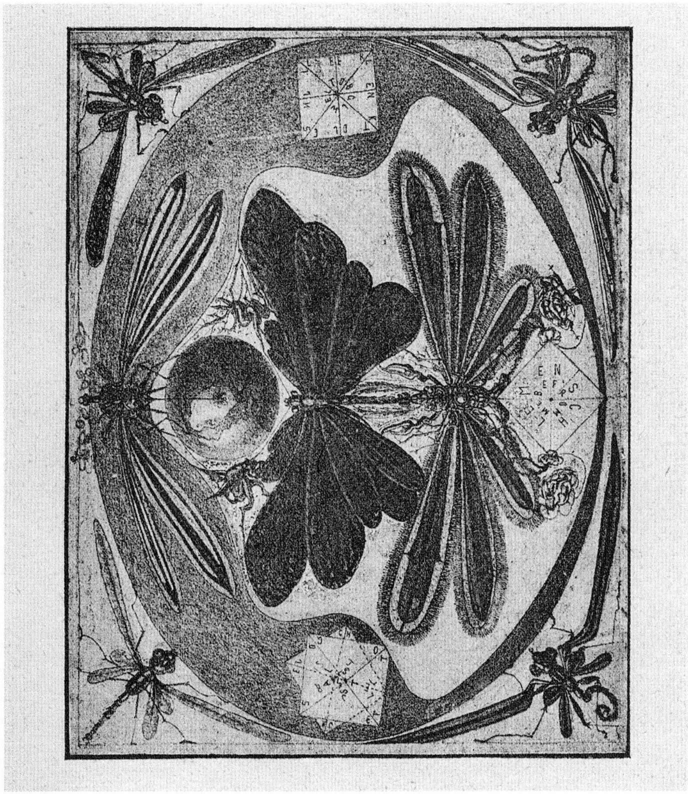

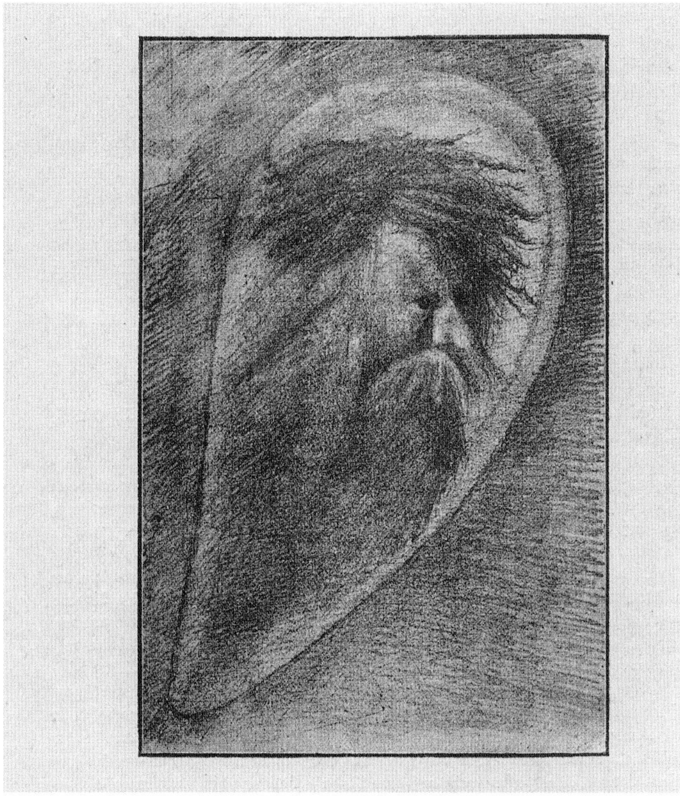

89 *Symbolic Representation of the Soul's Uplifting,* from Gustav Gessmann, "Somnambules Zeichnen," *Sphinx* 6, no. 32 (August 1888): 105.

90 *A Spirit Head*, from Gustav Gessmann, "Somnambules Zeichnen," *Sphinx* 6, no. 32 (August 1888): 104.

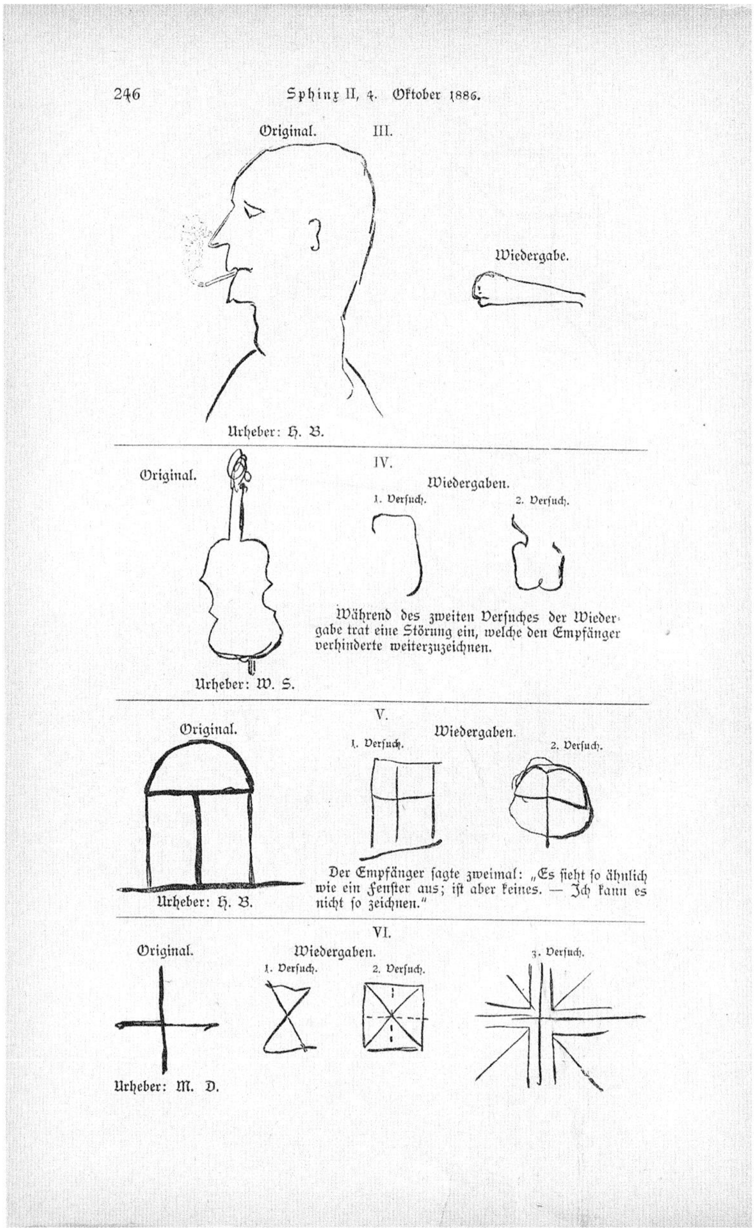

91 "Gedankenübertragung," *Sphinx* 2, no. 4 (October 1886): 246.

to the conscious mind, outside of the rational logic instilled by the systematic training of art academies, training that Munch knew extremely well.

In addition to these visual traces of somnambulistic creativity, *Sphinx* also published a series of drawing-based experiments into so-called thought transfers (*Gedankenübertragungen*), individual instances of telepathic communication, a phenomenon that aroused the interest of a number of Black Piglet members, including Strindberg and Dehmel.[53] These thought-transfer experiments typically involved controllers, usually called "agents," and mediums, called "percipients," who were often young women deemed especially susceptible to hypnosis and suggestion.[54] The agent, an "experienced experimenter" who was invariably male, would create a drawing out of sight of the percipient. Then, placing it behind the percipient's head, still out of the medium's sight, the agent would concentrate on the drawing with all his cognitive might. When the percipient signaled that she had received the mental suggestion, the drawing was removed from the room, and the percipient would then attempt to replicate it. Reporting on these experiments, *Sphinx* published the agent's "original" drawings next to the percipient's "reproductions" (*Wiedergaben*), implicitly claiming, regardless of how different the two drawings looked, that some essential formal aspect of the object image had been transferred telepathically from agent to percipient. The reproductions were almost always more schematic and childlike, and often bore almost no likeness at all to the originals. The drawing of a man with a cigarette, for instance, resembles the wormlike drawing labeled "reproduction" only by virtue of being placed next to it and labeled "original" (fig. 91). Nevertheless, the thought-transfer experiments implied that the two drawings existed within some larger genus, face, or, perhaps even more broadly, a category that might be composed of elongated, phallic things.

A number of Munch's drawings from the mid-1890s function similarly, reducing an object to its essential formal aspects and thereby inviting a thought-transfer-like aesthetic response. On the reverse of a more finished ink drawing, a representation of art nourished by male-female copulation (fig. 92), Munch inhabited the roles of both agent and percipient, producing a crayon and ink drawing of a man's head in profile that superimposes a controller original onto a mediumistic reproduction (fig. 93). Hesitant and schematic blue pencil lines are hardly if at all recognizable as a face without the tusche lines overtop that synthesize elements of the profile. Underneath the strokes of tusche, three lightly drawn noses suggest not so much that Munch changed his mind in drawing the picture as that he may have had his eyes closed and lifted the pencil a number of times as he drew the jerky lines, later to be smoothed over with ink. For those familiar with Munch's visual vocabulary from such works as *At the Roulette Table in Monte Carlo* and *Death in the Sickroom* (fig. 83 and fig. 86), the profile displays a strong chin and nose and distinctive hairline, recalling Munch's avatars. As agent, Munch has concentrated on a mental image of his own profile, and as percipient, he has re-created that image without looking. The process conjured up by the drawing seems designed to access perceptual and creative faculties beyond or before artistic training as part and parcel of a search for an essential formal language that will, eschewing naturalism and with a minimum of elements, communicate "man's head in profile" to the greatest number of viewers. In adding the tusche, Munch effectively restored an original to the reproduction, visually enacting the mental work necessary to synthesize the image. The presence of the two media encourages a back-and-forth between original and reproduction, thus instantiating a dynamic, dialectical mode of viewing that proclaims the existence of both essential form and mediumistic communication.

The Scream combines different elements of the mad and mediumistic creativity described by Gessmann and suggested by the thought-transfer drawings disseminated in *Sphinx*, providing a schema for both producing and receiving the 1893 painting as an instance of insane, somnambulistic, or mediumistic art. Functioning according to the expressive abstraction of *A Spirit Head* rather than the detailed realism of *Symbolic Representation*, *The Scream* invokes an irrational, anti-academic art-making practice rather than a delusional performance of artistic skill. While the shading of *A Spirit Head* produces a very different visual effect from Munch's brightly colored linear forms, the drawing's disembodied head, very summary evocation of facial features, and suggested orifices reveal structures analogous to Munch's partially formed, seemingly primitive embodiment of anxiety, fear, and insanity. The O that emerges toward the bottom of *A Spirit Head*'s teardrop-shaped face, and the dark, fuzzy area, presumably where one discerns eyes shaded by heavy brows, play with the visual conventions of anxiety and fear.

Likewise, the screaming figure of *The Scream* is more an evocation than a depiction of human form. Only cursory circles, dots, and small x's signify the sufficient

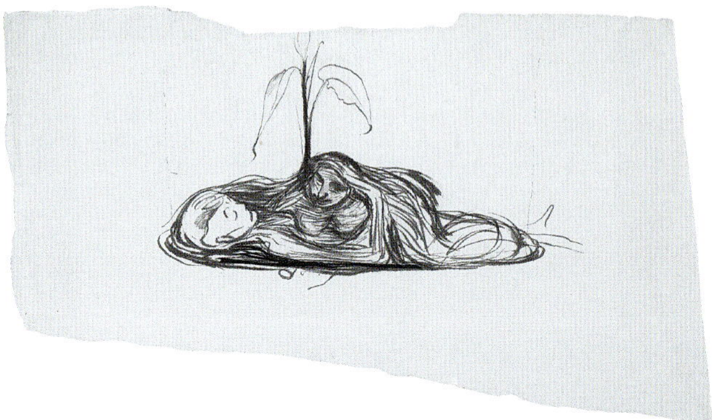

92 Edvard Munch, *Art,* ca. 1894–96. Tusche on paper, 22.5 cm x 44.5 cm. Munch Museum, Oslo, T408A.

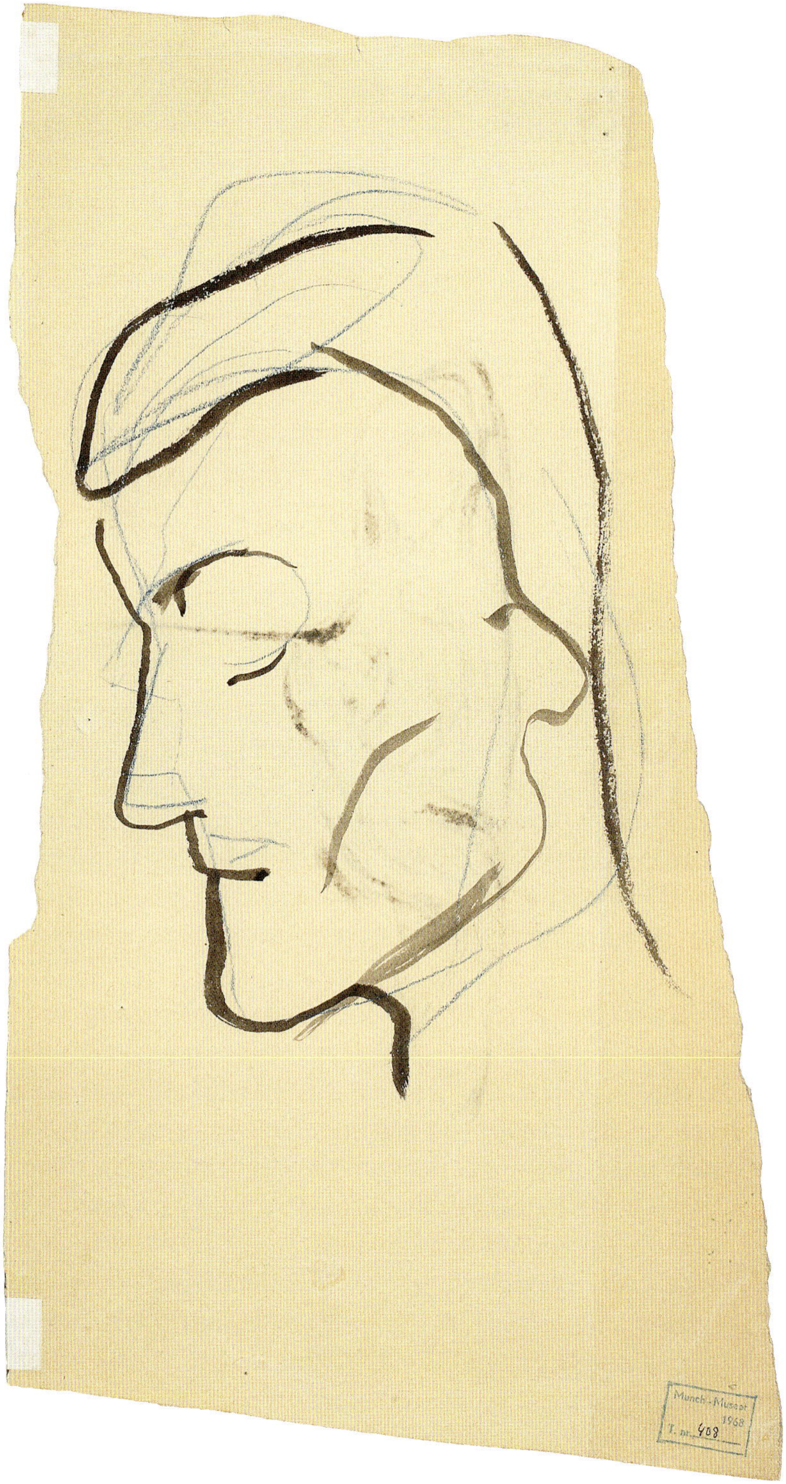

93 Edvard Munch, *Man's Head,* ca. 1894–96. Blue crayon and ink on paper, 44.5 × 22.5 cm. Munch Museum, Oslo, T408B.

orifices—eyes, nostrils, and mouth—needed to identify a humanoid face. Similar reductive processes appear to be at work in the thought-transfer reproduction of a fish drawing (fig. 94). It would not signify fish were the original not reproduced alongside it, but the schematic representation of a cell-like body, which also looks like a rudimentary head and three orifices, nevertheless conveys the idea of a primordial creature. Reversing the processes carried out by the thought-transfer experiments, *The Scream* invited its late nineteenth-century viewers to receive the mental suggestion of universal emotions such as fear and anxiety and then to re-create their own personal experiences. As percipient, Munch provided something to be read as essential, universal form, and asked his viewers to conjure up originals from a reproduction, ideally retrieving their own individual emotive experiences by engaging with the work.

Munch took enormous pains to signify *The Scream* as typical of insane expressivity, as anonymous and programmatically unimportant, as if it really "could only have been painted by a madman." He used seemingly undeliberated and often translucent strokes of oil and tempera, left the pastel lines largely unblended, and incorporated large areas of blank cardboard, including much of the main figure's face.[55] For its earliest intended viewers, these choices reiterated the ostensibly mad author's ignorance of artistic traditions and training, and displayed a professed lack of concern for permanence. Munch and his supporters immediately recognized that *The Scream* was a significant achievement; it was regularly exhibited almost from the moment of its creation, purchased by a Norwegian collector in 1909, and given to the Norwegian National Gallery the following year. And yet the painting looks as if it has been dragged through the mud, an impression emphasized by the slightly ridiculous gilt frame in which it usually hangs today. Flecks of dirt, paint, and other media are present all over the work, including a poetically placed tiny x in the left eye of the screaming figure. Splatters of a translucent white substance, possibly candle wax, appear in the lower

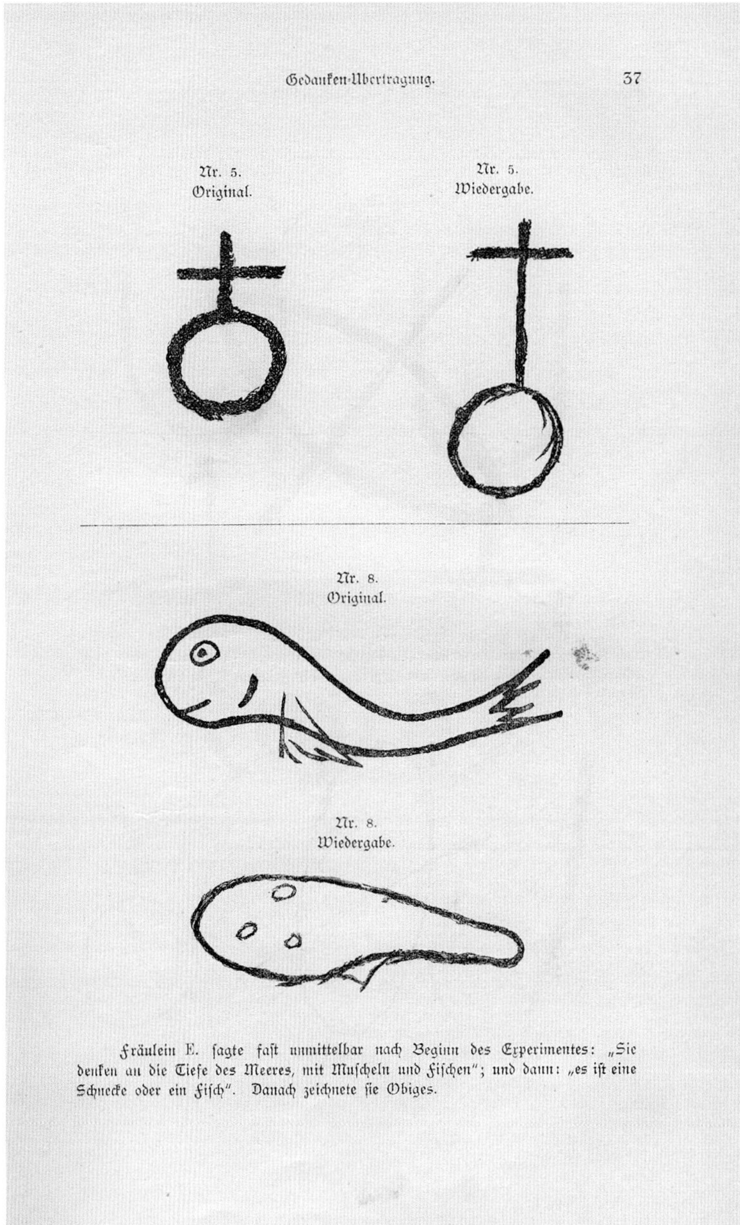
Gedanken-Übertragung. 37

Nr. 5. Original.

Nr. 5. Wiedergabe.

Nr. 8. Original.

Nr. 8. Wiedergabe.

Fräulein E. sagte fast unmittelbar nach Beginn des Experimentes: „Sie denken an die Tiefe des Meeres, mit Muscheln und Fischen"; und dann: „es ist eine Schnecke oder ein Fisch". Danach zeichnete sie Obiges.

94 "Gedankenübertragung," *Sphinx* 1, no. 1 (January 1886): 37.

right-hand corner and elsewhere. The cardboard is creased in a number of places, and a long horizontal scratch in the surface bisects the O of the screaming mouth. Perhaps one of the oddest elements is the long vertical incision in the cardboard running from the top to the bottom of the work about five centimeters in from the right edge of the painting. It borders one broad vertical stroke of translucent red paint, which makes it seem as if a strip has been added, as if some unknown person included a spacer so that the painting did not have to be custom framed. It is no small irony that many reproductions of *The Scream* crop this red strip out, as if it were somehow extraneous to the work.

In the early 1890s, the art of the mentally ill was rarely called art. It was valued primarily for diagnosis, for therapy, and, increasingly, as a curiosity. In one sense, *The Scream* was not in the end a painting but *only* a signifier of madness, a symptom and a site for diagnosis and experiment, useful only as a "soul preparation" that would enable putatively universal truths about humanity to be established. In another sense, however, *The Scream* was the ultimate avant-garde act, a painting so radical in its anonymity, in its antinaturalist form, and indeed in its fundamental experimentalism, as to establish itself as the first avant-garde performance of madness.

Munch and the Parisian Psychiatric Avant-Garde

If Munch's Berlin supporters most fully grasped and promoted his scientist-like ambitions, his Parisian critics most thoroughly recognized the connections between Munch's pathological subject matter and his pathological form. In fin-de-siècle Paris, there was remarkable agreement about the existence of the pathological in Munch's works, but the meaning, value, and function of pathology in his practice provoked stark differences of opinion. Less enthusiastic critics wrote disparagingly of his "horrors," his "inept drawing," "barbarous color," "childlike etchings," and "paintings of a sick mind."[56] But other critics proclaimed Munch's "feigned naïveté" the result of "a terrific mental effort" and insisted that works such as *The Scream* were the products of "an intelligent man," even an "intellectual," as Yvanhoé Rambosson wrote in a review accompanied by a reproduction of one of Munch's lithographs (fig. 95).[57] Alcanter de Brahm infused his criticism with the language of experimental psychology, highlighting Munch's subtle understanding of "natural hypnosis" and "multiple personality."[58] Writing in French, the critic William Ritter, having seen Munch's 1905 exhibition in Prague and struggling to grapple with art he instinctively abhorred, explicitly connected Munch's use of ugly technique to his depictions of the ugliness of modern life, calling him "a laboratory and hospital Velazquez" in a volume published by the press of the *Mercure de France.* "The great merit" of Munch's work, Ritter wrote, was that it "urgently affirm[ed] the demented nature of modern ugliness, adding the ugliness of the facture to the ugliness of the main subject matter, the ugliness of the presentation to the physical ugliness of the motif." "It is the cacophony of means," he concluded, "in the service of the cacophony of modernity."[59] For the critic and psychiatrist Marcel Réja, Munch's "clumsy, even puerile realism" was precisely what enabled the artist to achieve his symbolist syntheses. "It objectifies," Réja stressed, "what is subjective."[60]

In the late 1890s, Munch continued to explore the potential of a pathological approach to symbolist visual practice, especially in paintings and prints depicting institutionalized illness. In Paris, he befriended doctors and psychiatrists who gained him access to hospitals and asylums where he found numerous subjects in the patients and doctors he observed. Munch's Parisian circles included, in particular, psychiatrists who identified with a movement of asylum reform that justified itself by reiterating the basic assumption of the pathological method, that the normal and the pathological related to each other on a continuum, and that madness was therefore merely an exaggerated form of normal human psychological functioning. In paintings such as *Women in Hospital* (fig. 96), Munch politicized the pathological, using a pathological method to produce works that were aligned with the reforming rhetoric of an emerging psychiatric avant-garde.

After nearly three years of shuttling between Germany and Norway, Munch returned to Paris at the beginning of 1896 for a number of lengthy stays that extended into 1898.[61] By this time, the Black Piglet group had largely dispersed, many of its members, including Strindberg, having left Berlin to seek fame and fortune in the French capital. In Paris as in Berlin, Munch came to fraternize with an eclectic group of literary, musical, artistic, and scientific avant-gardes. These included the group centered around William Molard and Ida Ericson's apartment at 6, rue Vercingétorix, where Gauguin had rented a studio when he returned from Tahiti in 1893, and other groups associated with the symbolist journals *Mercure de France* and *La plume*.[62] During part of this period, Munch rented a studio on the edge of Montparnasse at 32, rue de la Santé, an address surrounded then, as now, by hospitals. Munch's neighbor

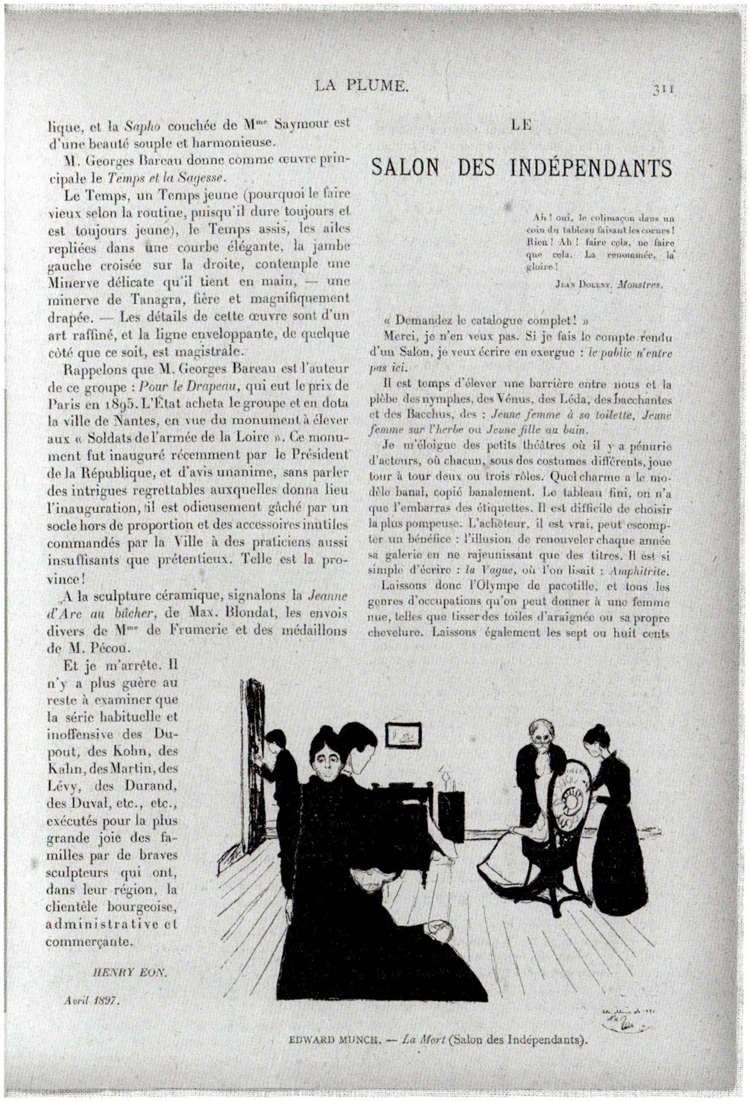

LA PLUME. 311

lique, et la *Sapho* couchée de Mme Saymour est d'une beauté souple et harmonieuse.

M. Georges Bareau donne comme œuvre principale le *Temps et la Sagesse*.

Le Temps, un Temps jeune (pourquoi le faire vieux selon la routine, puisqu'il dure toujours et est toujours jeune), le Temps assis, les ailes repliées dans une courbe élégante, la jambe gauche croisée sur la droite, contemple une Minerve délicate qu'il tient en main, — une minerve de Tanagra, fière et magnifiquement drapée. — Les détails de cette œuvre sont d'un art raffiné, et la ligne enveloppante, de quelque côté que ce soit, est magistrale.

Rappelons que M. Georges Bareau est l'auteur de ce groupe : *Pour le Drapeau*, qui eut le prix de Paris en 1895. L'État acheta le groupe et en dota la ville de Nantes, en vue du monument à élever aux « Soldats de l'armée de la Loire ». Ce monument fut inauguré récemment par le Président de la République, et d'avis unanime, sans parler des intrigues regrettables auxquelles donna lieu l'inauguration, il est odieusement gâché par un socle hors de proportion et des accessoires inutiles commandés par la Ville à des praticiens aussi insuffisants que prétentieux. Telle est la province!

A la sculpture céramique, signalons la *Jeanne d'Arc au bûcher*, de Max. Blondat, les envois divers de Mme de Frumerie et des médaillons de M. Pécou.

Et je m'arrête. Il n'y a plus guère au reste à examiner que la série habituelle et inoffensive des Dupout, des Kohn, des Kahn, des Martin, des Lévy, des Durand, des Duval, etc., etc., exécutés pour la plus grande joie des familles par de braves sculpteurs qui ont, dans leur région, la clientèle bourgeoise, administrative et commerçante.

HENRY EON.

Avril 1897.

LE

SALON DES INDÉPENDANTS

Ah! oui, le colimaçon dans un coin du tableau faisant les cornes! Rien! Ah! faire cela, ne faire que cela. La renommée, la gloire!

JEAN DOLENT. *Monstres.*

« Demandez le catalogue complet! »

Merci, je n'en veux pas. Si je fais le compte rendu d'un Salon, je veux écrire en exergue : *le public n'entre pas ici.*

Il est temps d'élever une barrière entre nous et la plèbe des nymphes, des Vénus, des Léda, des bacchantes et des Bacchus, des : *Jeune femme à sa toilette, Jeune femme sur l'herbe* ou *Jeune fille au bain.*

Je m'éloigne des petits théâtres où il y a pénurie d'acteurs, où chacun, sous des costumes différents, joue tour à tour deux ou trois rôles. Quel charme a le modèle banal, copié banalement. Le tableau fini, on n'a que l'embarras des étiquettes. Il est difficile de choisir la plus pompeuse. L'acheteur, il est vrai, peut escompter un bénéfice : l'illusion de renouveler chaque année sa galerie en ne rajeunissant que des titres. Il est si simple d'écrire : *la Vague*, où l'on lisait : *Amphitrite.*

Laissons donc l'Olympe de pacotille, et tous les genres d'occupations qu'on peut donner à une femme nue, telles que tisser des toiles d'araignée ou sa propre chevelure. Laissons également les sept ou huit cents

EDWARD MUNCH. — *La Mort* (Salon des Indépendants).

95 Yvanhoé Rambosson, "Le Salon des Indépendants," *La plume*, no. 194 (15 May 1897): 311, with reproduction of Edvard Munch, *Death in the Sickroom*, 1896. Lithograph. Bibliothèque nationale de France.

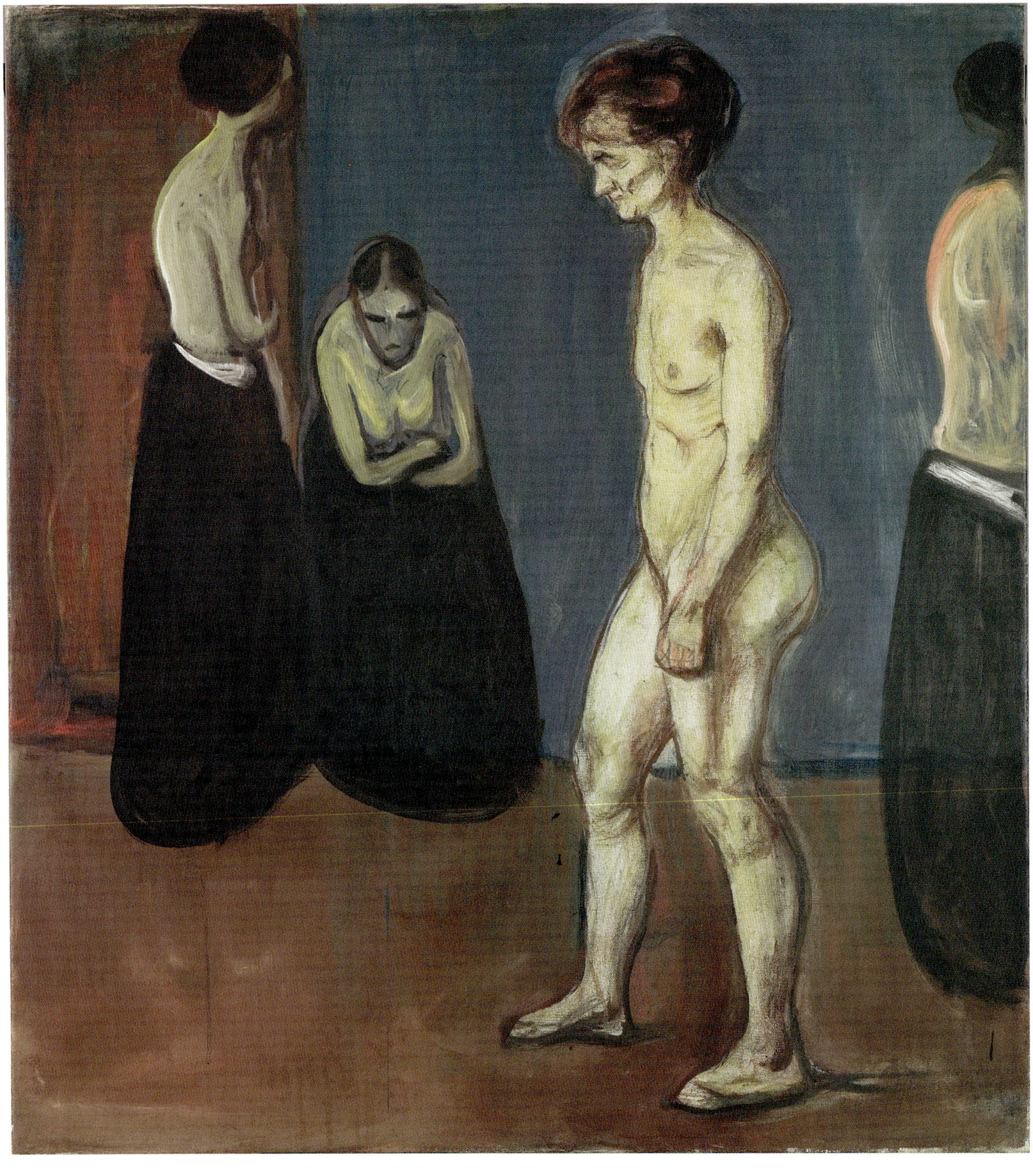

on the "street of health" was the enigmatic Dr. Paul Gaston Meunier, who published under the pseudonym Marcel Réja. A practicing psychiatrist at the Villejuif Asylum on the southern outskirts of Paris by day, and a symbolist poet and art critic, presumably by night, Réja would not only correct Strindberg's French for the autobiographical novel *Inferno* but would also write *The Art of the Insane* (*L'art chez les fous*, 1907). Réja's study, based on the collections of a number of French psychiatrists, is often considered the first critical assessment of the art of the mentally ill and a precursor to Hans Prinzhorn's *Artistry of the Mentally Ill* (1922), the best-known source of the avant-garde's interest in mental alterity.[63]

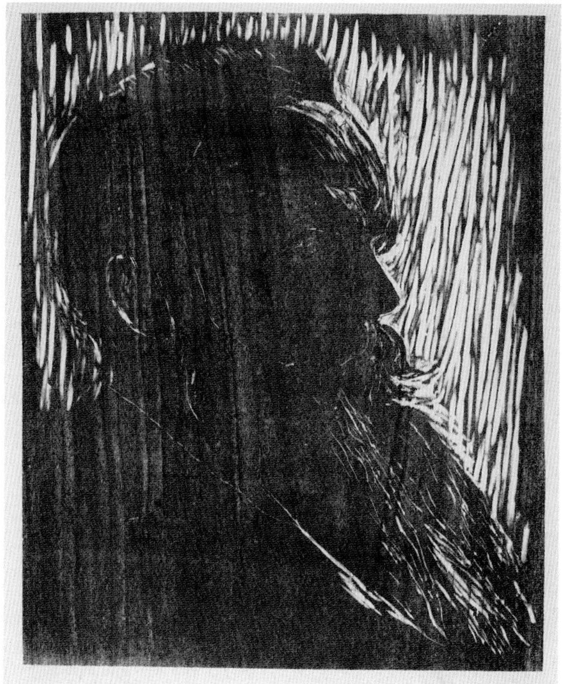

Judging by their correspondence, Réja's published praise for Munch's work, Munch's woodcut portrait of Réja (fig. 97), and the friendly dedications in Réja's books, conserved in Munch's library, the two were close during Munch's Paris period. Réja seems to have been Munch's main conduit to the many Parisian institutions that treated and housed the mentally ill at the fin de siècle, many of whom were suffering from the neurological pathologies associated with the dreaded tertiary stage of syphilis, a disease that greatly preoccupied Munch and his cohorts.[64] Munch's good friend Fritz Delius, who lived on the outskirts of Paris in Grez-sur-Loing, had been diagnosed with the secondary stage of syphilis in 1895, the same year that Munch's supporter Julius Meier-Graefe was falsely rumored to have taken to his bed with "a serious syphilis."[65] This was also the year that Gauguin probably contracted the disease, and the year that Strindberg was vehemently denying that the skin condition on his hands, for which he was being treated at the Hôpital Saint-Louis, was a sign that he too had been infected. There is some evidence that Munch visited both the Salpêtrière and the Hôpital Saint-Louis, but he probably also spent time at Sainte-Anne Asylum, the largest and most prestigious of the Parisian clinics for the mentally ill, a short walk from where both he and Réja were living.[66] In 1899, Réja would arrange for the artist Paul Herrmann, a close friend and colleague of Munch's, known in Paris at the time as Henri Héran, to sketch dancing women at Sainte-Anne's to publicize a fete at the asylum.[67] Such collaborations between psychiatrists and the symbolist avant-garde were hardly unusual. Réja's mentor at Villejuif, Dr. Auguste Marie, would soon invite the poet Gabriel Randon, better known as Jehan Rictus, to perform at Villejuif.[68] Rictus came into the Molard circle in 1896, exactly when Munch did; 1896 was also the year that Rictus made an anarchist splash with his scandalous recitation of the *Soliloquies of the Poor* at the Cabaret des Quat'z'Arts.[69]

96 (*opposite*) Edvard Munch, *Women in Hospital*, 1897. Oil and crayon on canvas, 109.5 × 99.5 cm. Munch Museum, Oslo.

97 Edvard Munch, *Marcel Réja*, 1896–97. Woodcut, 40 × 32.6 cm. Munch Museum, Oslo.

Munch himself claimed that an emotional visit to the Hôpital Saint-Louis was the basis for the painting now known as *Inheritance*, his "syphilis art," as he called it when it was exhibited at the Salon des Indépendants in 1903 (fig. 98).[70] He had martyred himself, Munch declared, in order to produce a "poster-like" image of a mother holding a syphilitic child in her lap. It was a "highly moral work" with a "true message of pain," in which he intended nonnaturalistic elements, including overtly religious iconography, to inspire in the beholder the same deep pathos he had felt in viewing the tragic scene.[71] A lithograph from 1896 (fig. 99) combines elements of *Inheritance* and the painting *Women in Hospital*, identifying the latter with Munch's category of "syphilis art."[72] The lithograph shows a doctor examining a woman's hand, looking perhaps for the rash characteristic of the secondary stage of syphilis. Dark spots on the figures in the background denote the chancres of syphilis, while the figures' varying states of undress, mussed hair, and caricatured expressions serve as visual signs of both mental illness and prostitution.

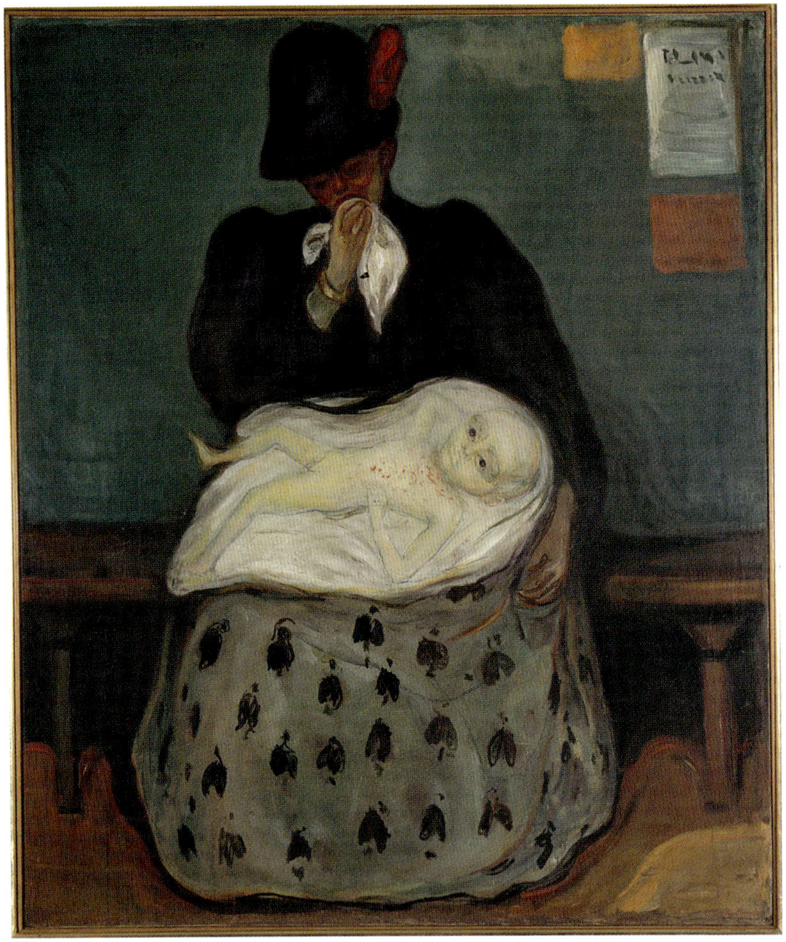

98 Edvard Munch, *Inheritance*, 1897–99. Oil on canvas, 141 × 120 cm. Munch Museum, Oslo.

The literary and pictorial tradition of representing the mad, especially the mad poor, in various states of undress dates at least as far back as the sixteenth century.[73] Undress, often to the point of nakedness, connoted degradation, otherness, irrationality, animality, and childishness—in short, the mad person's close proximity to nature. Examples abound, from the chained figures that graced the gates of Bethlem Hospital in eighteenth-century London to William Hogarth's *A Rake's Progress*, in which Tom Rakewell's increasing state of undress signifies his loss of material comforts and his progressively worsening syphilis, all of which eventually land him in Bedlam. When madwomen were shown in a state of undress, additional associations inhered between madness, unbridled sexuality, and prostitution, as in Tony Robert-Fleury's painting of Philippe Pinel supposedly liberating the patients of the Salpêtrière from their chains (fig. 100) and Albert Londe's photographs of hysterics, taken at the Salpêtrière under the auspices of Jean-Martin Charcot (fig. 101).[74]

In *Women in Hospital*, Munch deployed the familiar topos of the undressed mad, but he used a pathological means of painting to defy expectations about the madwomen depicted. The half-dressed women in the background are entirely painted but highly schematized, while the fully naked central figure is rendered naturalistically in crayon.

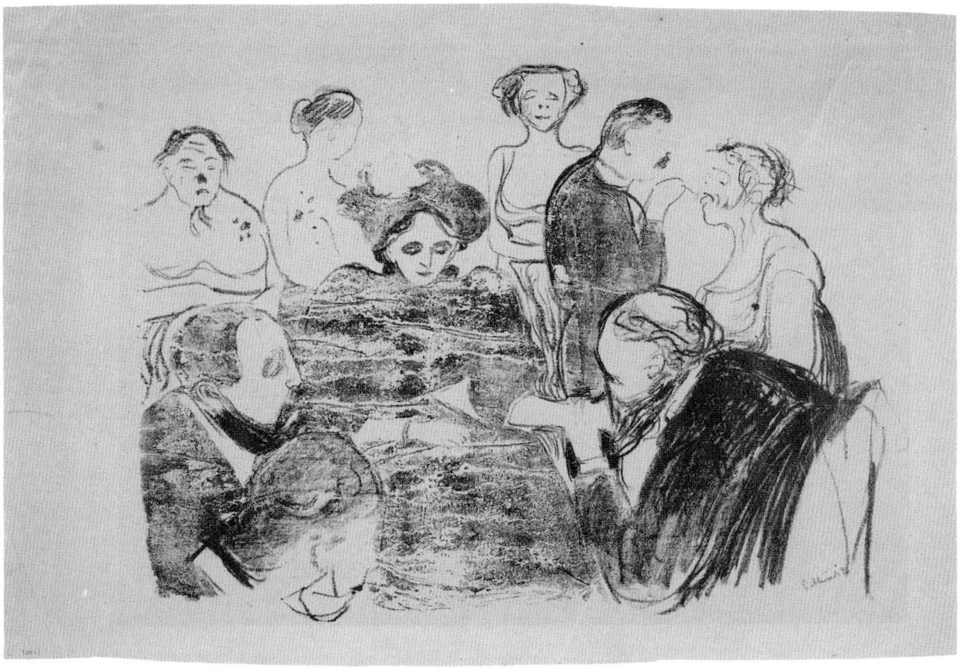

99 Edvard Munch, *Women in the Hospital*, 1896. Lithograph, 35 × 49 cm. Munch Museum, Oslo.

This main figure thus reads simultaneously as the most mad and the most human of the four, conveying a message in line with the moral stance taken by the specific psychiatric community to which Réja and his mentor Marie belonged. To call oneself a psychiatrist in Paris in 1896, as both Réja and Marie did, was to reject the label "alienist" and to embrace an emerging critique of the asylum system and its deeply institutionalized regimes of confinement and what was called the moral treatment.[75] French psychiatric reformers allied themselves with the *Revue de psychiatrie*, founded in 1896 by Édouard Toulouse, the most well placed of the self-declared reformers, who spent most of his career at Villejuif.[76] Before retiring from Sainte-Anne's in 1936, Toulouse claimed a final, albeit belated, reformist victory by having the words "asylum," "alienist," and "the insane" struck from the official lexicon; institutions such as Sainte-Anne's and Villejuif would henceforth be known as "psychiatric hospitals," their doctors as "psychiatrists," and their patients as "the mentally ill."[77] By the mid-1890s, Toulouse had taken a number of young reformers under his wing at Villejuif, including Réja, Marie, Paul Sérieux, and Nicolas Vaschide, some of the earliest collectors of the art of the insane, men who, nearly two decades before Prinzhorn, had begun to think about the artistic productions of their patients outside the contexts of curiosity and diagnostics. For these psychiatrists and others, collecting the work of patients and aligning themselves with literary, musical, and artistic avant-gardes were inextricably linked to their self-conscious identity as reformers.[78]

100 After Tony Robert-Fleury, *Philippe Pinel Releasing Lunatics from Their Chains at the Salpêtrière Asylum in Paris in 1795*, 1876–78. Photogravure, 24.4 × 30.5 cm. British Museum, London.

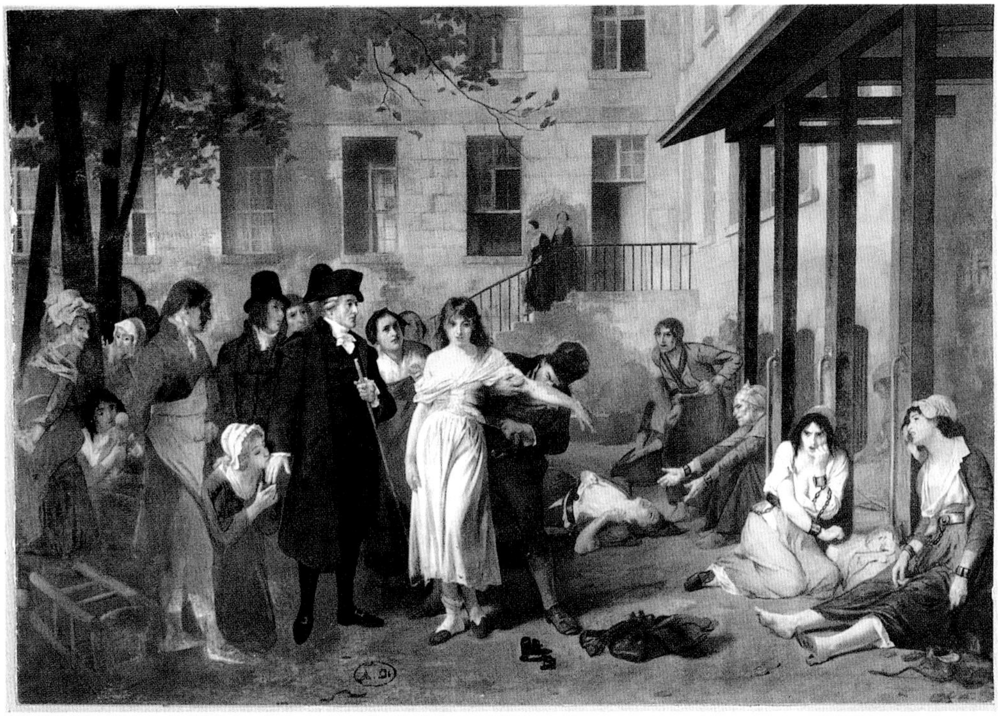

Évariste de Marandon de Montyel presented this new breed of psychiatrists as champions of the poor and politically disenfranchised, likening them to musicians whose genius can be realized only if their instruments—asylums—are properly tuned. Moreover, he compared the mentally ill to poets, in their heightened ability both to absorb their surrounding environment into their "state of soul" and to reflect that "state of soul" back onto their environment. Like artists, he argued, the mentally ill were particularly sensitive to their environment; a more liberated environment, he reasoned, therefore had the potential to free them from their disease. Similarly, when Réja's mentor Auguste Marie introduced his collection of patient art to the public, he maintained that it demonstrated the fundamental humanity of his patients. "The brain of the insane man is not that different from ours!" he insisted. "The only difference between him and us are exaggerations. . . . It is an all too common error to consider the insane as beings outside of humanity, to believe that their mental incoherence and disordered imagination are part of a world foreign to us." To underscore the point, Marie claimed that the majority of works in his collection were actually quite normal, or at least not nearly as "eccentric and baroque" as what one was likely to see at the Paris Salons.[79]

No doubt, Marie, like a number of Parisian critics, would have viewed Munch's work as "eccentric and baroque," but his protégé Réja proclaimed that Munch's "barbarous

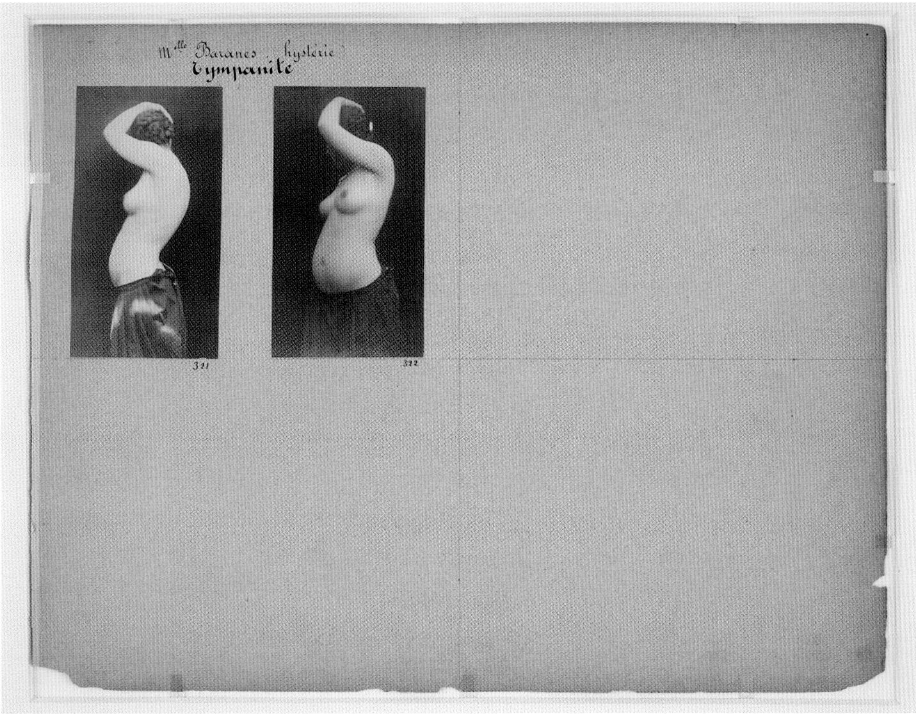

101 Albert Londe, *Mlle Baranes (hystérie): Tympanite*, ca. 1883. Photographs mounted on paper. Private collection, France.

brutality," although it might have appeared insane, in fact resulted directly from Munch's attempts to produce symbolist works of art.[80] Ideologically, it seems, Munch, Marie, and Réja were quite close, for *Women in Hospital* argued that even the maddest were individual human beings, not so different from "us," and therefore deserving of care and compassion. The three background figures, all half-dressed and highly schematized, are all partial, either doubled over, faceless, or cut off by the edge of the canvas. But they are all resolutely painted; their matter, though unevenly applied, is made self-evident. The saturated colors of their bare torsos—yellow and peach—draw attention to areas of thick paint, often single strokes, while in other areas, sketchily applied or thinned paint reveals the underlying blue of the background. The main female figure, however, is altogether different. An uncommon painted depiction of the elderly female body, she is about two-thirds life-size and dominates the canvas, head slightly tilted down, eyes apparently closed, in midstride or perhaps midshuffle. Munch chose not to paint her but to render her naturalistically, even individualistically, in colored crayon. Her sunken cheeks, visible collarbones, emaciated torso, and muscular legs are thoroughly realized, as if taken from the life. There could be no greater antithesis to Munch's otherworldly, youthful Madonnas. By virtue of their being half-dressed and painted, the three background women register as more mentally and materially present; in places, their bodies even protrude slightly

102 Edvard Munch, *In the Men's Clinic*, 1896. Lithograph, 33 × 55 cm. Munch Museum, Oslo.

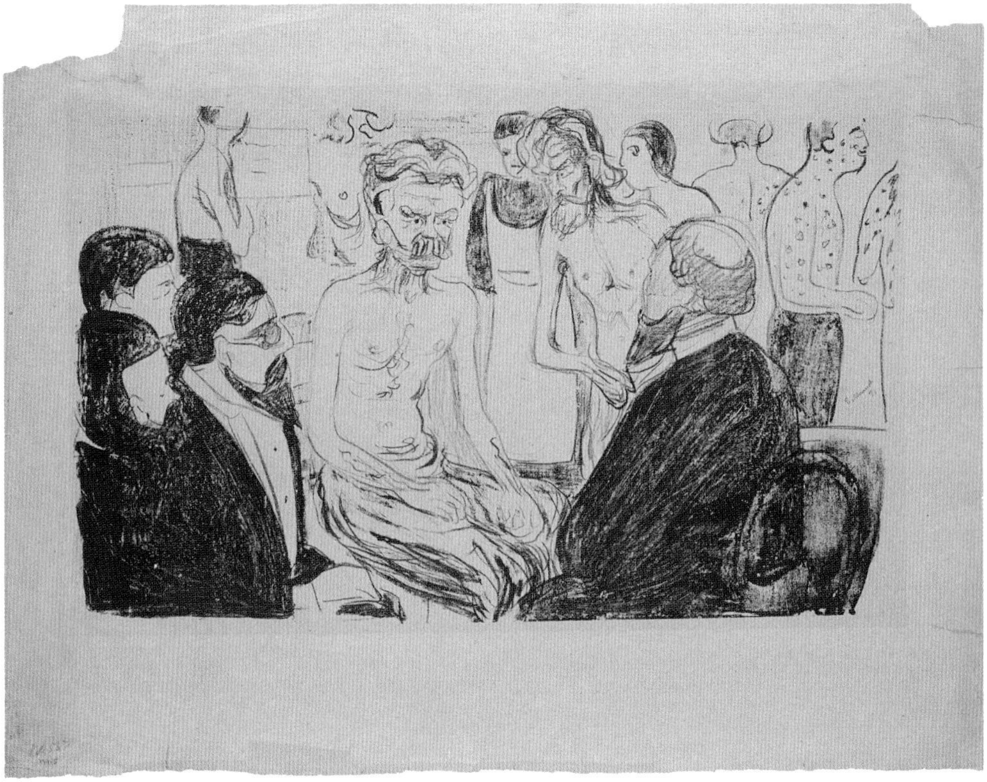

from the surface of the canvas. By contrast, the fully naked, unpainted figure is shallower, rendered before the matte background. Munch often began his painted compositions by sketching directly on canvas in crayon, pastel, or pencil.[81] Therefore, from the point of view of his working process, the figure can be read as a painterly other, unfinished or incomplete.

In this strategic return to naturalistic depiction, the central woman is at once the most diseased and least material, but also the most complete and particularized of the four figures depicted. That self, however, is disintegrating on the canvas. Especially between her legs and near her lower body, a chalky green medium, probably crayon or pastel, comes into view like a haze on the surface of the work, as if emanating from the woman's body. Using a much more friable medium than paint, Munch shows that the selfhood of this central woman is unstable, in the process of disappearing before our eyes. Her incomplete state derives not from the fact that she is not yet formed but that she was once formed and is now in the process of losing that form. Her liminal position between subjectivity and objectivity within the setting of a medical institution is what gives the painting its particular pathos. It is as if Munch painted the *process* of the subject becoming an object, the event of objectifying the subjective.

Women in Hospital highlights Munch's lifelong concern with the shifting subjectivities and objectivities in institutionalized settings of health and disease, but its message had a particular import for the fin-de-siècle Parisian psychiatric avant-garde. In this work, he posited the selfhood of the mentally ill as in the process of being lost, but even in their most vulnerable and fragile state as still connected to humanity. In *Women in Hospital*, Munch sought to provoke in the viewer the same emotional response he had experienced in the service of a moral message tied to asylum reform, which resulted in a posterlike piece of propaganda for the psychiatric avant-garde. Yet there nevertheless remained something clinical about the gaze he invited, an experimental gaze that rendered the individualized woman penetrable, her pathological body knowable. In the 1890s, despite his professed sympathy for the mentally ill, Munch identified more readily with doctors than with patients. His attitude, like that of psychiatric reformers such as Marie and Réja, remained paternalistic. In the 1896 lithograph *In the Men's Clinic*, Munch made this position explicit by including his avatar among the observing doctors (fig. 102).

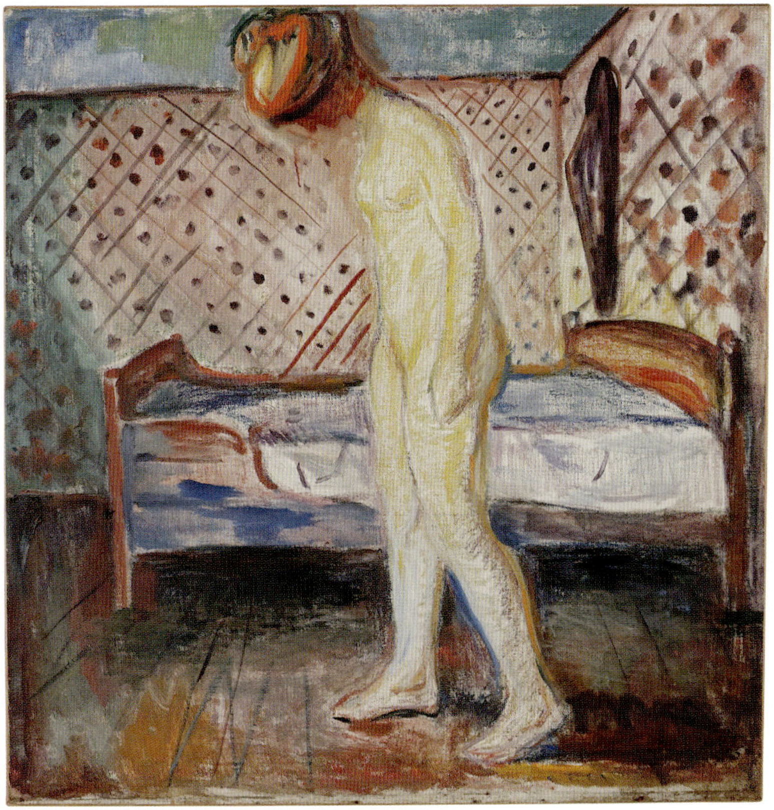

103 Edvard Munch, *Weeping Woman*, 1907. Oil on canvas, 121 × 119 cm. Munch Museum, Oslo.

The strategic alliance between psychiatrists and the avant-garde, between experimentalists and artists, would be stretched to the breaking point in the twentieth century, as the institutional critique brought forth by the avant-garde's so-called discovery of the art of "madmen, primitives, and children" was channeled into the antipsychiatry and anticolonialist movements of the 1960s. Munch registered this shift in subtle ways after the turn of the century, especially after his own voluntary stay at Dr. Daniel Jacobson's private clinic in Copenhagen in 1908–9. While he still often cultivated the authoritative gaze of the medical experimentalist, notably in his claims to be an anatomist, he also at times attempted to inhabit the much less certain role of the patient. Evidence of this shift appeared not for the first time in 1907, when Munch, in almost all of the media available to him at the time, revisited the figure of the striding or shuffling naked female, giving her an impenetrable, downturned face and an ambiguous contrapposto stance (fig. 103), in the process transforming her into a much less knowable entity. While this figure may derive from *Women in Hospital*, it is unclear whether she should be addressed as an object of eroticism, compassion, or even science. The relative certainty invited by the earlier painting, and indeed by the quantitative relationship between the normal and the pathological, and by the pathological method as a pathway to truth, had by then disappeared.

Dissecting the Soul

Later in life, Munch identified his work of the 1890s as a form of experimentation and stressed that his practice was intentionally engaged with the pathological.[82] "From 1894 to around 1915, I printed and experimented a lot," he wrote in a draft letter to an unidentified correspondent; "the main thing is that I experimented."[83] The *Frieze of Life*, the series of motifs that preoccupied Munch for much of his life, "dealt with inheritance as a curse," as he insisted in a letter from the 1930s. "It had something of an Osvald mood to it," he wrote, referring to the main character in Henrik Ibsen's play *Ghosts*, dying of hereditary syphilis.[84] Munch's use of the pathological, and in particular the psychopathological, was intertwined with a historically specific discourse of experiment. Nineteenth-century experimentalism positioned pathologies such as madness as privileged pathways for attaining knowledge about the most complex mechanisms of modern human psychology. The paradigmatic relationship between the normal and the pathological, the inextricable and functional links between what were imagined as two poles of being, encouraged Munch to chart the boundaries between health and disease, the normal and the pathological, and to play productively with reversing those poles with a view to making his work speak truth in universal terms. One irony of Munch's achievement, and indeed of the achievements of Denis, Vuillard, and Strindberg, was that they helped establish a new norm, that of symbolist simplification and distortion, which, in the accelerated logic of turn-of-the-century avant-gardes, would soon be superseded by even more radical experiments in abstraction. Perhaps this goes some way toward explaining what appears to be Munch's return to a form of naturalism in the twentieth century. Structurally, Munch's works needed the dialectical tensions of naturalism and symbolism, the normal and the pathological, that had implicitly operated in the term "psychic naturalism," Przybyszewski's extremely apt coinage of 1894.

Like Strindberg in the guise of Arvid Falk, Munch eventually wrote a manifesto on the usefulness of his artistic research into pathology, a compendium of notes initially titled *The Mad Poet's Diary*. And, also like Strindberg, he made clear that he thought of himself as a vivisectionist who dissected not bodies but souls in an experimental quest for truth. Although *The Mad Poet's Diary* began in the third person, Munch quickly dropped this pretense of objectivity, while still insisting that his aim had always been to study the soul scientifically, to objectify the subjective, and to do so through means such as "exaggeration" and the "use of others." "It is important for me to study the various inherited phenomena that form the life and destiny of a human being," he wrote, "especially the most common forms of madness. I am making a study of the soul, as I can observe myself closely and use myself as an anatomical soul preparation. The main thing is to make an artwork and a soul study, so I have changed and exaggerated, and have used others for these studies. . . . Just as Leonardo da Vinci studied the recesses of the human body and dissected cadavers, I try to dissect souls."[85] Munch never came closer than this to elucidating the role of madness in his art. Stressing the importance of an objective study of insanity, of the soul's pathologies, as well as of the use of exaggeration and alterity, he delivered his own version of Strindberg's *Madman's Defense,* the manuscript of which Strindberg, not insignificantly, gave him as a gift in the 1890s.[86] Munch positioned madness as the fundamental crux in a creative method aimed at achieving the most ambitious and idealistic of symbolism's goals: to determine and to represent objectively the mechanisms of the modern psyche. In the 1930s, as Max Nordau's diagnosis of the symbolists as mentally ill was increasingly extended to modernist artists, and to ever more horrific ends, Munch's pathologies were repathologized. When "Madness Becomes Method" appeared as a slogan to be ridiculed on the walls of the Nazis' Degenerate Art Exhibition in 1937,[87] the perverse reference to a central tenet of Munch's symbolist practice could only have been devastating.

CONCLUSION

The most modernist symbolist visual practices—the decorative deformations theorized by Maurice Denis, the automatisms and quasi-abstractions of Édouard Vuillard and August Strindberg, and the expressive distortions of Edvard Munch—were also the most indebted to naturalist precepts and notions of universal truth. Hence the currency, in the 1890s, of dialectic terms like "néo-traditionnisme," "interior impressionism," "naturalistic symbolism," and "psychic naturalism."[1] These coinages of Denis, Vuillard, Strindberg, and Stanisław Przybyszewski (describing the work of Munch) reveal symbolism at its most modernist to be dependent—paradoxically, it might seem—on naturalism, not only as a visual mode to reject but also as a truth-oriented idiom from which to draw and to adapt methods. Indeed, these artists and many others in the final decades of the nineteenth century, including Paul Cézanne, prolonged and cven intensified naturalism's search for truth.[2] They rejected what they imagined to be facile visual correspondences between the exterior, visible world and a painted world, and proclaimed the possibility of new visual forms deriving primarily from a search for knowledge of interior worlds.

Thus an artist like Vuillard could be at once among the most formally radical of the Nabis and the most reliant on naturalist tenets. And what is generally seen, in the cases of Denis, Vuillard, and Munch, as retrograde returns to naturalism, especially in the decades following World War I, in fact reveals these artists' fundamental underlying commitment to art as a form of knowledge production. As belief in the truth regimes of the Enlightenment was undermined in the twentieth century, as knowledge became suspect and the radically new a cherished avant-garde value,[3] the "retreat" to naturalism was for some a necessary shelter, especially in the face of modernist visual idioms that seemed strangely familiar—I am thinking here above all of the similarities between Denis's *Décor* and Henri Matisse's fauvist nudes, Vuillard's ink arabesques and Pollock's webs, Strindberg's and surrealism's automatisms, Munch and the avant-garde "discovery" of the art of the insane—yet radically unmoored from the epistemological regimes of the 1890s. Even a doubter like Vuillard, it seems, could not or would not disengage from an aesthetic position oriented toward truthfulness. It remains an open question whether Strindberg, who died in 1912, might have adjusted to the new logics of twentieth-century vanguardism, although his heteroclite practice, and the evidence that he came to regard pathology as something of a construction and avant-gardism with a certain suspicion,[4] make it tempting to imagine that Strindberg might have at least registered some uncertainty about the validity of the nineteenth-century beliefs that had structured his view of art.

I have argued throughout this book that experimentalism and avant-garde practices came to be aligned because the epistemology of experiment, especially as it had been applied to the new psychology, was especially suited to the goals of certain symbolist artists, those who took very seriously the dictates of symbolist theory to "objectify the subjective," and those who were conscious of doing so in ways that would signal their difference from the forms, if not the methods and assumptions, of naturalism. The authority of experiment, not just in its traditional domains of physics and chemistry but also in the life sciences and in newer scientific disciplines like psychology, was virtually unquestioned at the end of the nineteenth century. To seek knowledge meant to experiment. To establish truth increasingly meant to do so through experimentation. But in the second half of the century, the links between art as truth-producing and science as truth-producing were not necessarily, as Émile Zola discovered, easy to imagine or to justify until the advent of a scientific psychology with a hard-won purchase on experimentalism.[5]

Unlike in Germany or in England, the promise of a new scientific psychology in France depended above all on recourse to pathology, to those experiments that nature, so obligingly and in all her wisdom, offered up to the scientist so that he could seek and find truth. Pathology in its many forms thus became a sign of truth, or at least a sign that truth was being sought and might be found. To observe pathologies, to examine alterity, to experience something other than normality, was considered the very modern pathway to knowledge of the human psyche. Pathology might thus be said to be not only fundamental to modernity[6] but also determinative of modernism. Experimentalism, and in particular the enormous power invested in nature's experiments, not only resulted in some of the most formally radical works of the 1890s but also laid the epistemological foundations for the early twentieth-century avant-garde's manifold obsession with alterity.

This study, not explicitly a work of feminist art history but written by a feminist, has been haunted by a group of others, namely, women, overrepresented in the works discussed but largely absent as speaking subjects. "Women" is the missing term in Théodule

Ribot's privileged triad of psychological study, "madmen, primitives, and children." At once, in her white, bourgeois form, too similar to the apex of human normality—the white, bourgeois male—and too other to be contemplated as a norm, woman as sign, including as a sign of nature, has played between the lines of this book, but an analysis of gender has less often been foregrounded. Women like Anna Thibout and Marie Vuillard provided the bodies where form could be deformed and where the automatic arabesque could be transformed into the experimental arabesque. Woman was often the pathological object of study—the suspected adulteress Siri von Essen, the unnamed syphilitic mother, the mad *folle*, the talented medium. And even if womanhood itself is not as often explicitly named as pathological, the pathological was certainly feminized, and women were generally denied the opportunity to access truth through contemplating and cultivating the pathological. The self to be constructed, discovered, known, was not a female self. At a moment when concepts such as the conscious and the unconscious, exteriority and interiority, self and other, were gendered as well as classed and racialized, Vuillard, Strindberg, Munch, and to a lesser extent Denis embraced unstable selves and knowingly risked feminization for truth. But as a discourse of feminized pathologized artistic genius was already well established, even scientifically "proven,"[7] such risks were surely manageable, the dangers of feminization mitigated by adopting the position of the humble genius experimenter.

Symbolism, especially the symbolism that looks like modernism, is startling for its near absence of women artists, especially in comparison to impressionism and naturalism, movements in which women artists participated with some frequency and indeed some professional success. In most discussions of symbolism, which privilege gendered iconography such as the femme fatale, the idealized lover, and the tragic heroine, the absence of women practitioners seems to derive from, among other institutional factors, the preference for and valuing of exclusionary subject matter. But in my analysis of symbolist visual practice, the lack of women practitioners correlates to the gendered selves posited not only by symbolist theory but also by the institutionalized philosophies and the methods of experimental psychology that structured both the theory and many of the visual practices that allegedly applied that theory.

At the outset of this study, I admitted to being in thrall to many of the visual propositions offered by Denis, Vuillard, Strindberg, and Munch in their search for symbolist form, but for reasons that differed from those intended by the artists. In fact, I deliberately set out to avoid being complicit with the artists under discussion. I make no claim to know whether Munch's *The Scream*, for instance, embodies anxiety and madness. I do know that it was received as such in the 1890s and is received as such today, in our late modern moment. My intention has been to ask what visual aspects of the work made that reception possible and justifiable within the context of the conceptual possibilities of the late nineteenth century. This study has thus sought to historicize the most putatively innovative symbolist visual practices, to demystify the avant-garde value of experiment, and to expose the underlying positivist epistemology that enabled alterity in all its forms to become a source for antipositivist avant-garde creativity. In doing so, I have hoped to render experimental art not at all natural, but radically and usefully strange.

Notes

Introduction

1. Diderot, *De l'interprétation de la nature*, 177, 180, 217. See also the entry for *expérimental* (penned by Jean Le Rond d'Alembert), in Diderot and d'Alembert, *Encyclopédie*, accessed 11 June 2013. Despite Peter Bürger's dismissal of symbolism as the epitome of aestheticism and his focus on literature rather than visual art, I am borrowing his term "historical avant-gardes" to emphasize a historicized view of vanguardism. While Bürger, in his influential study, historicizes the new as an avant-garde value, he offers no such analysis of experiment. Bürger, *Theory of the Avant-Garde*, 60–63. For useful critiques of Bürger, see Buchloh, "Theorizing the Avant-Garde"; and Ward, *Pissarro, Neo-Impressionism*, 263–65.
2. Greenberg, "Avant-Garde and Kitsch," in Greenberg, *Collected Essays and Criticism*, 1:8.
3. The parody is summed up by the lyrics and video for the song "Experimental Film" by the American band They Might Be Giants, from the 2004 album *The Spine*.
4. Aurier, "Beaux-arts: Les symbolistes," 474.
5. For a fruitful study of symbolism and experience, which nevertheless conflates experience and experiment in ways critiqued in chapter 1, see Cándida Smith, *Mallarmé's Children*.
6. On the links between constructions of individual truth and universal truth in relation to both impressionism and symbolism, see Shiff, *Cézanne and the End of Impressionism*, 17, 21–26. In reality, this symbolist public was small, elite, and almost entirely male. On symbolism's gender biases, see Higonnet, "Imaging Gender," 152–53; Broude, *Impressionism: A Feminist Reading*, 144–65; and Mathews, *Passionate Discontent*.
7. Kahn, "Réponse des symbolistes." All translations are my own unless otherwise noted.
8. The epigraph to this section is from Bernard, "Étude sur la physiologie du coeur," in *Science expérimentale*, 366.
9. Bergson, "Philosophie de Claude Bernard," 257.
10. On the establishment of experimentalism in the physical sciences, see Shapin and Schaffer, *Leviathan and the Air-Pump*. Despite the fact that William Harvey's experiments on the circulation of blood date to the early seventeenth century, medical experimentation was contentious well into the nineteenth.
11. The distinction between empirical and experimental methods, however, remained far from clear, despite the efforts of Bernard and others to promote experimentalism by elevating it above empiricism. See Daston and Lunbeck, *Histories of Scientific Observation*, 3–4. There is a wealth of scholarship on Bernard and his *Introduction to the Study of Experimental Medicine*. See Olmsted and Olmsted, *Claude Bernard and the Experimental Method*; Canguilhem, *Idée de la médecine expérimentale*; Grande Covián and Visscher, *Claude Bernard and Experimental Medicine*; Grmek, *Raisonnement expérimental*; Michel, *Nécessité de Claude Bernard*; and Gendron, *Claude Bernard*.
12. Bernard, *Introduction*, 48.
13. Jay, *Songs of Experience*, 10. What follows is indebted to Jay's etymological analysis of *experience*, as well as the entries for *empirical* and *experience* in Williams, *Keywords*, 115–17, 126–29.
14. The notoriously slow publication schedule of the Academy's dictionaries, which track changes of at least half a century earlier, should be noted. According to the website of the current Academy dictionary, in progress, very few changes occurred between the seventh (1878) and eighth (1932–35) editions, which makes the changes in the entries for *expérience* between

these two editions all the more remarkable. See http://www.academie-francaise.fr/dictionnaire/index.html, accessed 31 May 2012. For other online dictionaries, see http://www.artfl-project.uchicago.edu/content/dictionnaires-dautrefois, accessed 31 May 2012.

15. Daston and Galison, *Objectivity*, 242–43.

16. German and Swedish dictionaries are slightly more revealing. A 1918 Swedish dictionary includes *experimental-roman*, crediting the term to Zola, as its only cultural usage. See the Swedish Academy's *Ordbok öfver Svenska Språket*. German, Swedish, and Norwegian all use a second word that is often translated as *experiment* (*Versuch*, *försök*, *forsøk*), but in relation to both scientific and cultural usage today, these are rarely favored over *Experiment* and *eksperiment*. Tellingly, German conserves two spellings for the adjectival form: *experimentell* for general and scientific contexts, and *experimental*, as in *Experimentalfilm*, for cultural contexts, suggesting that the latter derives from either French or English usage.

17. The claim that the avant-garde took up experiment allegorically is Renato Poggioli's, in *Theory of the Avant-Garde*, 139, and it informs Gombrich, "Experiment and Experience in the Arts."

18. Nietzsche, *Will to Power*, 29, citing Bernard, *Leçons sur la chaleur animale*, 391. See Canguilhem, *Normal and the Pathological*, 45.

19. Certainly, Francis Bacon's preternatural philosophy had long been superseded by studies of monstrosity understood as pathological rather than wondrous. On the role of monsters and other prodigies in sixteenth- and seventeenth-century philosophy, see Park and Daston, "Unnatural Conceptions"; and Daston and Park, *Wonders and the Order of Nature*. Jan Goldstein identifies an early version of the pathological method in the recommendations of Nicolas de Condorcet and Louis-Jacques Moreau de la Sarthe. Goldstein, *Console and Classify*, 240.

20. In the words of Comte, François-Joseph-Victor Broussais's "luminous maxim" states that "the phenomena of the pathological state are a simple prolongation of the phenomena of the normal state." Comte, *Système de politique positive*, 1:651–52, quoted in translation in Comte, *Essential Writings*, 191. See also Canguilhem, *Normal and the Pathological*, 49; and Nikolas Rose, "Normal," in Bennett, Grossberg, and Morris, *New Keywords*, 241–43.

21. Comte, *Essential Writings*, 244, 168. In a sense, this was Comte's response to the vitalists, whose rejection of experimental methods for living matter still held a great deal of sway. See Canguilhem, *Normal and the Pathological*, 53.

22. Canguilhem, *Normal and the Pathological*, 67. See also Beaune, "Notion de 'pathologique,'" 287–89.

23. Canguilhem, *Normal and the Pathological*, 67. A philosophy of error is Canguilhem's central concern and, in Foucault's view, the crux of his contribution to a history of the life sciences. See Foucault, "Introduction," ibid., 7–24.

24. Renan, *Avenir de la science*, 184. See also Canguilhem, *Normal and the Pathological*, 44–45. Renan's analogy between the physical and the psychological sciences, which relied upon the authority of the physical sciences, was a common one toward the end of the nineteenth century. On the experimental authority of the physical sciences at the fin de siècle, see Galison and Assmus, "Artificial Clouds, Real Particles," 264–69.

25. On Cousin's introspective method and its remarkable persistence, see Goldstein, "Advent of Psychological Modernism"; Brooks, "Philosophy and Psychology at the Sorbonne"; Brooks, *Eclectic Legacy*; Goldstein, *Post-Revolutionary Self*.

26. See Janet, "Chaire de psychologie expérimentale."

27. Société de psychologie physiologique, *Congrès international de psychologie physiologique*, 142, quoted in Carroy and Plas, "Origins of French Experimental Psychology," 74. As the following chapter reveals, however, metaphysical questions remained crucial to many in the field.

28. Carroy and Plas, "Méthode pathologique." Scientific models of psychology existed as early as the eighteenth century, particularly in English- and German-speaking contexts. Hatfield, "Remaking the Science of the Mind." But the mid-nineteenth century brought increasingly loud calls across Europe for a scientific psychology divorced from philosophy once and for all.

29. Taine, *De l'intelligence*, 1:16, 2–4. On Taine, see Kahn, *Science and Aesthetic Judgement*; Nordmann, *Taine et la critique scientifique*; Michaud and Le Pavec, *Taine au carrefour*; Seys, *Hippolyte Taine et l'avènement*; and Compagnon et al., *Relire Taine*. An early art-historical consideration of Taine can be found in Bouillon, "Histoire de l'histoire de l'art." See also Morton, "Naturalism and Nostalgia"; and Larson, *Dark Side of Nature*, 17–18, 40–41, 141–44.

30. Taine, *De l'intelligence*, 1:13–14. Zola, a great admirer of Taine, would take up this recommendation, submitting himself to Édouard Toulouse as an object of study. Toulouse, *Enquête médico-psychologique*. See Carroy, "'Mon cerveau.'" Zola's admiration for Taine, which waned in the 1880s, is evident in articles such as "M. H. Taine, artiste," in *Mes haines*, 201–32.

31. Ribot, *Psychologie anglaise contemporaine*, 35–36. "Morbid psychology" follows the terminology of Jacques-Joseph Moreau (de Tours). Ribot's reluctance to fully reject introspection may have been strategic, in that Cousinian eclecticism continued to dominate institutions of higher learning and the lycée curriculum well into the 1880s. Taine's academic career had been stymied by his vocal espousal of positivism, while Ribot, about a decade younger than Taine, had embarked upon a conventionally successful academic career.

32. Ibid., 36.

33. Compare the much less successful publishing record of Jean-Martin Charcot, who is often seen as a key representative of the new scientific psychology. While Charcot's impact was certainly large, it was earned primarily through his lectures and public demonstrations rather than through his writings.

34. See Delay, "Jacksonism and the Works of Ribot"; Faber, "Théodule Ribot and the Reception."

35. Ribot, *Psychologie anglaise contemporaine*, 25–26, 38. See also Carroy and Plas, "Méthode pathologique," 607–9.

36. Carroy and Plas, "Origins of French Experimental Psychology," 83.

37. Carroy, "Hypnose, psychothérapie et expérimentation"; Harris, "Murder Under Hypnosis." While French theories of the unconscious remained influential in France well into the twentieth century, they were largely ignored in post-Freudian histories of psychology and psychiatry until the work of Henri Ellenberger. See Ellenberger, *Discovery of the Unconscious*, 254–330.

38. On alterity as the construction of cultural others, see Taussig, *Mimesis and Alterity*.

39. See Heller, "Expressionism's Ancients"; Connelly, *Sleep of Reason*, 5, 11–36. Both authors make clear that the romantic view had its roots in the classical tradition.

40. *The Dark Side of Nature*, Barbara Larson's wide-ranging study of Redon's various engagements with scientific cultures, has suggested numerous avenues of research in developing the present work.

41. Michelle Facos offers a helpful introduction to symbolist painting as a historiographical problem in *Symbolist Art in Context*, 1–7.

42. *Redefining European Symbolism, 1880–1910* involved exhibitions, conferences, and a series of workshops. Jean-Paul Bouillon, following Pierre-Louis Mathieu, preferred to use the term "moment" rather than "movement." See Bouillon, "Moment symboliste"; and Mathieu, *Génération symboliste, 1870–1910*. To these fundamental texts should be added the special issues on symbolism of the *Art Journal* 45 (Summer 1985), edited by Sharon L. Hirsh, and the *Revue de l'art*, no. 96 (1992), edited by Jean-Paul Bouillon.

43. Michael F. Zimmermann proposed "definition" as a useful mode of inquiry during the final discussion of the "Redefining European Symbolism" conference held at the Musée d'Orsay, Paris, 13–14 April 2012.

44. In the early 1890s, the term *intellectual* was relatively new, its emergence coinciding with increasing opportunities and pressures for artists and other cultural figures to present themselves as thinkers. See Ward, *Pissarro, Neo-Impressionism*, 195; Datta, *Birth of a National Icon*.

45. Two previous studies, both addressing the work of Wassily Kandinsky and Piet Mondrian, artists with roots in symbolism, have elucidated

the relationship between symbolist aesthetics and abstraction: Cheetham, *Rhetoric of Purity*, and Reynolds, *Symbolist Aesthetics and Early Abstract Art*. This same problem—exactly how symbolism and modernism relate to each other—is noted in Facos and Mednick's introduction to *Symbolist Roots of Modern Art*, 1–7, and is addressed in different ways by a number of that volume's contributors.

46. See Burhan, "Vision and Visionaries"; Silverman, *Art Nouveau in Fin-de-Siècle France*; Zimmermann, *Seurat and the Art Theory*; Gamboni, *Potential Images*; Rapetti, *Symbolisme*; and Larson, *Dark Side of Nature*, 133–56.

47. My approach might be compared to that of Jonathan Crary, both in his short essay "Seurat's Modernity" and in his *Suspensions of Perception*, and that of Brain, *Pulse of Modernism*. Crary describes a wide-ranging pan-European concern for attention as a particularly pressing problem of modernity for psychologists, aestheticians, critics, and artists, while I am interested in more localized usages of experimentalism and the pathological method. Brain focuses on the "experimental systems" (Hans-Jörg Rheinberger's term) and technics of physiology, rather than experimental method, to make arguments about modernism not incompatible with my own. Tellingly, we both use the work of Strindberg and Munch, although our readings of specific works, Munch's *The Scream*, for instance, differ significantly.

48. See Heller, "Art Work as Symbol" and "Concerning Symbolism."

49. Clark, *Farewell to an Idea*, 165.

50. Michael Leja's work, which examines how abstract expressionists and their critics instituted a system of representation that successfully rendered the unconscious and the primitive in post–World War II United States, has been a useful model for this study. See especially Leja, *Reframing Abstract Expressionism*, 1–17.

51. On symbolist journals and the communities surrounding them, see Genova, *Symbolist Journals*.

52. On *La revue blanche*, see Jackson, *Revue blanche, 1889–1903*; Comes, "*Revue blanche* et le mouvement"; Datta, "*Revue blanche*"; and Bourrelier, *Revue blanche*. On the relationship between *La revue blanche* and artists, see Waller and Seiberling, *Artists of* La revue blanche; and Bernier, *Revue blanche, ses amis, ses artistes*.

53. Natanson, "Correspondance de Kristiania." Munch had been mentioned in *La revue blanche* the month before, in an article by Henri Albert, one of the earliest French translators of Nietzsche. See Albert, "Lettres allemandes," 381.

54. Strindberg, "Exposition d'Edward [*sic*] Munch." Apart from his own articles in *La revue blanche*, Strindberg had been the subject of a lengthy appraisal in the journal by Henri Albert not long after his arrival in Paris. See Albert, "Auguste Strindberg."

55. See Comes, "'Mercure de France' dans l'évolution"; and Décaudin, "*Mercure de France*."

56. Goldstein, "Neutralizing Freud," 53–59.

57. On the Nabis' lycée philosophy training, see especially Burhan, "Vision and Visionaries," 42–45; Bouillon, "Denis, Taine, Spencer," 295–97.

58. On *La revue blanche* and the *Mercure de France* and the construction of the intellectual in France, see Datta, *Birth of a National Icon*. On artistic *enquêtes* as a form of art criticism at the fin de siècle, see Méneux, "Enquêtes artistiques."

59. Bélugou, "Quelques opinions sur l'oeuvre de Taine," 266. For a comparison of Vuillard and Barrès, see Chastel, *Vuillard, 1868–1940*, 59.

60. Quoted in Bélugou, "Quelques opinions sur l'oeuvre de Taine," 269. In many ways, Denis's project would also be to reconcile Tainean positivism with Catholicism. See Bouillon, "Denis, Taine, Spencer," 302–4.

61. Bélugou, "Quelques opinions sur l'oeuvre de Taine," 274. On Boutroux and Bergson, see Grogin, *Bergsonian Controversy in France*, 14–16. On Boutroux and Durkheim, see Lukes, *Émile Durkheim*, 57–58.

62. Bélugou, "Quelques opinions sur l'oeuvre de Taine," 275, 283, 273, 295.

63. Before World War I, Bélugou (1865–between 1942 and 1948) was a member of the Stendhal Club, along with Maurice Barrès, Paul Bourget, and Paul Signac. His correspondence

with Edith Wharton, to whom he became close in the first decade of the twentieth century, is published in Wharton, *Edith Wharton*. After World War I, Bélugou worked for mining companies in French Indochina. See Léon Bélugou Papers, Archives nationales, 176 AQ (1997 076). There is also extant correspondence between Bélugou and Ribot, and Bélugou and Marcel Proust.

64. Bélugou, "Chronique philosophique" (November 1894), 458.
65. Bélugou took assiduous notes in Ribot's classes from late 1887 until 1901. For more on these notes, see chapter 3.
66. Bélugou, "Chronique philosophique" (1 February 1895), 182.
67. Bélugou, "Philosophie," 304–5.
68. Bélugou, "Chronique universitaire" (February 1898).
69. Bélugou, "Chronique universitaire" (June 1899).
70. Bélugou, "Chronique universitaire" (February 1902), 502–3.
71. Among the many investigations of Bouvard and Pécuchet imagined by Flaubert, the two men meditate on a candle flame. See Goulet, *Optiques*, 1–2, 13.
72. Bélugou, "Pouvoir de l'imagination," 136, 140.
73. Bélugou, "Chronique universitaire" (February 1902), 502.
74. Another alternative might be found in a microhistorical approach, such as the one I offer in the article "*Musée de la folie*."
75. Scc Shiff, *Cézanne and the End of Impressionism*, xv–xvi.
76. See Jumeau-Lafond, "Symbolismes."
77. Foucault, *Archaeology of Knowledge*, 29.

Chapter 1

1. The spiritualist reading is exemplified by Caro, *Matérialisme et la science*, and the positivist reading by André-Nuytz's refutation of Caro, "Spiritualisme et l'école expérimentale." See also Hirst, *Durkheim, Bernard, and Epistemology*, 19–32, 43–57. In France, spiritualism connotes metaphysical philosophy and belief in the afterlife of the soul, while in Anglo-American contexts, spiritualism refers more specifically to the nineteenth-century movement, which is denoted here, following Lynn L. Sharp, by spiritism. See Sharp, *Secular Spirituality*, xi.
2. Paul, "Debate over the Bankruptcy of Science." One exception, where the authority of the experimental method is held up for some ridicule, is Paul Bourget's novel *Le disciple* (1889). See Larson, *Dark Side of Nature*, 182.
3. Bernard, "Étude sur la physiologie du coeur," 366.
4. The essay first appeared in September 1879 in the Saint Petersburg journal *Le messager de l'Europe*, and the following month appeared in Paris in *Le Voltaire*, along with the first fascicule of *Nana*. In 1880, Zola included the essay in a collection of theoretical and polemical texts; see Zola, *Roman expérimental*, 1–53.
5. Zola, "Roman expérimental," in *Oeuvres complètes*, 9:337.
6. Zola discusses hereditary degeneration only briefly in "The Experimental Novel." There is an extensive literature on the theme of degeneration in Zola's novels and more generally in late nineteenth-century French culture. As starting points, see Barrows, *Distorting Mirrors;* Nye, *Crime, Madness, and Politics;* Chamberlin and Gilman, *Degeneration;* and Pick, *Faces of Degeneration*. On degeneration and symbolism, see Larson, "Microbes and Maladies," 386.
7. Brunetière, "Revue littéraire," 939, 946, 936.
8. See de Pontmartin, "M. Émile Zola," 162. Benjamin Aubé made a very similar critique in "La littérature et la science," *Journal officiel*, 5 November 1880, quoted in Zola, *Correspondance*, 4:136n3.
9. Ferdas, *Physiologie expérimentale*, 6, 9–10. Zola referred only obliquely to vivisection in his text. The perceived need to defend Bernard from Zola's travesty persisted well into the twentieth century. See, for example, Voorhoeve, "Contre Zola."

10. Zola, "Roman expérimental," in *Oeuvres complètes*, 9:334.

11. It is telling, perhaps, that Alphonse Daudet criticized Benjamin Aubé's article (see note 8 above) and proceeded to defend not "The Experimental Novel" but another essay, "Naturalism at the Theater," which appeared in the *Journal officiel* in 1880. See Daudet, "La vérité au théâtre," *Journal officiel*, 29 November 1880, quoted in Zola, *Correspondance*, 4:135–36. At least one of Zola's correspondents might have found solace in an apparent private retreat from dogma. See Zola to unknown, 26 November 1879, *Correspondance*, 3:411–12.

12. Céard had previously studied medicine. See Verne, "Une lettre de M. Henry Céard"; Burns, "Henry Céard," 317–18; and Chantal Pierre-Gnassounou, introduction to Zola, "Roman expérimental," in *Oeuvres complètes*, 9:317.

13. Flaubert to Maupassant, 21 October 1879, in Flaubert, *Oeuvres complètes*, 16:317. We have no recorded opinion from Edmond Duranty, an author very close to Zola beginning in the mid-1860s and the most disposed to link both naturalism and impressionism with scientific methods. This may be because Duranty died less than six months after the essay appeared and before the most vituperative critiques were published. See below.

14. Becker, Gourdin-Servenière, and Lavielle, *Dictionnaire d'Émile Zola*, 289.

15. "Within the narrowed discourses and restricted ranges of nineteenth-century criticism, silence continued to be a powerful form of expression." Ward, *Pissarro, Neo-Impressionism*, 14.

16. Such analyses were inaugurated by Butor, "Émile Zola, romancier expérimental," and have continued with Guedj, preface to Zola, *Roman expérimental*; Pagès, "En partant de la théorie"; Mitterand, "Textes en intersection"; Cabanès, "Zola et le modèle bernardien"; and Bourdieu, *Rules of Art*, 116–17.

17. Zola, "Mon Salon," *L'événement* (April–May 1866), in Zola, *Écrits sur l'art*, 85–135. See also Broude, *Impressionism: A Feminist Reading*, 122–23.

18. Zola, "Roman expérimental," in *Oeuvres complètes*, 9:328.

19. Ibid., 9:339.

20. This formula first appeared in Zola, "Proudhon et Courbet," *Le salut public* (26 July and 31 August 1865), in Zola, *Écrits sur l'art*, 44. Kahn's statement has often been read as declaring a break with naturalism and impressionism, but, as Richard Shiff has argued, it should be understood as strategically dramatizing their opposition. See Shiff, *Cézanne and the End of Impressionism*, 39–52.

21. Pagès, "En partant de la théorie"; Pierre-Gnassounou, introduction to "Roman expérimental," in *Oeuvres complètes*, 9:319.

22. Zola, "Roman expérimental," in *Oeuvres complètes*, 9:339.

23. Guieu, "Caricatures antidreyfusards," 439–40.

24. Louis Leroy, "L'exposition des impressionnistes,"*Le Charivari*, 25 April 1874, 79–80, quoted in Berson, *Reviews*, 25. There is some debate over which painting of the Boulevard des Capucines Monet exhibited, but generally the painting now in the Pushkin Museum in Moscow is favored. See Kennedy, "Monet's Boulevard des Capucines."

25. See Ward, *Pissarro, Neo-Impressionism*, 263–65.

26. Gombrich, "Experiment and Experience in the Arts," 162–65; Isaacson, "Observation and Experiment"; Druick and Zegers, "Scientific Realism: 1873–1881"; Shiff, "Morality, Materiality, Apples," 169.

27. It is discussed only briefly in Marcussen and Olrik, "Le réel chez Zola," and in Mitterand, "Zola et le réalisme artiste."

28. Shiff, *Cézanne and the End of Impressionism*, 14–26. Norma Broude revealed the gender stakes involved in inserting impressionism into a dominant masculinizing narrative of scientific progress in *Impressionism: A Feminist Reading*, 110–43.

29. Cugini, *Er sieht einen Fleck*. See also the forthcoming work of Laura Anne Kalba on impressionism and chemistry, *Color in the Age of Impressionism*.

30. Quoted in Gombrich, "Experiment and Experience in the Arts," 145. Neither this nor a related statement on Claude Lorrain ("The

greatest masters considered their best efforts but as experiments, and perhaps as experiments that had failed when compared with their hopes, their wishes, and with what they saw in Nature") appeared in Constable's statements that were translated into French and published in the *Magasin pittoresque*, later excerpted in the short-lived journal *Réalisme*, edited by Duranty (*Réalisme*, no. 2 [15 December 1856]: 29–30). See Constable, "Pensées d'un paysagiste"; Duranty, "Idées de Constable." See also Isaacson, "Constable, Duranty, Mallarmé," 428–29.

31. On art and experiment in the seventeenth century, see Hunter, *Wicked Intelligence*.

32. Gombrich's position is informed by his particular blind spot regarding abstraction and his difficulties in accounting for shifts between illusionistic and nonillusionistic forms of art making in the modern period. His own view of avant-garde experimentation, however, did not stop him from trying to redeem, or at least to reiterate the connection between, the avant-garde and experimentation by titling the final chapter of *The Story of Art* "Experimental Art: The Twentieth Century." Gombrich's understanding of the avant-garde was, he readily acknowledged, heavily indebted to the work of Renato Poggioli. For a less woeful view of experimentation and twentieth-century art, see Bann, *Experimental Painting*.

33. A similar problem is found in Joel Isaacson's suggestive but ultimately inconclusive attempt, in "Observation and Experiment," to use Bernard's *Introduction* to illuminate aspects of Monet's early working practices.

34. For a critique and analysis of this discursive practice, see Broude, *Impressionism: A Feminist Reading*. My argument in this chapter for the most part squares with Broude's argument that the art-historical relationship between impressionism and science is both overdrawn and based on a failure to distinguish between "normal science"—the dominant tradition of mechanistic, materialist science—and the more romantic traditions of *Naturphilosophie*, which persisted into the nineteenth century.

35. Fry, "The Philosophy of Impressionism" (1894), in *Roger Fry Reader*, 12–20.

36. Fry, "Post Impressionism," *Fortnightly Review*, 1 May 1911, ibid., 100.

37. Brettell, *Impression*, 11, 13.

38. Melot, *Impressionist Print*, 181. Druick and Zegers translate Degas's catalogue text as "experiments and states of plates." "Scientific Realism: 1873–1881," 199.

39. Quoted in Druick and Zegers, "Scientific Realism: 1873–1881," 199.

40. Cézanne to Geffroy, 26 March 1894, in *Paul Cézanne Letters*, 237.

41. Cézanne to Vollard, 2 April 1902, ibid., 283.

42. Quoted in Stevens et al., *Émile Bernard, 1868–1941*, 11.

43. See entry no. 14 in the catalogue for the Fifth Impressionist Exhibition, reprinted in Moffett, *New Painting*, 311.

44. Duranty, "Sur l'outillage dans l'art."

45. Cézanne to Geffroy, 26 March 1894, in *Paul Cézanne correspondance*, 299. A new translation of Cézanne's letters by Alex Danchev renders "tentatives" as "efforts." Cézanne, *Letters of Paul Cézanne*, 256.

46. Cézanne to Vollard, 2 April 1902, in *Paul Cézanne correspondance*, 360–61. Again, Danchev's new translation tries to remain closer to Cézanne's original meaning, rendering "recherches" literally as "researches." Cézanne, *Letters of Paul Cézanne*, 319.

47. Retté, "Septième exposition des artistes indépendants," 297.

48. By my count, "tentative" or "tentatives" appcars at least seventy-seven times, while "expérience" and its cognates appear thirty-two times. No instances of the neologisms *expérimentateur, expérimentation, expérimentaliste,* or *expérimentalisme* are recognized by Littré in 1863 or Larousse in 1870. More surprisingly, there are also no instances of the adjectival or adverbial forms *expérimental* or *expérimentalement*.

49. Elie de Mont, "L'exposition du boulevard des Capucines," *La civilisation*, 21 April 1881, 2, quoted in Berson, *Reviews*, 361.

50. Paul Adam, "Peintres impressionnistes," *La revue contemporaine: Littérature, politique et*

philosophique 4 (April 1886): 4, quoted in ibid., 427. This phrase comes near the beginning of Adam's review, in which he distinguishes the impressionists (not the soon to be named neoimpressionists, with which his text is largely concerned) from previous schools of art. In contrast with other schools that rely on the "always uncertain data of experience" added to "the perceived sensation" (*la sensation perçue*), the impressionists, he writes, are concerned to "reproduce the pure phenomenon, the subjective appearance of things" (*reproduire le phénomène pur, l'apparence subjective des choses*). Mallarmé wrote that the impressionists "have attained a praiseworthy result, to make us understand when looking on the most accustomed objects the delight we should experience could we but see them for the first time." Stéphane Mallarmé, "The Impressionists and Edouard Manet," trans. George T. Robinson, *Art Monthly Review and Photographic Portfolio*, 30 September 1876, quoted in ibid., 97. The original French version of this text has not been located.

51. Ph. B. [Philippe Burty], "L'exposition des artistes indépendants," *La République française*, 16 April 1879, 3; J.-K. Huysmans, "L'exposition des indépendants en 1880," in *L'art moderne* (Paris: Charpentier, 1883); Henry Fouquier, "Chronique," *Le XIXe siècle*, 27 April 1879, 3; Émile Blavet, "Avant le Salon: L'exposition des réalistes," *Le Gaulois*, 31 March 1876, 1, all cited in ibid., 210, 286, 220, and 62, respectively.

52. Duranty's pamphlet is reprinted in Moffett, *New Painting*, 477–84. For a close reading of this text, see Armstrong, *Odd Man Out*, 73–100. Broude took issue with Marianne Marcussen's reading of *The New Painting* as a text proposing impressionism as scientific, objective, and analytical, an argument Marcussen based in part on the contents of Duranty's library. See Marcussen, "Duranty et les impressionnistes II"; and Broude, *Impressionism: A Feminist Reading*, 126. For a very different take on Duranty's text, one attentive to the kind of modernity Duranty was keen to reflect, see Fried, *Manet's Modernism*, 259–61.

53. Duranty, "La quatrième exposition faite par un groupe d'artistes indépendants," *La chronique des arts et de la curiosité*, 19 April 1879, quoted in Berson, *Reviews*, 219.

54. Outside France, Diego Martelli accepted the invitation to justify impressionist practice by comparing it to physiological research. Martelli, "Gli impressionisti," in *Scritti d'arte*, 106–8. See Ward, "Camille Pissarro in the 1880s," 258–64; Broude, *Impressionism: A Feminist Reading*, 127–33. My own analysis of Strindberg and Munch below suggests that fertile ground for ideas like Duranty's might be found in Scandinavia, and I suspect that Øystein Sjåstad's work on the naturalist painting of Christian Krohg will bear this out.

55. Henry Havard, "L'exposition des artistes indépendants," *Le siècle*, 27 April 1879, 3, cited in Berson, *Reviews*, 222. Druick and Zegers provide a different reading of Havard's citation of Duranty. See their "Scientific Realism: 1873–1881," 202.

56. Farfadet, "La semaine en quatrains: Jeudi," *Le tintamarre*, 5 March 1882, 1, quoted in Berson, *Reviews*, 387.

57. Pierre Veron, "Chronique parisienne," *Le journal amusant*, no. 1076 (14 April 1877): 2, quoted in ibid., 193.

58. Ferdinand Brunetière, "L'*impressionnisme* dans le roman," 15 November 1879, reprinted in Brunetière, *Roman naturaliste*, 75–104. Brunetière held Alphonse Daudet up as the exemplar of a "true" naturalism over and against the "false" naturalism of Zola, the Goncourt brothers, and company.

59. Zola, "Nouvelles artistiques et littéraires," *Le messager de l'Europe*, July 1879, in Zola, *Écrits sur l'art*, 397–405. A passage from Zola's review was translated from Russian and published in the *Revue politique et littéraire* on 26 July 1879, with the name Monet rather than Manet. The following day, *Le Figaro* picked up the story, with the commentary "M. Zola vient de rompre avec M. Manet," which led Zola to write immediately to his friend to assure him of his ongoing support. See Zola to Édouard Manet, 27 July 1879, in Zola, *Correspondance*, 3:355–56. In his review of the Salon of 1880, however, Zola reiterated the critique

of impressionism that he had introduced in the 1879 review. See "Le naturalisme au Salon," *Le Voltaire*, 18 and 22 June 1880, in Zola, *Écrits sur l'art*, 423–27. On the effects of Zola's decision to publish significant pieces of writing, including much of his art criticism, in Russia, see Hilton, "*Messager de l'Europe*." Although Hilton is reluctant to make the point explicitly, the publishing of "The Experimental Novel" in Russia must have had some relation to the rhetoric surrounding the 1896 *Exhibition of Experiments*, in light of the rarity of the discursive conjunction between impressionism and experimentation discussed in this chapter. See Hilton, "Exhibition of Experiments."

60. That view is surely overstated. On *The Masterpiece*, and on its reception by Cézanne and other painters in Zola's circle, see Rewald, *Cézanne, sa vie, son oeuvre*; Niess, *Zola, Cézanne, and Manet*, 78–86; and D'Souza, "Paul Cézanne, Claude Lantier."

61. Laforgue, "Émile Zola," *La guêpe*, 2 November 1879, in Laforgue, *Oeuvres complètes*, 3:204.

62. Shiff, *Cézanne and the End of Impressionism*, 22–23, 28; Hannoosh, "Poet as Art Critic"; Broude, *Impressionism: A Feminist Reading*, 134–37; Simpson, "Bourget, Laforgue, and Impressionism's Inside Story."

63. Laforgue to Charles Henry, ca. 10 December 1883, in Laforgue, *Oeuvres complètes*, 1:847. Gustav Theodor Fechner's law states that the difference between two stimuli has a proportional relationship to the quantity of the stimuli. Laforgue once addressed Henry as "mon cher (scientifiquissme)." Laforgue to Henry, [1 May 1885], ibid., 2:757.

64. Laforgue specifically referred to Thomas Young's work on the three fibrils of the eye, also taken up by Helmholtz and James Clerk Maxwell. On the importance of Helmholtz's theories for Seurat, see Foa, "On the Spaces of Painting and Perception"; Foa, *Georges Seurat*, 20–28, 49–50, 74–76, 87–91, 121–22.

65. Laforgue, "L'impressionnisme" (1883), in *Oeuvres complètes*, 3:329–36. On the links between anarchism and symbolist vanguards, see Leighten, *Liberation of Painting*, 130–34.

66. Hannoosh, "Poet as Art Critic," 561.

67. See Hartmann, *Philosophie des Unbewussten*. Laforgue would have read the 1877 French translation. See also Ellenberger, *Discovery of the Unconscious*, 209–10. As many scholars have pointed out, Hartmann conceives of the unconscious in much the same way that Hegel conceives of the absolute idea. Writing to Kahn in February 1884, Laforgue insisted that Hartmann's unconscious explained absolutely everything. Laforgue to Kahn, [28 February 1884], in *Oeuvres complètes*, 2:694.

68. Laforgue, "Notes d'esthétique sur Taine et Renan" (ca. 1880–82?), in *Oeuvres complètes*, 3:358.

69. According to Michèle Hannoosh, Laforgue also knew Ribot's *Philosophie de Schopenhauer* (1874), *Psychologie allemande contemporaine* (1879), and, probably, *Psychologie anglaise contemporaine* (1870). See ibid., 3:1143.

70. These ideas were initially laid out in Spencer's 1857 essay "Progress: Its Law and Cause," later incorporated into Spencer, *First Principles*.

71. Laforgue, "Un carnet de notes," ca. 1880–81, in *Oeuvres complètes*, 3:378–79.

72. Laforgue to Ephrussi, December 1883, ibid., 1:850.

73. Laforgue, "Notes d'esthétique sur Taine et Renan," ibid., 3:357.

74. Laforgue's idealizing of Bernard probably stems from his reading of Elme Caro. Laforgue mentions Caro's upcoming course at the Sorbonne in "Chronique parisienne," 2 November 1879, for *La guêpe*, and he read Caro's *Pessimisme au XIXe siècle* carefully in conjunction with Hartmann's work and Ribot's text on Schopenhauer. See Laforgue, "Mélanges philosophiques," in *Oeuvres complètes*, 3:1130–36, 1140–44, and the introduction to these notes by Michèle Hannoosh, 1123–27. One of the draft plans for the poem "Angoisses" mentions a dedication to Caro. Laforgue, *Oeuvres complètes*, 1:1986, 290.

75. Laforgue, "Notes d'esthétique sur Taine et Renan," in *Oeuvres complètes*, 3:357; Péladan, "Idéalisme expérimental," 196. Péladan's text began with a modified version of the familiar Comte-Bernard three-step evolution. In 1884–85, Charles Henry published three articles on

Leonardo in the *Revue de l'enseignement secondaire et de l'enseignement supérieur*, which Laforgue noted with appreciation in *La chronique des arts et de la curiosité*, 31 January 1885, 39, in *Oeuvres complètes*, 3:293.

76. Laforgue, "Notes d'esthétique sur Taine et Renan," in *Oeuvres complètes*, 3:359–60. See also Hannoosh, "Poet as Art Critic," 561–62.

77. Laforgue, "L'art moderne en Allemagne," in *Oeuvres complètes*, 3:338–39.

78. Laforgue, "Un carnet de notes," ibid., 3:385.

79. Laforgue, "Notes sur l'art en Allemagne," ibid., 3:353–54; Laforgue to Gustave Kahn, [28 February 1884], ibid., 2:694. Laforgue surely also intended to render obsolete Eugène Véron's 1878 *L'esthétique*, a text he despised, calling it "pédant et enfantin." Laforgue to Charles Ephrussi, [31 December 1883], ibid., 1:854.

80. Berthelot, "La science idéale et la science positive—à M. E. Renan," November 1863, in *Science et philosophie*, 1–40. See also Paul, "Debate over the Bankruptcy of Science," 310–11.

81. Zimmermann, *Seurat and the Art Theory*, 229. Zimmermann's text provides crucial context for Henry's ideas and their influence within neoimpressionist and symbolist circles. See also Argüelles's earlier monograph on Henry, *Charles Henry*.

82. For neoimpressionism as fundamentally scientific, see Homer, *Seurat and the Science of Painting*.

83. For Seurat, idealism, and Wagnerian aesthetics, see Smith, *Seurat and the Avant-Garde*, 105–56.

84. For Seurat's interest in Henry's theories of dynamogeny, see Zimmermann, *Seurat and the Art Theory*, 227–300. On Fénéon's scientific rhetoric and on neoimpressionism's engagement with psychophysiological aesthetics from the perspective of vanguardism, see Ward, *Pissarro, Neo-Impressionism*, 57–59, 124–46. See also Crary, *Suspensions of Perception*, 149–280; and Roslak, *Neo-Impressionism and Anarchism*. For a recent consideration of the intersections between neoimpressionism and symbolism, especially in the criticism of Paul Adam and Gustave Kahn, see Rapetti, *Symbolisme*, 124–32. For a discussion of "divisionism" as a visual language in relation to symbolism, based on the writings of Paul Signac, see Facos, *Symbolist Art in Context*, 93–94. For a study on Seurat's understanding of vision, with an emphasis on the optics of Helmholtz, see Foa, *Georges Seurat*.

85. Hennequin, "Beaux-arts." On Véron, see Jacqueline Lichtenstein's preface to Véron's *Esthétique*, 7–19; Colrat, "Eugène Véron"; and Hennequin, "Esthétique de Wagner." Hennequin (1858–1888) was the literary critic for *La revue indépendante*. See also LeGouis, *Positivism and Imagination*, 34–41, 59–86, 122–41; Bouillon and Méneux, *Promenade du critique influent*, 226–27.

86. Kahn, "Chronique de la littérature," 303–4. For a much less enthusiastic response to Hennequin's book, see Lecomte, "Estopsychologie [*sic*]."

87. Kahn, "De l'esthétique du verre polychrome," 54, 61–62.

88. Kahn, "Réponse des symbolistes."

89. Kahn, "Difficulté de vivre," 1. See also Ward, *Pissarro, Neo-Impressionism*, 62.

90. Quoted in Huret, *Enquête sur l'évolution littéraire*, 405. In his response, Kahn expressed his continued faith in the symbolist movement as a necessary corrective to naturalism but offered little praise for writers claiming to be symbolists, with the exceptions of Paul Verlaine, Laforgue, and Arthur Rimbaud. On Kahn's development of *vers libre* in dialogue with Henry's scientific aesthetics, see Brain, *Pulse of Modernism*, 150–65.

91. Caillié, "Spiritualisme expérimental," 20. This, of course, was nothing new. French spiritists since Allan Kardec had enlisted science and scientists to their cause. See Kardec (pseudonym of Hippolyte Léon Denizard Rivail), *Spiritisme expérimental*.

92. Quoted in Caillié, "Études cosmologiques," 157. Although Caillié attributes the statement to Bernard, I have been unable to locate its source.

93. Dévoluy, "Lettre ouverte," 156. Bois was a key figure in occult symbolist circles. See Bergman-Carton, "Medium Is the Medium"; Keshavjee, "'*Art inconscient*,'" 186–247.

94. Art historians have also linked Schuré's text with specific works by Paul Gauguin and Paul Ranson. See Jirat-Wasiutyński, "Paul Gauguin's 'Self-Portrait'"; and Bitker and Genty, *Paul Ranson, 1861–1909*, cat. nos. 28, 92. See also Mauner, *Nabis*, 3; Burhan, "Vision and Visionaries," 30.
95. Schuré, *Grands initiés*, v.
96. Ibid., x.
97. Burhan, "Vision and Visionaries," 18–20. Gombrich makes this common error; see his "Experiment and Experience in the Arts," 166.
98. Aurier, "Beaux-arts: Les symbolistes," 475.
99. Aurier, "Ratiocinations familières," 37.
100. For "false science" and "banal and petty naturalism," see Denis, preface to *L'École de Pont-Aven*.

Chapter 2

1. Ward, *Pissarro, Neo-Impressionism*, 128. The epigraph to this chapter is from Denis, "Cézanne—II," 276. Denis's text originally appeared in September 1907 (see Denis, "Cézanne"). Fry rendered Denis's plural "*déformations*" as the singular "deformation."
2. Denis, "Définition du néo-traditionnisme," 10, 7 (citations of this essay are from *Ciel et l'Arcadie*, 5–21).
3. By 1895, Denis had acknowledged that his promotion of the term *néo-traditionnisme* had failed, although he was always careful to situate his own work among "symbolist" and "synthetist" tendencies, in contrast to "neo-idealist" ones. See Bouillon, *Maurice Denis*, 38.
4. Denis, "Définition du néo-traditionnisme," 5.
5. Quoted in Retté, "Septième exposition des artistes indépendants," 300.
6. In 1892, Denis gave the painting to Bonnard, who seems to have sold it sometime after 1916. See Whitfield and Elderfield, *Bonnard*, 114. No photographs of *Décor* are known, not even in the archives of Eugène Druet, which is surprising given how active Druet was in documenting both Denis's and Bonnard's works. It is tempting to see this lack of photographic documentation as deliberate on Denis's part.
7. Junod, *Transparence et opacité*, 221.
8. Shiff, *Cézanne and the End of Impressionism*, 37–38.
9. For other examples of this modernist elision, see Ockman, *Ingres's Eroticized Bodies*, 85–124.
10. Concerns over European cultural degeneration culminated in the publication of Max Nordau's *Entartung* in 1892, translated into French in 1894. For the sociopolitical dimensions of degeneration in fin-de-siècle France, see Nye, *Crime, Madness, and Politics*; Larson, *Dark Side of Nature*, 70–78.
11. Deleuze, *Francis Bacon*, 59, 34.
12. On the debates over morality and abstraction centered on Cézanne, see Shiff, "Morality, Materiality, Apples."
13. Lemaître, "Ferdinand Brunetière," in Lemaître, *Contemporains*, 247–48.
14. Zola to Antony Valabrègue, 18 August 1864, in Zola, *Correspondance*, 1:375.
15. Zola, "Lettres de Paris: L'école française de peinture à l'exposition de 1878," in Zola, *Écrits sur l'art*, 376. This point is made in Lethbridge, "Homme et *L'Oeuvre*," 135–36. As Lethbridge argues, the history of deformation as an operational term and as a concept for modern literature and painting remains to be written.
16. Philippe Burty, "Exposition des oeuvres des artistes indépendants," *La République française*, 10 April 1880, 2, quoted in Berson, *Reviews*, 269.
17. Hennequin, "Beaux-Arts."
18. Véron, *Esthétique*, 289–305. For more on Véron, see Lichtenstein's introduction to Véron, *Esthétique*, 7–19; and Colrat, "Eugène Véron."
19. Symptomatic of this new and ambiguous use of the term *deformation* is the first English translation of Véron's text (1879), in which *déformation* appears variously as "irregularity," "malformation," "modification," and "contortion." Véron, *Aesthetics*, 250–63. Taine tended to downplay the role of individual temperaments in favor of collective ones determined by race, milieu, and moment. As a result, in both *On Intelligence* (1870) and *Philosophy of Art* (1865),

he made very little if any recourse to the term *déformation* but briefly considered what he called the *altérations* of artists.

20. Émile Hennequin, "Notes d'art: L'exposition des artistes indépendants," *La vie moderne*, 11 September 1886, 581–82, in Bouillon and Méneux, *Promenade du critique influent*, 285–87.
21. Moréas, "Symbolisme," 39. On literary deformation, see Stead, *Monstre, le singe et le foetus*, 29.
22. Daston and Galison, *Objectivity*, 191–251.
23. Poe's original, from *Silence: A Fable* (1837), reads, "Now there are fine tales in the volumes of the Magi—in the iron-bound, melancholy volumes of the Magi." Baudelaire's translation was published in 1857 in *Nouvelles histoires extraordinaires*.
24. Quoted in Retté, "Septième exposition des artistes indépendants," 300. On Retté, see Dessy, *Écrivains et les Nabis*, 71–73.
25. On *Décor* as a possible study for a decorative project, see Cogeval, Denis, and Barruel, *Maurice Denis, 1870–1943*, 301. On decoration, symbolism, and the Nabis, see Groom, *Beyond the Easel*; and Kuenzli, *Nabis and Intimate Modernism*.
26. On the École des beaux-arts as a "lieu de concours," see Vaisse, *Troisième République et les peintres*, 82. On Denis's student records, see Vincent, "Maurice Denis," 134. In charcoal, on the recto, appears the inscription "Denis M, élève de MM Lefebvre Balla and Doucet," and, in red pencil, "Mention." A red stamp on the verso reads "École des Beaux-Arts Concours d'émulation."
27. On the naturalist models Denis rejected, see Thomson and Thomson, "Maurice Denis's 'Définition.'"
28. Denis later claimed that this exchange occurred in 1888. The anecdote appears twice in his writings, in the "Définition du néo-traditionnisme," 13, and then in "L'enseignement du dessin" (1917), in Denis, *Nouvelles théories sur l'art*, 96. I have combined the two versions of the anecdote here, but it is significant that the first version foregrounds nonnaturalistic color, while the second is concerned with drawing.
29. The address on the sheet is 54, rue de Paris. A Françoise Anna Thibout was born 6 May 1870 at 66, boulevard Davout in the largely working-class twentieth arrondissement, close to the Porte de Montreuil. Nearby are two rues de Paris, one in Montreuil and the other to the north in Les Lilas. Anna Thibout would have modeled for the artist at 28, rue Pigalle, in the studio rented by Bonnard in 1890, which he shared with Denis, Vuillard, and Aurélien Lugné-Poë through 1891. Letters from Vuillard to Denis from the fall of 1892 suggest that the two artists were in the habit of sharing models. See MSS 12080 and 12081, unpaginated, Musée départemental Maurice Denis.
30. Denis to Lugné-Poë, undated, in Lugné-Poë, *Sot du tremplin*, 261, 246.
31. See Morowitz, "Anonymity, Artistic Brotherhoods," and scholarship on the Nabis, including Mauner, *Nabis*; Frèches-Thory and Terrasse, *Nabis*; Boyle-Turner, *Nabis*; Frèches-Thory and Perucchi-Petri, *Nabis: 1888–1900*; Cogeval, Frèches-Thory, and Genty, *Time of the Nabis*; and Groom, *Beyond the Easel*.
32. Ockman, *Ingres's Eroticized Bodies*, 3.
33. Denis, a committed Catholic, subscribed to an idealized vision of heterosexual love, as evidenced in his journal and in his numerous maternity scenes and family group pictures depicting his own wives and children. The overt sensuality of some of the *Décor* sketches is at odds, however, with much of Denis's work, and it seems plausible that in this overtness he deliberately sought the attention of heterosexual male symbolists. For more on the gender politics at play in symbolist visual practice, see Mathews, *Passionate Discontent*. The gender and sexual identities of members of the homosocial Nabi brotherhood and the school of Pont-Aven have received surprisingly little attention. The work of Damien Delille might begin to rectify this. See Delille, "Troisième genre."
34. See Garb, *Sisters of the Brush*, 88–91; and Benjamin, "Expression, Disfiguration," 76, 90.
35. These are part of the so-called glued sketches (*croquis collés*), a gathering of sketches that Denis glued to large folio sheets, usually to demonstrate the development of a particular motif, although the organizing logic is not always

clear. I am grateful to Claire Denis and Fabienne Stahl for facilitating my study of these sketches and other unpublished material.

36. On *Le voyage d'Urien*, see Vaughan, "André Gide and Maurice Denis"; and Christin, "Livre double."

37. On deformation and caricature, see Thomson, *Art of the Actual*, 177–83.

38. Jumeau-Lafond, "Symbolismes," 128.

39. Dauberville and Dauberville, *Bonnard: Catalogue raisonné*, 396; Whitfield and Elderfield, *Bonnard*, 114; Cogeval and Cahn, *Bonnard: Peindre l'Acadie*, 152. On *The Mantelpiece* as part of Bonnard's transition to classicism, including a discussion of the figure's pose as related to that of the *Dying Niobid*, see Newman, "Bonnard, 1900–1920," 102–3.

40. The color scheme of Denis's oil sketches and Bonnard's rendering of *Décor* recalls the palette of Denis's *Sunlight on the Terrace*, 1890, oil on cardboard, 24 × 20.5 cm, Musée d'Orsay, Paris.

41. For a recent discussion of Bonnard's oil study, see McDermott and Brettell, *Reflections*, 82.

42. Harrison, *Painting the Difference*, 208. Although Harrison does not discuss it, a pencil sketch in which Bonnard's head appears reflected in the lower right part of the mirror supports his reading—Bonnard's *The Model and the Artist Reflected in a Mirror* (study for *The Mantelpiece*), 1916, pencil and black chalk on paper, 31.1 × 24.8 cm, private collection, reproduced in Easton, *Snapshot: Painters and Photography*, 75.

43. See Pollock, *Avant-Garde Gambits, 1888–1893*, 16–21; Duncan, "MoMA's Hot Mamas."

44. Paul Gauguin, *Copy of Manet's Olympia*, 1891, oil on canvas, 89 × 130 cm, private collection.

45. *Nude Woman*, 130 × 162 cm, location unknown, possibly destroyed. Catalogue no. 140 in Rewald, *Paintings of Paul Cézanne*, 119–20. Fred Leeman first suggested this comparison to me. For more on the painting's reception, see Geelhaar, "Painters Who Had the Right Eyes," 285.

46. On Cézanne's early work as a harbinger of psychological modernity, and indeed as an early modernist attempt to exteriorize the interior, see Dombrowski, *Cézanne, Murder, and Modern Life*.

47. Émile Bernard, "Julien Tanguy, dit le Père Tanguy,'" *Mercure de France* 69 (16 December 1908), in Bernard, *Propos sur l'art*, 1:166.

48. Émile Bernard, "Souvenirs sur Paul Cézanne et lettres inédites I," *Mercure de France* 64 (1 October 1907), ibid., 1:144; Maurice Denis, "L'époque du symbolisme," *Gazette des beaux-arts* 12 (March 1934), in Denis, *Ciel et l'Arcadie*, 208.

49. Denis, "Aventure posthume de Cézanne," 194.

50. The three Émile Bernard paintings were *Bathers*, ca. 1889, present whereabouts unknown; *Bathers with a Red Cow*, 1887, oil on canvas, 92.7 × 72.4 cm, Musée d'Orsay, Paris; and *Bathers with Water Lilies*, ca. 1889, oil on canvas, 92.7 × 72.4 cm, Josefowitz Collection. See Luthi, *Émile Bernard: Catalogue raisonné*, nos. 67 and 68; Stevens et al., *Émile Bernard, 1868–1941*, 159–64; and Leeman, *Émile Bernard (1868–1941)*, 144–46. The frieze was in place by 1890, when Bernard completed a self-portrait with one of the canvases in the background: *Portrait of the Artist*, 1890, oil on canvas, 55.5 × 46 cm, Musée des beaux-arts, Brest (Luthi 252). See Stevens et al., *Émile Bernard, 1868–1941*, 165–67. Bernard notes a visit from Denis to his new studio in Asnières, possibly the first, on 12 February 1891. See Bernard to Eugène Bloch, 13 February 1891, in Bernard, *Émile Bernard: Les lettres d'un artiste*, 133.

51. See especially the figure at far right with the buttock shoulders in Cézanne's *Large Bathers*, ca. 1904–6, oil on canvas, Philadelphia Museum of Art. See Clark, *Farewell to an Idea*, 153–54.

52. *The Loss of Virginity* was probably purchased by Antoine de La Rochefoucauld at the February 1895 sale of Gauguin's works. See Brettell et al., *Art of Paul Gauguin*, 195. Gauguin's figural deformations were often provoked by unusual points of view and sometimes seemed to derive from the actual physical state of the subject. *Nude Breton Boy* (1889, oil on canvas, 93 × 73.5 cm, Wallraf-Richartz Museum, Cologne), for example, deploys a bird's-eye view similar to one of Denis's *Décor* sketches, but it also depicts a ruddy face and hands that have led scholars to describe the boy as sickly. See ibid., 151.

53. As James Kearns has argued, Aurier's article on Gauguin fractured rather than unified symbolists on the question of symbolist form. See Kearns, *Symbolist Landscapes*, 1–29.

54. Antoine, "Impressionnistes et synthétistes," 370–71.

55. Aurier, "J.-F. Henner," *Le moderniste illustré* 2–3 (13–20 April 1889), in Aurier, *Textes critiques, 1889–1892*, 90. See also Shiff, *Cézanne and the End of Impressionism*, 33.

56. Aurier, "Symbolisme en peinture," 159, 162.

57. Quoted in Retté, "Septième exposition des artistes indépendants," 300. For his own account of symbolism, see Retté, *Symbolisme, anecdotes et souvenirs*.

58. Quoted in Retté, "Septième exposition des artistes indépendants," 300.

59. For a detailed reading of *Catholic Mystery*, see Bouillon, *Maurice Denis*, 30–31. For Retté's account of the meeting between Denis and Verlaine, see Retté, *Symbolisme, anecdotes et souvenirs*, 50–51. For more on Retté's reception of *Catholic Mystery* in relation to neoimpressionism, see Young, "Death of Georges Seurat."

60. Retté, "Septième exposition des artistes indépendants," 299–300.

61. Verhaeren, "Salon des Indépendants," 111.

62. Leclercq, "Aux Indépendants," 299.

63. Marx, "Petits salons—Les Indépendants"; Geffroy, "Les Indépendants" (1891), in Geffroy, *Vie artistique*, 1:310–12; Saunier, "Exposition des Indépendants"; Antoine, "Critique d'art."

64. Bouillon, *Maurice Denis*, 32–33.

65. On Germain, see Ward, *Pissarro, Neo-Impressionism*, 108, 202–10; Marlais, *Conservative Echoes*, 171–81; Bouillon and Méneux, *Promenade du critique influent*, 340–41; Lucbert, *Entre le voir et le dire*, 49–51, 101–4; and Dessy, *Écrivains et les Nabis*, 69–71.

66. Germain, "Du symbolisme dans la peinture."

67. Germain, "Du temperament peintre," 619. Denis responded publicly, under his pseudonym, Pierre Louis, to Germain's letter. Denis, "À M. Alphonse Germain."

68. "Je désire vivement faire votre connaissance, un peintre écrivant, esthétisant comme vous, c'est si rare!" Germain to Denis, 13 November 1890, MS 4531, unpaginated, Musée départemental Maurice Denis.

69. Germain to Denis, 28 November 1890, MS 4532, unpaginated, ibid.

70. Denis to Lugné-Poë, undated, in Lugné-Poë, *Sot du tremplin*, 246.

71. "Vos peintures d'un si doux mysticisme m'avaient transporté dans l'au- delà." Germain to Denis, [2 March 1891], MS 4534, unpaginated, Musée départemental Maurice Denis.

72. Germain, "Modernisme et le beau."

73. Germain, "À l'exposition des Indépendants."

74. "Comme il me lâchait le nom de néo-traditionnistes comme titre de l'article . . . j'ai fait la grimace et l'ai prié de ne rien faire avant que nous n'ayez donné votre avis. . . . Il m'a prié de vous demander aussi si vous aviez qq. changement[s] dans vos théories depuis celles exprimées l'année dernière. . . . Vous me disiez ces jours derniers qu'il était probable que vous ne seriez pas aussi catégorique si c'était à refaire." Vuillard to Denis, [ca. August 1891], MS 12159, unpaginated, Musée départemental Maurice Denis.

75. Germain, "Théorie des déformateurs." For more on Germain's essay, see Shiff, *Cézanne and the End of Impressionism*, 233n18; and Ward, *Pissarro, Neo-Impressionism*, 209.

76. Rambosson, "3ème exposition des peintres," 532. On Rambosson, see Bouillon and Méneux, *Promenade du critique influent*, 401. On Denis's later depictions of "healthy" bodies in the context of neoclassicism and calls for regeneration, see Larson, "Curing Degeneration," 166–67, 177–79.

77. Denis met Marthe Meurier in the fall of 1890. Notes in Denis's journal from early 1892 suggest that their relationship became sexual around this time, with Denis expressing his shame and renewing his pledge to remain chaste until marriage. See Denis, *Journal*, 1:91–92. Denis's conceptions of ideal love, spirituality, and family life were deeply intertwined with

his art practice, as Bouillon argues in his 1993 monograph, *Maurice Denis*.

78. In 1890, apart from the "Définition," Denis published only the brief response to Germain and a short exhibition review. As far as is known, he published nothing in 1891, three short texts in 1892, and then nothing again in 1893 and 1894. In March 1895, Gauguin remarked upon how little Denis had published, thinking the "Définition" had been his only text prior to the review of the Séguin exhibition. Gauguin to Denis, [early March 1895], in Gauguin, *Lettres à sa femme et à ses amis*, 306.
79. Denis, "À propos de l'exposition d'A Séguin," 118. Here, Denis implicitly referred to Germain's *Pour le beau: Essais de kallistique* and *Du beau moral et du beau formel*.
80. Ibid.
81. Denis, "Notes sur la peinture religieuse," *L'art et la vie*, no. 54 (October 1896), in Denis, *Ciel et l'Arcadie*, 36.
82. Filiz Burhan briefly mentions the notebook in "Vision and Visionaries," 86, but it was analyzed in greater depth in Bouillon, "Denis, Taine, Spencer," 295–300, which is foundational for my account. Bouillon dates Denis's period of study for the baccalaureate to between 16 October 1887 and November 1888. An earlier discussion of Denis's philosophical background, including the Tainean echo in the opening salvo of the "Définition," appears in Bouillon, "Denis: Du bon usage," 61.
83. On the philosophy curriculum's inclusion of the new psychology, and in particular the pathological method, see Goldstein, "Neutralizing Freud," 47–59, and the introduction, above.
84. Émile Boirac (1851–1917), editor of Leibniz and co-author, with A. Magendie, of *Leçons de psychologie appliquée à l'éducation* (1902), was conversant with the latest psychological and psychical research, which he would later write about in *La psychologie inconnue* (1917) and *L'avenir des sciences psychiques* (1918).
85. "La vérité est le but de la pensée. Bien penser c'est former des idées vraies. . . . Une idée est toujours réelle, vraie" (1); "Le but de la Science est de prouver d'une façon claire et incontestable la vérité à tout le monde" (21); "Théorie de l'induction de Cl. Bernard: Induire, c'est déduire dans des conditions particulières: c'est faire une déduction provisoire qui réclame la vérification expérimentale" (27). Denis, *Cahier philosophique*, 1887–88, private collection.
86. Denis, "Paul Sérusier, sa vie, son oeuvre," in Sérusier, *ABC de la peinture* (1942), 40–41. For more on Sérusier's rigorous philosophical education, his attentive study of Neoplatonism, and his later theosophical beliefs, see Boyle-Turner, *Paul Sérusier*, 24–25. On Sérusier's Neoplatonist theories as related to the emergence of abstraction, see Cheetham, *Rhetoric of Purity*, 24–35.
87. One might also choose to refer to this crisis as one of style or technique. See Shiff, *Cézanne and the End of Impressionism*, xvii. Using the term *method*, however, points to the appeal of scientific methods among some symbolist avant-gardes. It was also Denis's preferred term. *The Talisman* would begin to play a role in Denis's writings only in 1903. See Denis, "L'influence de Paul Gauguin," *L'Occident*, no. 23 (October 1903), in Denis, *Ciel et l'Arcadie*, 74.
88. Sérusier to Denis, undated ("Jour de Venus," 1889),in Sérusier, *ABC de la peinture* (1995), 27–28. For Sérusier's concerns over the role of nature in the artwork and the creative process, see Boyle-Turner, *Paul Sérusier*, 21, 24.
89. Sérusier to Denis, undated, in Sérusier, *ABC de la peinture* (1995), 28–30.
90. Denis, "Les arts à Rome, ou la méthode classique," *Le spectateur catholique* 4 (July–December 1898 [1899?]), in Denis, *Ciel et l'Arcadie*, 64.
91. Denis, "De la gaucherie des primitifs," *Les arts de la vie*, July 1904, in Denis, *Théories, 1890–1910*, 173. In the same essay, Denis wrote of critics who accused Ingres of "vouloir rebalbutier," another of Germain's characteristic phrases. As he had done in 1895, he repeated the critique and let it stand, effectively revaluing what Germain had denigrated.
92. Denis, "À propos de l'exposition d'A. Séguin," 119. Certainly, Denis oversimplified Aurier's position on deformation.
93. Denis, "De Gauguin et de Van Gogh au classicisme," *L'Occident*, no. 90 (May 1909):

187–202, in Denis, *Ciel et l'Arcadie*, 155–74. As Bouillon notes, this essay summed up Denis's own understanding of his theoretical and artistic journey from 1890.

94. Denis, *Journal*, 2:101. Denis saw both works in Ivan Morosov's collection in Moscow in early 1909.

95. Denis, "De Gauguin et de Van Gogh au classicisme," in Denis, *Ciel et l'Arcadie*, 163. In 1890, Aurier had allied Van Gogh's pathology with his sincerity, but the view of Van Gogh as mad is largely a product of the early twentieth century. See Zemel, *Formation of a Legend*.

96. Denis, "La peinture," *L'Ermitage*, no. 11 (15 November 1905), in Denis, *Ciel et l'Arcadie*, 86. In 1912, Denis retitled this article "De Gauguin, de Whistler et de l'excès des théories."

97. Daston and Galison, *Objectivity*, 210, 234–51.

98. Fry, "Post Impressionism," *Fortnightly Review*, 1 May 1911, in *Roger Fry Reader*, 101.

99. Sickert, "Post-Impressionists," *Fortnightly Review*, January 1911, in Sickert, *Complete Writings on Art*, 276.

100. Quoted in Christopher Reed, "Forming Formalism: The Post-Impressionist Exhibitions," in Fry, *Roger Fry Reader*, 52. Reed highlights Fry's hesitation over the role of Cézanne's will in producing distortion. This association between deformation and abstraction continued to inflect the British reception of abstraction into the 1920s and beyond. See Lewison, "Jackson Pollock," 206–7.

101. Greenberg, "Cézanne and the Unity of Modern Art," *Partisan Review* (May–June 1951), in Greenberg, *Collected Essays and Criticism*, 3:85.

102. Shiff, *Cézanne and the End of Impressionism*, 55–152.

103. Denis to Vuillard, 15 February 1898; Vuillard to Denis, 19 February 1898; Denis to Vuillard, 22 February 1898, all in Denis, *Journal*, 1:134, 136–37, 139, respectively. For a close reading of these letters, see Bouillon, "Vuillard et Denis."

104. Salomon and Cogeval, *Vuillard: Le regard innombrable*, 1:134. See also the very similar seated nude by Bonnard, illustrated on the same page, *Seated Nude*, 1891, oil on cardboard, private collection.

105. This claim is made most explicitly in Cogeval, *Édouard Vuillard*, 100.

106. Denis, "Pierre Bonnard," 4.

107. On Denis's critiques, see Bouillon, "Denis: Du bon usage," 66; and Benjamin, *Matisse's "Notes of a Painter,"* 87–99. Benjamin also makes clear how Matisse responded to Denis's critiques. See also Clark, *Farewell to an Idea*, 91–94; Wright, *Matisse and the Subject of Modernism*, 68–70, 81–83; Clark, "Madame Matisse's Hat"; and Shiff, "Dream of Abstraction," 54.

108. Denis, "La peinture," in Denis, *Ciel et l'Arcadie*, 96–98.

Chapter 3

1. Chu, "Dormeurs et rêveurs," 9–14.

2. For the caricature's context and a reading of the painting that supports my emphasis on violence, see Berry, "Bedrooms," 133–34.

3. The French word *carnets* denotes both sketchbook and notebook. The two earliest *carnets*, numbered I.1 and I.2, are held at the Bibliothèque de l'Institut de France, along with many of Vuillard's journals, all deposited after the artist's death. For the purposes of this chapter, I have relied mostly on the original documents and published excerpts, but I have also consulted for comparison two unpublished transcriptions: Court, "Vuillard: Les années de jeunesse"; and Alexandre, "Édouard Vuillard, carnets intimes." See also Alexandre, "Carnets de Vuillard." I am indebted to a number of scholars who have worked on parsing and analyzing Vuillard's journals, but I want to acknowledge as especially influential the work of Cogeval, Easton, and Groom, who have all brought a particularly critical regard to the journals and to Vuillard's attempts to theorize his practice.

4. Benjamin, "Decorative Landscape."

5. Denis to Vuillard, 22 February 1898, in Denis, *Journal*, 1:138. On the importance of these letters for Denis's turn to classicism in and around 1898, see Bouillon, "Vuillard et Denis."

6. Jacques Salomon claimed that Vuillard never showed his journals to anyone. See Salomon and Cogeval, *Vuillard: Le regard innombrable*, 1:6. But a letter from Denis to Vuillard, cited in Thomson, *Vuillard*, 28, suggests that Denis may have been privy to the journals at an early date.
7. Denis to Vuillard, 22 February 1898, in Denis, *Journal*, 1:140.
8. Ibid., 1:120. Denis was apparently still worrying about Vuillard's inability to finish works in 1906. See Thomson, *Vuillard*, 73. Elsewhere, Denis disparaged Vuillard's "sensualism" as part and parcel of a "Semitic" rather than a "Latin taste." In 1899, in the midst of the Dreyfus Affair, Denis divided the core Nabis into two groups, Vuillard, Bonnard, and Vallotton representing the "Semitic taste," and himself, Sérusier, and Paul Ranson representing the "Latin taste." See Denis, *Journal*, 1:150. Vuillard, Bonnard, and Vallotton had been especially closely associated with *La revue blanche*, an important forum for assimilated Jewish intellectuals who publicly supported Dreyfus. See Datta, "Dreyfus Affair and Anti-Semitism." On Vuillard's support for Dreyfus and his views on anti-Semitism, see Cogeval, *Édouard Vuillard*, 33–35. On Vuillard's Jewish patrons, see Brown and Brettell, *Édouard Vuillard*. On Denis and anti-Semitism, see Kuenzli, "Aesthetics and Cultural Politics."
9. Denis to Vuillard, 23 November 1897, in Denis, *Journal*, 1:124; Denis to Vuillard, 15 February 1898, ibid., 1:133–34.
10. Vuillard to Denis, 19 February 1898, ibid., 1:136–38; Denis to Vuillard, 22 February 1898, ibid., 1:139.
11. Goldstein, *Post-Revolutionary Self*, 165–71, 184–94.
12. Goldstein, "Neutralizing Freud," 53–59.
13. For a detailed look at Vuillard's lycée education, and in particular his aptitude for mathematics, see Easton, "Édouard Vuillard's Interiors of the 1890s," 16–30.
14. Vuillard's highly personal approach to volition was undoubtedly also informed by his early education from the Marist Brothers. See Alexandre, "Carnets de Vuillard," 22–24.
15. Ribot, *Essai sur l'imagination créatrice*; Flournoy, *Des Indes à la planète Mars*.
16. For a study of naturalism as much more varied and complex, see Thomson, *Art of the Actual*.
17. "Impressionisme intérieur." Vuillard, *Carnet* I.2, fol. 57r, [1 July 1895]. This term suggests a play between the psychological interior and the domestic interior that has inspired a lively area of research. See especially Silverman, *Art Nouveau in Fin-de-Siècle France*; Sidlauskas, *Body, Place, and Self*; Krämer, *Das unheimliche Heim*.
18. Vuillard, *Carnet* I.1, fol. 10r.
19. Vuillard to Denis, 4 October 1888, MS 12071, unpaginated, Musée départemental Maurice Denis. An undated visiting card with a handwritten note in MS 12142 suggests that Vuillard and Denis had renewed their acquaintance from the Lycée Condorcet in the summer or early fall of 1888, either at the Académie Julian or at the École des beaux-arts.
20. Cogeval has similarly described the journal entries as "automatic," "telegraphic," and "metronomically obsessive." *Édouard Vuillard*, 6–8, 53. Additional sketchbooks from the 1890s are known to be in private hands. Vuillard may well have kept other journals during this time, some of which may have been destroyed. Ibid., 6–7. The dating of the journals in this chapter relies on that provided in Salomon and Cogeval, *Vuillard: Le regard innombrable*, and Cogeval, *Édouard Vuillard*. Cogeval imposes no punctuation on transcriptions of the text. In my translations, I have added a few punctuation marks for legibility but have translated the texts conservatively, favoring accuracy, even literalness at times, over easy comprehension.
21. For Denis's mythologizing account of the "revelation" of *The Talisman*, see above all his "Paul Sérusier, sa vie, son oeuvre," in Sérusier, *ABC de la peinture* (1942), 37–121. For a refutation of Denis's insistence that Sérusier revealed the painting in October 1888, see Bouillon, "Denis, Taine, Spencer," 294; Bouillon, *Maurice Denis*, 122. Vuillard was certainly associating with the Nabis in 1889, and a letter from Sérusier to Denis, written in the early summer of 1889,

mentions the "nouveau frère que Jahvé a dirigé vers nous," which has been interpreted as a reference to Vuillard. See Thomson, *Vuillard*, 18; Cogeval, *Édouard Vuillard*, 8. Groom, however, believes that Vuillard only committed himself to the Nabi group the following year, in the spring of 1890, which is also entirely plausible. See Groom, *Édouard Vuillard: Painter-Decorator*, 8.

22. Vuillard's first public showings of his work, especially compared to Denis's ten entries at the 1891 Indépendants, remained somewhat tentative. Vuillard exhibited only two paintings and two unidentified pastels at the Château of Saint-Germain-en-Laye in September 1891, and two unidentified works at the gallery of Le Barc de Boutteville from December 1891 to February 1892. He had previously shown a large *conté* drawing of his grandmother at the 1889 Salon.
23. "Soir immense trottoir mur à droite arbres roux poussiéreux grande femme toute noire soir effet de nuit." Vuillard, *Carnet* I.1, fol. 13v.
24. "Travaille dans la pâte avec entrain, œil dilaté." *Carnet* I.1, fol. 31v, quoted in Salomon and Cogeval, *Vuillard: Le regard innombrable*, 1:27.
25. "Effet dans l'œil effet gris poussiéreux." Ibid., fol. 14r.
26. "Mauvais état d'esprit"; "énervement considérable." *Carnet* I.1, fol. 19r.
27. "Désordre dans ma cervelle," *Carnet* I.2, fol. 47v (2 August 1894); "déroute intérieure. . . . Nerveux," *Carnet* II.5, fol. 59r (19 January 1912), quoted in Salomon and Cogeval, *Vuillard: Le regard innombrable*, 2:1119.
28. "Gueule de bois sensations naturelles activité des yeux ciel clair bleu et blanc." *Carnet* I.1, fol. 10r.
29. Cogeval, *Édouard Vuillard*, 53; and Salomon and Cogeval, *Vuillard: Le regard innombrable*, 1:6.
30. "En travaillant perds l'impression . . . trouvée en ébauchant lentement ne pas fixer un seul objet." *Carnet* I.1, fol. 11v.
31. Taine, *Philosophie de l'art*, 1:28–29.
32. "Si l'appareil cérébral n'est pas en état de saisir ces rapports, de les garder un moment et de les reporter comme un somnambule sur le papier ou la toile inutile de dépenser son temps"; "Faire nature c'est vous émotionner illusionner." *Carnet* I.1, fol. 12v, 12r. Vuillard's reference to working like a somnambulist has proved very difficult to contextualize, coming so early in the journals, before Vuillard was thought to have been exposed to symbolist theory. Cogeval argues that Vuillard is already advocating somnambulism as a method (*Édouard Vuillard*, 243), and elsewhere has suggested a Wagnerian connection (Salomon and Cogeval, *Vuillard: Le regard innombrable*, 1:69–70). Similarly, Gamboni reads the passage as above all prescriptive. Gamboni, "Vuillard and Ambiguity," 407. A number of scholars have pointed out that aspects of Vuillard's early writings seem to anticipate the arguments in Bergson, *Essais sur les données immédiates*. See Gamboni, "Confondre avec méthode," 54. Certainly, this is the case for Vuillard's understanding of the incompatibility of perception as an instantaneous act and painting as a durational activity.
33. Doran, *Conversations avec Cézanne*, 170.
34. "Essayé de peindre de mémoire, travaillé avec Mimi." *Carnet* I.1, fol. 32v.
35. Easton, "Édouard Vuillard's Interiors of the 1890s," 61.
36. "It is better to paint from memory, such that your work will be yours; your feeling, your intelligence, and your soul will thus survive for the amateur's eye." Morice, *Paul Gauguin*, 213, quoted in translation in Chu, "Lecoq de Boisbaudran and Memory Drawing," 289n42. See also Cheetham, *Rhetoric of Purity*, 4–5; and Sidlauskas, *Body, Place, and Self*, 6–19.
37. See, for example, Van Gogh's *Memory of the Garden at Etten (Ladies of Arles)*, 1888, oil on canvas, 73 × 92 cm, Saint Petersburg, State Hermitage Museum. On this point, see Cheetham, *Rhetoric of Purity*, 12–13; and Druick et al., *Van Gogh and Gauguin*, 197–205.
38. "Pas de changement dans les peintures toujours la même chose (!)." *Carnet* I.1, fol. 33v.
39. "Commence à travailler de mémoire," *Carnet* II.2, fol. 12r, quoted in Cogeval, *Édouard Vuillard*, 475; "petites salissures de mémoire," *Carnet* I.2, fol. 77v. This phrase has been translated variously as "little daubs" (Thomson, *Vuillard*, 28) and "little bits of nothing" (Groom, *Édouard*

Vuillard: Painter-Decorator, 8). Vuillard often made these autobiographical notes on or around his birthday (11 November). See Salomon and Cogeval, *Vuillard: Le regard innombrable*, 1:4.

40. "C'est l'idée de travail d'après nature qui me poursuit." *Carnet* I.2, fol. 31r [March 1891].

41. "Année de Sérusier" [1905?]. *Carnet* I.2, fol. 77v. In another chronology, in *Carnet* II, fol. 12, written 11–12 November 1908, Vuillard recalled that he met Sérusier in 1890, the year that his name first appears in the journal (entry for 9 or 13 January 1890, *Carnet* I.2, fol. 42r). But it is likely that Vuillard met Sérusier earlier, either at the Académie Julian or at one of the Nabi gatherings in 1889.

42. Vuillard to Denis, undated [summer 1890], MS 12146, unpaginated. Musée départemental Maurice Denis.

43. For example, Vuillard's *Women in the Garden, "Le cantique des cantiques"* (1891, 74 × 51 cm, private collection, United States), was long ascribed to Paul Ranson. See Cogeval, *Édouard Vuillard*, 113; Salomon and Cogeval, *Vuillard: Le regard innombrable*, 1:185–87.

44. "La même arabesque dessinée par 2 caractères ~~différents~~ donnera 2 expressions différentes . . . pas s'inquiéter de Sérusier et la bande." *Carnet* I.1, fol. 55r, quoted in part in Cogeval, *Édouard Vuillard*, 54.

45. "L'œuvre d'art partie d'une sensation motivante est par elle-même doit être choyée par l'artiste comme œuvre présente non comme renouvellement de sensation passée comme décorant un bout de papier de toile." *Carnet* I.1, fol. 55r (12 July 1890).

46. "Concevons bien un tableau comme un ensemble d'accords, s'éloignant définitivement d'idée naturaliste." *Carnet* I.1, fol. 71r. See Easton, "Édouard Vuillard's Interiors of the 1890s," 78.

47. Taine, *Philosophie de l'art*, 2:334–35. Bouillon discusses the similarity of this statement with Denis's opening line of the "Définition"; see Denis, *Ciel et l'Arcadie*, 5n3.

48. Taine, *Philosophie de l'art*, 2:333–34. As Denis knew well, having cited Charles Henry's scientific aesthetics in the "Définition," Henry would attempt to codify Taine's expressivity of line by developing the principles of dynamogeny.

49. Easton, *Intimate Interiors of Édouard Vuillard*, 17.

50. "Influence toujours de Denis"; "les habitudes d'esprit idées naturalistes pourtant bien attaquées, sont un obstacle; le travail avec soin sa nécessité dans une œuvre . . . préparer ses couleurs, sa palette, et que les formes soient précises qui doivent les contenir: ce qu'il y a sur la toile superficiel et seul affectant notre regard . . . mais il faut que ce . . . soit conçu complètement formes et couleurs (ancienne idée de <u>voulu</u>); l'application, l'exécution devient alors une affaire de grande patience et de soin et c'est tout l'œuvre d'art. Déjà mises en pratique il y a 6 mois (portrait de Mimi en Jaune)." *Carnet* I.1, fols. 55v–56v [1–2 September 1890], quoted in Cogeval, *Édouard Vuillard*, 63.

51. Salomon and Cogeval argue that if Vuillard had meant to refer to this work, he would have written "Marie en ochre" or "Marie en Bistre," and that therefore the "portrait de Mimi en Jaune" must be a lost work. *Vuillard: Le regard innombrable*, 1:76.

52. "Idée basse mais qui me tient, idée de pélichtim." *Carnet* I.2, fol. 24r.

53. Groom translates both words as "methodology" to help overcome the confusion. *Édouard Vuillard: Painter-Decorator*, 8.

54. "Il faut donc avoir une méthode pour la production dont on ne peut <u>connaître</u> par avance le résultat. Entendons-nous, je dois <u>imaginer</u>, voir, les lignes, les couleurs que je pose et ne rien fairc au hasard, cela est parfaitement juste. Je dois réfléchir toutes mes combinaisons. Mais justement pour la possibilité de ce travail, il me faut une méthode dont je sois convaincu . . . théorie décidément assise qui me délivre du souci vulgaire d'originalité. Une théorie bellement établie peut seule me donner le repos d'esprit et me permettre de me développer." *Carnet* I.2, fols. 22v–23r, quoted in part in Salomon and Cogeval, *Vuillard: Le regard innombrable*, 1:157–58.

55. "L'œuvre n'a qu'une seule méthode, l'ensemble de toutes les actions. + Donc tous les ouvrages auront la même méthode + Cette

méthode dépend de l'esprit, de l'harmonieux développement des facultés. . . . L'esprit qui devrait dans notre cas être toujours le plus fort devient l'esclave des sens." *Carnet* I.2, fol. 30r [March 1891].

56. Easton, *Intimate Interiors of Édouard Vuillard,* 9.

57. "C'est l'idée de travail d'après nature qui me poursuit." *Carnet* I.2, fol. 31r [March 1891].

58. "On observer, on ne peut vouloir observer. L'émotion continue (l'observation) est la première condition d'une œuvre d'art avant l'esprit de méthode et l'intelligence pratique qui naissent de l'âme même qui subit cette émotion. . . . L'observation pure et simple est un acte de vie simple et c'est cette simple et primitive observation qui est nécessaire, qui est." *Carnet* I.2, fol. 33r [2 April 1891]. For a different reading of Vuillard's practices and theories of observation, see Easton, "Édouard Vuillard's Interiors of the 1890s," 55–56, 87–91.

59. "Faire attention simplement ou aller se promener et faire ses pieds—Voilà un bon système." *Carnet* I.2, fol. 20v [6 September 1890].

60. Bélugou's notes can be found in Nouvelles acquisitions françaises (hereafter NAF), MSS 18173–18185, Département des manuscrits, BnF.

61. NAF, MS 18173, fol. 22, ibid.

62. "Je n'ai confiance dans les idées et raisonnement que contrôlés par des impressions directes de la nature, il me faut nourrir de nouvelles observations. . . . Je l'ai essayé et je suis étonné de la perspective de découvertes que cela me découvre. Car cette observation qui s'applique à ce cas particulier, s'étend à toutes sortes de faits. Mais celui-ci justement à cause de ce besoin de contrôle par l'observation directe et que je n'ai jamais pu voir . . . qu'en caricatures par suite de toute mon éducation." *Carnet* I.2, fols. 46v–47r (27 July 1894), reproduced in Salomon and Cogeval, *Vuillard: Le regard innombrable,* 1:316, 388.

63. "Se servir d'un modèle pour soutenir son imagination. L'imagination généralise toujours." *Carnet* I.2, fol. 47v (2 August 1894), quoted in part in Cogeval, *Édouard Vuillard,* 304.

64. NAF, MS 18173, Département des manuscrits, BnF, fols. 1–4, 46.

65. Easton, "Édouard Vuillard's Interiors of the 1890s," 58–61.

66. "Exprimons ce que je sens (c'est une simple désignation, un fait pur et simple de designer la chose que j'ai dans la tête): une expression d'attendrissement causé par tel objet. . . . Puis je trace sur le papier d'un coup (comme on parle mal quand une idée se présente) la ligne ou la tache imaginée et voulue." *Carnet* I.2, fol. 24r.

67. See Perucchi-Petri, "Nabis et le Japon."

68. Denis, "Définition du néo-traditionnisme," 13.

69. Jacques-François Blondel, "Arabesque ou Moresque," in Diderot and d'Alembert, *Encyclopédie,* accessed 11 June 2013.

70. The French Academy noted the infrequent use of *arabesque* as a noun in the sixth edition of 1835, but Émile Littré's 1872–77 dictionary gave equal weight to the adjective and noun forms.

71. Quoted in Woźniakowski, "De l'arabesque romantique," 59, 69. See also Gamboni, *Potential Images,* 52.

72. Frank, "Arabesque, Cipher, Hieroglyph," 147–48.

73. One of the earliest musical arabesques, if not the very earliest, is Robert Schumann's *Arabeske in C,* op. 18 (1839). The use of the word *arabesque* in ballet dates from 1828. See, both by Falcone, "Arabesque" and "Evolution of the Arabesque."

74. Gamboni, *Potential Images,* 86–104.

75. Bhogal, *Details of Consequence,* 64–114. Bhogal's suggestive distinction between "arabesque in theory" and "arabesque in practice" (75) is echoed here.

76. See Bouillon, "Arabesques"; Kuenzli, *Nabis and Intimate Modernism,* 69–103.

77. This explains the equal, in fact, sometimes greater, place given to Vuillard's drawings, pastels, and lithographs in recent exhibitions and in Salomon and Cogeval, *Vuillard: Le regard innombrable.*

78. On the Nabis' involvement with the theater, and in particular with the Théâtre de l'Oeuvre, see Boyer, *Artists and the Avant-Garde Theater;* and Cahn et al., *Théâtre de l'Oeuvre.* Cogeval has most thoroughly incorporated Vuillard's theater work into the artist's oeuvre in Salomon and Cogeval, *Vuillard: Le regard innombrable,* 1:155–215; and

Cogeval, *Édouard Vuillard.* On Vuillard, theater, and caricature, see Thomson, *Art of the Actual,* 143–83.

79. This is consistent with a point made by Kuenzli, who argues that the Nabis' interpretation of the Wagnerian *Gesamtkunstwerk* nevertheless retained painting at the top of a hierarchy of the arts. Kuenzli, *Nabis and Intimate Modernism,* 149–83.

80. "Les moyens expressifs de la peinture peuvent rendre une analogie et non une photographie impossible d'un instant. Oh la chose différente d'un instantané et l'Image." *Carnet* I.2, fol. 20v. Second part cited in Cogeval, *Édouard Vuillard,* 241–42.

81. "Je la développe, si besoin et je compose fait simple en lui-même composer et toute ma patience (Oh alors cela est facile) est absorbée par le souci de bien faire et alors possibilité de travail manuel vraiment très prolongé." *Carnet* I.2, fol. 24r.

82. "Si donc nous étions des êtres simples et non gâtés par des préjugés et des habitudes, nos œuvres seraient facilement belles parce qu'il n'y aurait pas de lutte pour distinguer la vérité du mensonge. . . . Le beau alors serait seulement subjectif? Non pourtant si nous croyons à l'existence objective, si nous croyons à Dieu." *Carnet* I.2, fol. 33r (2 April 1891). On the crucial religious context of this statement, see Alexandre, "Carnets de Vuillard," 24.

83. *Woman Sleeping* was exhibited at Le Barc de Boutteville in November 1892, and then at the Salons de La Dépêche from May to June 1894. The portrait of Lugné-Poë was exhibited at Le Barc de Boutteville from July to September 1893 in the *Portraits du vingtième siècle* exhibition.

84. Cogeval, Patry, and Guégan, *Van Gogh, Gauguin, Cézanne,* 196.

85. On the scientific study of dreams in France, see Carroy, *Nuits savantes.*

86. James, *Dream, Creativity, and Madness,* 153–57, 234–49; Kroker, *Sleep of Others,* 147–48. For the broader context of scientific studies of sleep in France, see Fay, "Stories of the Sleeping Body," 28–66.

87. Ribot, *Maladies de la personnalité,* 4–14, 40–56. See also Janet, *Automatisme psychologique;* and Carroy and Plas, "How Pierre Janet Used Pathological Psychology."

88. Both *Woman Sleeping* and *Sleeping Woman* are thought to depict Vuillard's sister Marie. Salomon and Cogeval, *Vuillard: Le regard innombrable,* 1:144–46. The more caricatured *In Bed,* which Vuillard never exhibited during his lifetime, is by far the most extensively discussed of the sleeping figure paintings. See, most recently, Perucchi-Petri, "Au lit, im Bett."

89. Many contemporary critics and art historians have compared Vuillard's decorations to tapestries, especially in the use of pattern to create an overall effect. See, for instance, Groom, *Édouard Vuillard: Painter-Decorator,* 93–94, 127–30.

90. Jules Lemaître was known to have described him as such. See Cogeval, *Édouard Vuillard,* 12.

91. See Daston and Galison, *Objectivity,* 191–251.

92. Gamboni, "Confondre avec méthode"; Gamboni, *Potential Images,* 100–104. Gamboni argues that Vuillard's uncertainty became part of his method insofar as he courted the viewer's uncertainty through strategies such as ambiguity and using schemata similar to caricature and puzzle pictures.

93. "Pas de force dans la tête." *Carnet* I.1, fol. 25r.

94. "Énervé ne peux me concentrer sur un point une idée. . . . Esclave de mes sensations et d'un orgueil condamnant à l'impuissance. À l'atelier paresse et orgueil." *Carnet* I.2, fol. 55v (November 1894), quoted in part in Gamboni, "Vuillard and Ambiguity," 407.

95. "Pourquoi l'esprit résiste-t-il? Probablement fatigue physique et morale. . . . Mais comment l'effort vient-il? De la vitalité? Et alors le plus ou moins d'effort? De la conscience de cet effort. Qu'est-ce qu'il y a sous ces mots efforts, envie, désir! Confusion des mots et des idées qu'ils représentent!" *Carnet* I.1., fols. 72v–73r (31 August 1890).

96. Ribot, *Maladies de la volonté,* 3, 59.

97. NAF, MS 18181, fol. 188, Département des manuscrits, BnF.

98. "Je ne pourrai avoir de tranquillité en pensant au travail qu'en m'étant délivré . . . par habitude du travail machinal, que mes longues

flemmes n'ont pu encore détruire. La plupart du temps, fatigue de mon oisiveté d'esprit, mon désir de travail n'est qu'une espérance de flemme rendue facile qui se contente. Peut-être l'habitude rend-elle des services, mais l'habitude d'une activité." *Carnet* I.2, fol. 44v (23 July 1894).

99. "Esclave de certains dimensions, de certaines matières (carton, huile) quand l'idée me vient de travailler. Dérouté devant les sensations ou impressions nouvelles au moment de leur appliquer la même méthode de travail qu'aux anciennes. . . . Essai de retour au travail de 90–91." *Carnet* I.2, fol. 41v (10 July 1894).

100. Bernard, *Introduction*, 86.

101. Walter Sickert, "The International Society," *English Review*, May 1912, in Sickert, *Complete Writings on Art*, 314.

102. See chapter 2.

103. "Mon seul désir maintenant est la vie de mon âme. Le travail est le signe en même temps le produit de cette vie. . . . La vérité est la vie, la vie est la vérité, la vérité est une ou ne serait pas la vérité. Les dogmes ne sont que des méthodes d'arriver à la vérité; toute méthode contemplée par l'Esprit offre un spectacle captivant c'est un ensemble harmonieux d'idées plus ou moins pures ce qui fait la plus ou moins grande et vaste Beauté du dogme. Le dogme n'est pas une fin, le dogme est une œuvre d'art, du moins c'est ainsi que je l'entends, c'est le résultat d'expériences pour une aspiration finale qui est et ne peut être que Dieu." *Carnet* I.2, loose sheet 84 (September 1891).

104. Gooding, Pinch, and Schaffer, "Introduction," in *Uses of Experiment*, 4–5.

105. See Sidlauskas, "Contesting Femininity."

106. On "critical cross-dressing," see Showalter, "Critical Cross-Dressing"; and Goldstein, "Uses of Male Hysteria," 156–57.

Chapter 4

1. Maria Weyr to Rudolf Weyr, ca. 1893–94, in Strindberg, *Marriage with Genius*, 139–40, 145–46.

2. Strindberg to Leopold Littmansson, [4? August 1894], in Strindberg, *Brev*, no. 2885, 10:192; English translation in *Strindberg's Letters*, no. 325, 2:500; Strindberg to Richard Bergh, 26 June 1894, in *Brev*, no. 2828, 10:106. Where they exist, I have relied on authoritative English translations of Strindberg's writings as starting points for my own translations, and have therefore included, where appropriate, references to both the original and the translation. My own translations often differ, however, from the published translations for reasons discussed below.

3. Strindberg to Leopold Littmansson, [14? July 1894], in *Brev*, no. 2847, 10:130; *Strindberg's Letters*, no. 314, 2:480.

4. The foundational study on Strindberg's experimental method is Sprinchorn, "'Zola of the Occult.'" See also Gavel Adams, "Strindberg som Ockultismens Zola"; and Grimal, "Intuition du tout."

5. Quoted in Puech, "Zola et l'Académie." Puech also interviewed Maurice Maeterlinck and Georg Brandes, among others, for this piece. See also Strindberg's original letter to Puech, 1 May 1894, in *Brev*, no. S405, 22:40–42.

6. Varvara, "Justice de l'étranger," 357–58.

7. Jaspers, *Strindberg und van Gogh*, 46–48. My approach in this chapter is similar to the Foucault-inspired work of Ulf Olsson. See Olsson, "Simulacres"; "Going Crazy"; and *Jag blir galen*.

8. Strindberg to Zola, 29 August 1887, in *Brev*, no. 1440, 6:262.

9. As with English translations of impressionist criticism and Cézanne's letters, English translations of Strindberg's works often liberally translate words such as *försök*, as in *Försöksteater*, and *essai*, as in "Essai sur le rationalisme mystique," as *experiment* or *experimental*. Nearly all of Strindberg's works are searchable online at http://www.litteraturbanken.se and http://www.spraakbanken.gu.se/konk/.

10. On the Black Piglet circle, see Lathe, "Group Zum schwarzen Ferkel"; and Heller, "'Schwarze Ferkel.'"

11. Strindberg's engagement with the visual has long been central to analyses of his work. For an overview, see Szalczer, "Strindberg and the Visual Arts."

12. Strindberg to Helena and Carl Rupert Nyblom, 24 January 1882, in *Brev*, no. 402, 2:357;

Strindberg's Letters, no. 47, 1:88. For an important early account of Strindberg's engagements with Zola, see Ahlstedt and Morizet, "Strindberg et Zola," 34.

13. Zola, *Naturalisme au théâtre*, 58.
14. Strindberg to Zola, 29 August 1887, in *Brev*, no. 1440, 6:262.
15. Strindberg to Edvard Brandes [12? June 1885], ibid., no. 1079, 5:100; *Strindberg's Letters*, no. 124, 1:185.
16. Zola, *Naturalisme au théâtre*, 16.
17. Strindberg to Helena and Carl Rupert Nyblom, 24 January 1882, in *Brev*, no. 402, 2:357; *Strindberg's Letters*, no. 47, 1:88.
18. Strindberg, "Det icke medtagna förordet till Tjänstekvinnans son," in *Samlade Verk*, 20:373. English translation in Strindberg, *Son of a Servant*, 23. This preface, written in mock interview form, was intended for the original 1886 publication of the novel but was published later in Strindberg's life.
19. On Strindberg's studies in psychology and psychiatry, see the work of Madsen, "Impact of the French Naturalists"; and Hansen, "Strindberg und die dynamische Psychiatrie."
20. Lindström, *Strindberg och böckerna*, 1:23, 52–54.
21. Strindberg to Verner von Heidenstam, 5 October 1886, in *Brev*, no. 1314, 6:78; *Strindberg's Letters*, no. 150, 1:217.
22. Léouzon Le Duc, "Nuits du Somnambule," 90.
23. See Strindberg to Gustaf af Geijerstam, [29 February 1884], in *Brev*, no. 766, 4:64; *Strindberg's Letters*, no. 74, 1:132.
24. Strindberg to Edvard Brandes, [5?] April 1886, in *Brev*, no. 1212, 5:306; *Strindberg's Letters*, no. 133, 1:196.
25. Strindberg to Albert Bonnier, 25 April 1886, in *Brev*, no. 1218, 5:314; *Strindberg's Letters*, no. 134, 1:197; Strindberg to Gustaf af Geijerstam, [13?] March 1886, in *Brev*, no. 1207, 5:295; *Strindberg's Letters*, no. 132, 1:195; Strindberg to Edvard Brandes, 27 May 1886, in *Brev*, no. 1236, 5:339; *Strindberg's Letters*, no. 137, 1:202. See also Michael Meyer's critical review of Evert Sprinchorn's translation of *The Son of a Servant*, "Young Strindberg," and Sprinchorn's response, "Vulgar Style."
26. Strindberg to Edvard Brandes, 27 May 1886, in *Brev*, no. 1236, 5:339; *Strindberg's Letters*, no. 137, 1:202.
27. Strindberg, "Det icke medtagna förordet till Tjänstekvinnans son," in *Samlade Verk*, 20:374; English translation in Strindberg, preface to *Son of a Servant*, 24.
28. Zola to Strindberg, 14 December 1887, in Berendsohn, *Briefe an Strindberg*, 50.
29. On using Zola's influence to court André Antoine, see Nietzsche to Strindberg, 27 November 1888, ibid., 51.
30. Strindberg to Gustaf af Geijerstam, 15 November 1888, in *Brev*, no. 1679, 7:157; *Strindberg's Letters*, no. 215, 1:290.
31. Strindberg to Georg Brandes, 29 November 1888, in *Brev*, no. 1707, 7:183–84; *Strindberg's Letters*, no. 216, 1:291.
32. Strindberg, preface to *Fröken Julie*, in *Samlade Verk*, 27:101, 109–10; English translation in Strindberg, *Miss Julie and Other Plays*, 56, 64.
33. Strindberg, preface to *Fröken Julie*, in *Samlade Verk*, 27:111, 113; Strindberg, *Miss Julie and Other Plays*, 65–66, 68. Strindberg first encountered impressionist painting in Paris in 1876, during a period in which he wrote art criticism for Swedish daily newspapers. Strindberg, "'Salonen' och den franska konsten." See Berman, "Strindberg, Painter, Critic, Modernist"; Usselmann, "Strindberg et l'impressionnisme."
34. Strindberg, "Om modern drama," 36; English translation in Strindberg, *Selected Essays*, 78.
35. Strindberg, preface to *Fröken Julie*, in *Samlade Verk*, 27:103, 109; Strindberg, *Miss Julie and Other Plays*, 58, 64.
36. See Strindberg to Joseph Seligmann, 16 October 1888, in *Brev*, no. 1663, 7:144; *Strindberg's Letters*, no. 214, 1:289; Strindberg to Verner von Heidenstam, [9? October 1889], in *Brev*, no. 1894, 7:377; *Strindberg's Letters*, no. 241, 1:322; Strindberg to Karl Otto Bonnier, 30 December 1888, in *Brev*, no. 1735, 7:212.
37. Strindberg to Gustaf af Geijerstam, 12 April 1890, in *Brev*, no. 1949, 8:25; *Strindberg's Letters*, no. 246, 1:326.

38. Strindberg to Per Hasselberg, [10?] September 1892, in *Brev*, no. 2388, 9:63; *Strindberg's Letters*, no. 282, 1:368.
39. Strindberg to Leopold Littmansson, 22 July 1894, in *Brev*, no. 2860, 10:150; *Strindberg's Letters*, no. 317, 2:485. For a stimulating set of essays on Strindberg and the other, see Houe, Rossel, and Stockenström, *August Strindberg and the Other*.
40. Strindberg to Isidor Bonnier, 25 April 1887, in *Brev*, no. 1399, 6:191; *Strindberg's Letters*, no. 163, 1:229; Strindberg to Verner von Heidenstam, 9 January 1887, in *Brev*, no. 1361, 6:142.
41. Strindberg to Claës Looström, 20 January 1888, in *Brev*, no. 1523, 6:377; *Strindberg's Letters*, no. 186, 1:262.
42. Strindberg to Isidor Bonnier, 25 April 1887, in *Brev*, no. 1399, 6:191; *Strindberg's Letters*, no. 163, 1:229.
43. Strindberg to Albert Bonnier, 6 February 1887, in *Brev*, no. 1371, 6:155; *Strindberg's Letters*, no. 158, 1:225. Strindberg wrote two essays in Swedish in the 1880s on the "woman question," and in 1895 he published "Vivisection: De l'infériorité de la femme" in *La revue blanche*. See Fahlgren, "Strindberg and the Woman Question."
44. Strindberg to Alexander Kielland, 29 November 1886, in *Brev*, no. 1341, 6:110; *Strindberg's Letters*, no. 152, 1:218; Strindberg to Pehr Staaff, 13 August 1887, in *Brev*, no. 1433, 6:249; *Strindberg's Letters*, no. 168, 1:236. The psychological investigation of Siri would also result in *Le plaidoyer d'un fou*, an account of his first marriage.
45. Strindberg to Pehr Staaff, [5 August 1887], in *Brev*, no. 1429, 6:242–43; *Strindberg's Letters*, no. 167, 1:234–35.
46. Strindberg to Albert Bonnier, 20 September 1886, in *Brev*, no. 1304, 6:63; English translation in *Strindberg's Letters*, no. 149, 1:216. On Strindberg's use of vivisection as a metaphor for a penetrative antimimetic dramatic strategy, see Holzapfel, "Strindberg as Vivisector."
47. See Plas, "De la vivisection morale"; and Carroy, "'Mon cerveau,'" 202.
48. Strindberg, "Hjärnornas kamp," in *Samlade Verk*, 29:24–49; Strindberg, *Selected Essays*, 25–46.
49. Strindberg to Isidor Bonnier, 25 April 1887, in *Brev*, no. 1399, 6:192; *Strindberg's Letters*, no. 163, 1:230.
50. Strindberg, "Mystik—tills vidare," in *Samlade Verk*, 29:71; Strindberg, *Selected Essays*, 62.
51. Strindberg, "Själamord (Apropå Rosmersholm)," in *Samlade Verk*, 29:79; Strindberg, *Selected Essays*, 70.
52. See Rugg, "August Strindberg," 8–9.
53. Strindberg to Verner von Heidenstam, 25 May 1888, in *Brev*, no. 1612, 7:92; *Strindberg's Letters*, no. 201, 1:278.
54. Strindberg to Brandes, 12 April 1890, in *Brev*, no. 1950, 8:26; *Strindberg's Letters*, no. 247, 1:327. Strindberg and Nietzsche had a brief correspondence in late 1888, just before Nietzsche collapsed in Turin and entered a clinic. Exactly what role Nietzsche's work played in Strindberg's development is much debated. See Milochevitch, *Nietzsche et Strindberg*; and Dahlkvist, "*By the Open Sea*."
55. Gilman, "The Nietzsche Murder Case; or, What Makes Dangerous Philosophies Dangerous," in *Difference and Pathology*, 59–75.
56. Strindberg to Bernhard Meijer, 11 December 1890, in *Brev*, no. S304, 21:311–12; *Strindberg's Letters*, no. 256, 1:335.
57. See Daston and Galison, *Objectivity*, 191–251.
58. See chapter 5.
59. Max Asch, a physician, and Carl Ludwig Schleich, a radical neurologist, were the main sources of Strindberg's scientific encouragement within Zum schwarzen Ferkel. Dierig, "Con Sordino for Piano and Brain." For a wide-ranging discussion of Strindberg's and Munch's scientific interests in protoplasm and monism, see Brain, "How Edvard Munch and August Strindberg Contracted Protoplasmania"; and Brain, *Pulse of Modernism*, 174–200.
60. Richard Dehmel to Strindberg, 16 July 1893, Strindberg Papers, Ep S53b, Kungliga Biblioteket, Stockholm.
61. Schleich, *Erinnerungen an Strindberg*, 14, quoted in translation in Dierig, "Con Sordino for Piano and Brain," 437.

62. On Strindberg's participation in Parisian occult circles, see Gavel Adams, "Strindberg et l'occultisme en France."
63. Kerner, *Kleksographien*. *Sphinx* reproduced the Klecksographs in volume 11 (January 1891 and February 1891). For more on the Klecksographs, see Hofmann and Praeger, "Bilder aus Klecksen"; Gisbourne, "Spiritisme chez Victor Hugo"; and Gamboni, "'Fabrication of Accidents.'"
64. See also Strindberg's discussion of Kerner's inkblots in *En blå bok I*, in *Samlade Verk*, 65:205.
65. In a series of essays on the Jardin des Plantes in 1896, Strindberg included an article titled "La tête de mort (Acheronita atropos): Essai de mysticisme rationnel," in *Samlade Verk*, vol. 35, translated into English as "The Death's Head Moth: An Experiment in Rational Mysticism," in *Selected Essays*, 149–54.
66. Strindberg to Per Hasselberg, [10?] September 1892, in *Brev*, no. 2388, 9:62; *Strindberg's Letters*, no. 282, 1:366–67. On Strindberg's paintings, see Söderström, *Strindbergs Maaleri*; Söderström, *Strindberg och bildkonsten*; Schmidt, *Strindbergs måleri*. More recently, see these exhibition catalogues: Hedström, *August Strindberg: Painter and Photographer*; and Granath et al., *August Strindberg: Painter, Photographer, Writer*. See also Feuk, *August Strindberg: Paradisbilder*; Rosenblum, *Paintings of August Strindberg*; and Gamboni, "'Dieses Schillern der Eindrücke freute mich.'"
67. Strindberg's first intensive period of painting began in 1869 and continued into the 1870s, the decade in which he regularly wrote art criticism for a Swedish newspaper. The paintings produced between 1892 and 1894 constituted his second intensive campaign, and there was a third between 1901 and 1905.
68. Heller, "Concerning Symbolism," 150.
69. Anna Westerståhl Stenport argues that Strindberg "sought to challenge an aesthetics of simplistic symbolism without material components." Stenport, *Locating August Strindberg's Prose*, 112.
70. Strindberg had written on Courbet's use of the palette knife. See Granath et al., *August Strindberg*, 15.
71. Strindberg to Richard Bergh, 26 November 1894, in *Brev*, no. 3010, 10:314; *Strindberg's Letters*, no. 339, 2:519.
72. Strindberg, "Sensations détraquées."
73. Strindberg to Leopold Littmansson, 31 July 1894, in *Brev*, no. 2875, 10:177; *Strindberg's Letters*, no. 322, 2:494. See also Robinson, "*New Arts, New Worlds*"; Gamboni, *Potential Images*, 176–77.
74. See chapter 2.
75. Quoted in Hübbe-Schleiden, "Esoterische und exoterische Naturen"; Schuré, *Grands initiés*, xiii.
76. Strindberg, "Des arts nouveaux," 268.
77. Przybyszewski, *Werk des Edvard Munch*, 35–37. See also Heller, "Making a Painting Scream."
78. See the exchange of letters between Strindberg and Gauguin, published in *L'éclair*, 15–16 February 1895, as well as in the preface to Gauguin's February 1895 sale catalogue.
79. On the contradictory nature of Strindberg's radicalism and vanguardism, see Stenport and Szalczer, "Strindberg and Radicalism."
80. Jan Goldstein makes a similar point in arguing that psychological modernism, rather than constituting a complete break with psychologies of the nineteenth century, emerged directly out of positivist modes of thinking. See Goldstein, "Advent of Psychological Modernism."
81. Sprinchorn, "'Zola of the Occult,'" 258.
82. Strindberg to Torsten Hedlund, 21 June 1896, in *Brev*, no. 3305, 11:219. Quoted in English translation in Sprinchorn, "'Zola of the Occult,'" 255.
83. Strindberg to Leopold Littmansson, [25 July 1894], in *Brev*, no. 2864, 10:161; *Strindberg's Letters*, no. 320, 2:492.
84. Strindberg to Mathilde Praeger, 12 June 1896, in *Brev*, no. 3297, 11:210; *Strindberg's Letters*, no. 369, 2:556.
85. Strindberg, *Götiska rummen*, in *Samlade Verk*, 53:37; English translation in Sprinchorn, "'Zola of the Occult,'" 258–59.
86. Strindberg, *Svarta fanor*, in *Samlade Verk*, 57:219; English translation in Robinson, "Finding a New Language," 206.

87. André Breton made a direct connection between surrealism and Strindberg's theory of "automatic art" in the early 1960s after encountering Strindberg's paintings. See Stenport, *Locating August Strindberg's Prose*, 112.

Chapter 5

1. My reading relies on Reinhold Heller's dating of the self-portrait in *Munch: His Life and Work*, 110–11, which has been bolstered by technical analyses discussed in Thurmann-Moe, "Roßkur und Firnis," 118. On the controversies over dating this work, see Mørstad, "Edvard Munch under Medusas hode."
2. There remains debate about the timing and extent of Paul Gauguin's influence on Munch, but certainly by 1891 Munch was responding to synthetism. See Heller, *Munch: His Life and Work*, 229n13; and Arne Eggum, "Importance des deux séjours de Munch en France en 1891–1892," in Eggum and Rapetti, *Munch et la France*, 106, 116–19. For the argument that Munch might have seen Gauguin's work at the Volpini exhibition in 1889, see Dorra, "Munch, Gauguin, and the Norwegian Painters," 175–76.
3. Christian Krohg, 1891 review, published in both *Dagbladet* and *Verdens Gang*, quoted in Heller, *Munch: His Life and Work*, 82.
4. Mørstad, "Edvard Munch under Medusas hode"; Müller-Westermann, *Munch by Himself*, 32–33.
5. See Bonillo, *Riviera de Charles Garnier*.
6. Aubert, "Høstudstillingen." See also Heller, "Edvard Munch's Night," 80–81, and *Munch: His Life and Work*, 69. For more on Munch and neurasthenia, see Berman's essays "Edvard Munch's *Self-Portrait with Cigarette*," esp. 632–33, and "Edvard Munch's 'Modern Life of the Soul,'" esp. 51n59.
7. Mørstad, "Responding to *Self-Portrait with Cigarette*."
8. Monica Solem quoted in Gibbs, "Museum Reopens."
9. Heller, "'Could Only Have Been Painted by a Madman'"; Berman, "Edvard Munch's 'Modern Life of the Soul.'" See also Yarborough, "Exhibition Strategies and Wartime Politics"; Clarke, *Becoming Edvard Munch*; Ydstie and Guleng, *Munch Becoming "Munch"*; and Clarke, "Art Equals Life?"
10. Munch to Eberhard Grisebach, 17 February 1913, quoted in Stang, *Edvard Munch*, 262.
11. Especially via the Danish poet Emanuel Goldstein. See Heller, *Munch: His Life and Work*, 58–62, 66–68; Frydenberg-Flaatten, "Edvard Munch's 'Skrik,'" 59–60, 66–80.
12. MM T 2771, Munch Museum, dated 1890–91, Sketchbook, fol. 2r, *Edvard Munch's Writings, Digital Archive*, http://www.emunch.no/. On this rich digital archive, see Guleng, *eMunch.no—Text and Image*. On the billiard motif, see Svenaeus, *Universum der Melancholie*, 13–15.
13. There is no documentary evidence that Munch saw *The Night Café*, but he was certainly familiar with Van Gogh's work by this time, having visited the Salon des Indépendants of both 1889 and 1890, where a number of Van Gogh's paintings were shown. Munch may also have helped Erik Werenskiold organize the exhibition *Les XX* in Brussels in 1890, which included works by Van Gogh. With reference to the billiard notes, Heller explores the impact of Van Gogh's work, including *The Night Café*, on Munch, comparing its exaggerated perspective with that in *Night in Saint-Cloud*. See Heller's essays "Edvard Munch's 'Life Frieze,'" 65–67, 117–18; "Edvard Munch's Night"; and "'In Part Myself,'" 66–67.
14. Vincent van Gogh to Theo van Gogh, 8 and 9 September 1888, translation slightly amended from *Vincent van Gogh: The Letters*.
15. MM T 2760, Munch Museum, dated 1891–92, Sketchbook, fol. 16r, *Edvard Munch's Writings, Digital Archive*.
16. See chapter 2; and Heller, "Edvard Munch's 'Life Frieze,'" 67.
17. The following section revisits material from Morehead, "'Are There Bacteria.'"
18. Skredsvig, "Edvard Munch i Monte Carlo," 114. On Munch and Skredsvig's time together on the French Riviera, see Thomsen, "Skredsvig og Munch på Rivieraen."

19. Munch's shift mirrors a nineteenth-century shift from viewing gambling as a metaphor for sociohistorical structures, and the target of moral concerns over the crumbling of those structures, to a fascination with the psychopathological dimensions of betting on chance. See Reith, *Age of Chance*, 3–8.
20. Praed, "Casino at Monte Carlo." This text accompanied the reproduction of Béraud's painting in the *Graphic*.
21. Offenstadt, *Jean Béraud, 1849–1935*, 188–89. Gösta Svenaeus first connected Béraud's painting with Munch's depictions of roulette, pointing to the woman seen from the back as a similar element in all three paintings. See Svenaeus, *Universum der Melancholie*, 15–17. Munch could have seen Béraud's painting at the Salon du Champ-de-Mars in 1890, but it is just as likely that he saw the large print of the painting published in *The Graphic*, perhaps in the well-stocked reading rooms of the casino itself. See Aurier, "Salon du Champ-de-Mars," 248.
22. Benjamin addresses gambling in a number of his works. See especially Benjamin, *Arcades Project*, 488–515.
23. A number of scholars have noted the similarity between Munch's and Dostoevsky's writings on gambling, most recently Sivert Thue. Thue points out how popular Dostoevsky's work was in Norway, especially *The Gambler*, which was translated into Norwegian in 1889. Knut Hamsun was even accused of plagiarizing Dostoevsky's novella in his short story *Hazard* (1889). See Thue, "From Types on the Promenade," 119.
24. MM N 24, Munch Museum, dated 1891–92, Note, pp. 3–4, *Edvard Munch's Writings, Digital Archive*.
25. Becker, "Games and Gamesters," 83; Wells's luck spawned the song "The Man Who Broke the Bank in Monte Carlo," written in 1892 by Fred Gilbert and popularized by the British music hall performer Charles Coborn.
26. Gabriel Charmes, *Journal des Débats*, 17 February 1884, cited in Visitor to the Riviera, *Monte Carlo and Public Opinion*, 184–86.
27. MM N 1, Munch Museum, dated 1891–92, Note, pp. 1–4, *Edvard Munch's Writings, Digital Archive*.
28. Scholars have long assumed that as the largest and most fully realized version of the roulette theme, *At the Roulette Table in Monte Carlo* was shown at Munch's fall 1892 solo exhibitions in Kristiania and at various venues in Germany, but research now indicates that only the two smaller versions were probably exhibited. See Krisch, *Die Munch-Affäre*, 66n266. For a slightly dissenting view, see Pettersen, "Berlin 1892," 185–86.
29. Gary Saul Morson, "Introduction: Writing Like Roulette," in Dostoevsky, *Gambler*, xi–xliii.
30. MM N 1, Munch Museum, dated 1891–92, Note, pp. 1–4, *Edvard Munch's Writings, Digital Archive*.
31. In his analysis of Dostoevsky, Sigmund Freud linked the "passionate play of hands" in gambling to masturbation. Freud, "Dostoevsky and Parricide" (1928), in Wellek, *Dostoevsky*, 98–111.
32. On Munch and caricature, see, among others, Mørstad, "Linguaggio formale." In a letter to Rolf Stenersen, Munch argued that profiles revealed inherited and racial characteristics, while frontal views revealed more individual traits. See Heller, "Edvard Munch's 'Life Frieze,'" 135n11. On caricature and physiognomy more generally, see Sheon, "Caricature and the Physiognomy of the Insane."
33. An earlier version of this figure might appear in the upper left background of the second roulette painting. On Munch's use of "avatars," see Endresen, "Serial Experiments," 145, 150–53, 157–59.
34. Heller, "Concerning Symbolism," 149–50.
35. MM T 2760, Munch Museum, dated 1891–92, Sketchbook, fol. 56r, dated "Nizza 22/1 92," *Edvard Munch's Writings, Digital Archive*. English translation adapted from Francesca M. Nichols's translation, accessed 12 December 2013, at http://www.emunch.no/english.xhtml; and Heller, *Scream*, 107.
36. The work of Patricia Berman and Shelly Wood Cordulack has been foundational for my understanding of Munch's engagements with science, especially beginning in the Berlin

period. See Berman, "Edvard Munch's 'Modern Life of the Soul'"; Berman, "From Munch's Laboratory"; and Cordulack, *Edvard Munch and the Physiology of Symbolism*. On the scientific context of At the Black Piglet, see Dierig, "Con Sordino for Piano and Brain"; Dierig and Brink, "Beyond the Temples of Science"; and Brain, *Pulse of Modernism*, 180–84.

37. Przybyszewski, "Psychischer Naturalismus," 150.

38. On Przybyszewski's studies, see Uggla, "Przybyszewski and Strindberg," 201.

39. See Czarnocka, "Nietzsche, Przybyszewski," 44.

40. This surprising fact and its implications are discussed in Heller, "Edvard Munch's 'Life Frieze'"; and Lathe, "Group Zum schwarzen Ferkel," 13.

41. Quoted in Przybyszewski, *Werk des Edvard Munch*, 74. Pastor's review first appeared as "Edvard Munch" in *Frankfurter Zeitung*, 24 January 1894.

42. Quoted in Przybyszewski, *Werk des Edvard Munch*, 37; see also Heller, "Making a Painting Scream," 28.

43. Robert M. Brain tentatively makes a similar claim about *The Scream*, asserting that "it might be said to function like an experiment in psycho-physiological aesthetics," and suggests that the attitude of the central figure recalls subjects of those experiments as seen in photographs. Brain, "How Edvard Munch and August Strindberg Contracted Protoplasmania," 21; Brain, *Pulse of Modernism*, 185–86.

44. Czarnocka, "Nietzsche, Przybyszewski," 43.

45. See Servaes's essay in Przybyszewski, *Werk des Edvard Munch*, 48, quoted in English translation in Heller, "Edvard Munch's 'Life Frieze,'" 208.

46. "Kan kun være malt af en gal mand!" This text, noted as early as 1904, is controversial. Some scholars argue that Munch wrote it, others that it was a later addition by an unknown "critic." Regardless, as Heller points out, Munch retained the text, a choice that argues for its significance. See Heller, "'Could Only Have Been Painted by a Madman,'" 17. For a summary of the argument against Munch as the author of the text, see Woll, *Edvard Munch: Complete Paintings*, 1:316. The most thorough account of the painting remains Heller, *Scream*.

47. Heller, "'Could Only Have Been Painted by a Madman,'" 27. My debts to Heller are many, but his suggestion that Munch must have had some conception of the art of the mentally ill prior to painting *The Scream* has driven much of my research on Munch. I am grateful to him for sharing these insights with me at an early stage of my research.

48. MacGregor, *Discovery of the Art of the Insane*, 151–60.

49. Strindberg apparently got halfway and then decided against the visit, hopping off the train at Charlottenburg and returning immediately to Berlin. Paul, *Min Strindbergsbok*, 46n, cited in Meyer, *Strindberg: A Biography*, 258. Munch would later recommend that his companion Tulla Larsen consult Asch for her health problems. For more on Asch, see Czarnocka, "Nietzsche, Przybyszewski," 44; and Dierig, "Con Sordino for Piano and Brain," 423.

50. Hübbe-Schleiden, "Sphinx, Aufruf und Vorwort." On *Sphinx*, see Treitel, *Science for the Soul*, 33, 83–85; and Wolffram, *Stepchildren of Science*, 54–55, 60–61.

51. Evers, "Psychischer Naturalismus." Evers wrote to Munch in 1932, recalling their Berlin years together and saying how much joy the artist's etchings had brought him over the years. See Franz Evers to Edvard Munch, 10 December 1932, Munch Papers, Munch Museum Library, Oslo.

52. Gessmann, "Somnambules Zeichnen."

53. Strindberg to Dehmel, 4 July 1893, in *Brev*, no. 2578, 9:229; Dehmel to Strindberg, 1 July 1893, in Berendsohn, *Briefe an Strindberg*, 79; and Dehmel to Strindberg, 7 July 1893, Strindberg Papers, Ep S53b, Kungliga Biblioteket, Stockholm. The following four paragraphs incorporate material from Morehead and Otto, "Representation in the Age of Mediumistic Reproduction."

54. See the *Sphinx* articles "Experimente übersinnlicher Gedanken-Übertragung" and

"Übersinnlichen Gedanken-Übertragung." The Society for Psychical Research experiments were soon repeated by the Munich and Berlin experimental psychology groups and duly published in *Sphinx*.

55. Tempera was specifically linked in Munch's mind with the chemical experiments of Black Piglet member Carl Schleich, which are discussed in a letter from Richard Bergh to Munch, 15 June 1893, Munch Papers, Munch Museum Library, Oslo. Schleich's experiments extended to members of the Black Piglet group itself, on whom he supposedly conducted experiments to assess the effects of excessive alcohol consumption on the "brains of geniuses." Servaes, "Schwarze Ferkel," cited in Lathe, "Group Zum schwarzen Ferkel," 37.

56. Nanteuil, "Trois expositions à l'Art nouveau" ("horrors"); Camille Mauclair ("inept drawing," "barbarous color," "childlike etchings"), quoted in Rapetti, "Munch face à la critique française," 28; Chapoy, "Choses d'art en 1896" ("paintings of a sick mind"). On Munch's reception at the Salon des Indépendants in 1896 and 1897, see Van Dijk, "International Artists at the Salon," 44–47.

57. Jourdain, "Petites expositions"; Rambosson, "Salon des Indépendants."

58. See de Brahm, "Un côté inconnu" and "Salon des Indépendants."

59. Ritter, "Un peintre norvégien," 82, 85. Ritter sent this volume to Munch with the following inscription: "Au poëte que je ne sais s'il faut louer ou éxécrer des tares physiques et de la laideur contemporains, en temoignage au moins de réelle bonne volonté à chercher à comprendre." For a new Norwegian edition of Ritter's text on Munch, see Ritter, *Edvard Munch*. On Munch's attempts to paint ugly paintings in relation to Van Gogh, see Heller, "'In Part Myself,'" 66–71.

60. Réja, "Symbolisme pictural," 10–11.

61. On Munch's time in France, see Torjusen, "Mirror"; Eggum and Rapetti, *Munch et la France;* and Zarobell, "Year in Paris."

62. See Millroth, *Molards Salong;* Deryng, "Variations Popoffsky."

63. On Réja in the context of symbolism, see Rapetti, *Symbolisme*, 293–99. On Réja and the art of the mentally ill, see Thévoz, "Marcel Réja"; and MacGregor, *Discovery of the Art of the Insane*, 161–84.

64. On discussions of syphilis in Munch's circles in the 1890s and their wider medical context, see Ydstie, "Edvard Munch's Painting *Inheritance*," 208–12.

65. *Delius: A Life in Letters*, vol. 1; Richard Dehmel to Henri Albert, 7 April 1895, in Krahmer, *Deutsch-französische Brieffreundschaft*, no. 64; Strindberg to Anders Eliasson [1 April 1895?], in *Strindberg's Letters*, no. 349, 2:532.

66. Réja later worked at Sainte-Anne's. See Salmon, *Souvenirs sans fin, 1903–1940*, 207.

67. These were produced as lithographs. See Singer, *Graphische Werk des Maler-Radierers*, cat. nos. 63 and 64. According to Singer, these images of dancing women were done at the request of the psychiatrist Dr. Dide. Héran made a total of eight prints after sketches at Sainte-Anne's.

68. Auguste Marie to Jehan Rictus [Gabriel Randon], 26 May 1901, NAF, MS 24566, Département des manuscrits, BnF.

69. William Molard and Ida Ericson to Jehan Rictus [Gabriel Randon], NAF, MS 24568, ibid. On Rictus, see Coon, "Building of the Rictus Legend."

70. Munch to Dr. Max Linde (1903), in Lindtke, *Edvard Munch—Dr. Max Linde*, quoted in Stang, *Edvard Munch*, 288n19. An unidentified clipping from a French newspaper at the Munch Museum (marked "*Paris encyclopédique* 17 Sept? 1898?") identifies the location of the scene as the Hôpital Saint-Louis. However, a pencil drawing (T2357, Munch Museum, Oslo), usually dated 1893–94, suggests that Munch had the motif in mind earlier. The Norwegian painter Alfred Hauge, with whom Munch shared a Parisian studio in 1897, wrote to a friend that Munch had specifically sought opportunities to draw syphilitics, enlisting a doctor friend to help him gain access to the hospital. See Woll, "Graveur," 259. On symbolism and syphilis, including a discussion of Munch's *Inheritance*, see Høifødt, "Kvinnen, Kunsten, Korset," 78–83; Larson, "Microbes

and Maladies," 390–92; Berman and Nimmen, *Munch and Women*, 175; Hirsh, *Symbolism and Modern Urban Society*, 129–44; and Ydstie, "Edvard Munch's Painting *Inheritance*."

71. MM T 2800, Munch Museum, dated 1908, Sketchbook, p. 39, *Edvard Munch's Writings, Digital Archive*.

72. Gerd Woll connects the lithograph to *Inheritance* in *Edvard Munch: The Complete Graphic Works*, 100.

73. See Kromm, *Art of Frenzy*, 55, 79–83; Andrews, "(Un)dress of the Mad Poor."

74. Kromm, *Art of Frenzy*, 74–75, 83.

75. See Fauvel, "Aliénistes contre psychiatres."

76. Morel, *Dictionnaire biographique de la psychiatrie*, 235.

77. Fauvel, "Mots des sciences."

78. See Morehead, "*Musée de la folie*." The following paragraph revisits material from this essay.

79. Marie, "Musée de la folie," 353–54. For a recent reconsideration of Marie's collections and Réja, see Nelson, "Timeknots: Science and Reform," 204–29.

80. Réja, "Symbolisme pictural," 10.

81. See, for example, *Vision* and *The Scream*. On Munch's process, see Topalova-Casadiego, "Technical Aspects of Munch's Paintings."

82. The word *experiment* and its cognates appear thirty-six times in Munch's archived writings, but rarely if ever before 1908–9, judging by the texts that can be confidently dated. In the 1890s, Munch tended to use the more generic term *forsøg*, which, like the Swedish *försök*, is often translated into English as *experiment*.

83. MM N 2644, Munch Museum, undated draft of a letter to an unidentified man, fol. 1r, *Edvard Munch's Writings, Digital Archive*.

84. MM N 37, dated 1933–34, draft of a letter to Jens Thiis, fols. 1v, 2v, ibid. Munch would design sets for Ibsen's *Ghosts* in the early twentieth century. See Woll, *Edvard Munch: Complete Paintings*, 2:700–708.

85. MM T 2734, *Edvard Munch's Writings, Digital Archive*. This project, intended to bring together notes, drawings, and prints, would eventually be titled *The Tree of Knowledge of Good and Evil*. It would remain unpublished. See Woll, "Tree of Knowledge."

86. Flugsrud, "En dåres försvarstal."

87. See, most recently, Lüttichau, "'Crazy at Any Price,'" 36.

Conclusion

1. To these might be added J.-K. Huysmans's term "naturalisme spiritualiste."

2. The interrelations of naturalist and symbolist tenets have been explored most productively in relation to Cézanne in Shiff, *Cézanne and the End of Impressionism*. André Dombrowski is less explicit about how Cézanne's work relates to symbolist aesthetics, but his study *Cézanne, Murder, and Modern Life* offers a useful prehistory of Cézanne's engagement with new models of psychological interiority that resulted in work that is both dependent on and radically different from naturalist models.

3. Bürger, *Theory of the Avant-Garde*, 60–63.

4. Stenport, *Locating August Strindberg's Prose*, 100.

5. This link had been possible, however, before the early nineteenth century, when painting and science had existed together under the rubric of natural philosophy.

6. Gordon, "From Charcot to Charlot," 124.

7. Mathews, *Passionate Discontent*, 46–63.

Bibliography

Manuscript Sources

Archives nationales, Paris
Léon Bélugou Papers, 176 AQ (1997 076)
Bibliothèque de l'Institut de France, Paris
Édouard Vuillard, *Carnets*, MS 5396 (I.1, I.2), MS 5397 (II.2, II.5)
Bibliothèque nationale de France, Paris (BnF), Département des manuscrits
Léon Bélugou, notes taken from Théodule Ribot's courses at the Sorbonne and the Collège de France, 1887–94, Nouvelles acquisitions françaises, MSS 18173–18185
Gabriel Randon Correspondence, Nouvelles acquisitions françaises, MSS 24566, 24568
Kungliga Biblioteket, Stockholm
August Strindberg Papers
Munch Museum Library, Oslo
Edvard Munch Papers
Musée départemental Maurice Denis, Le Prieuré, Saint-Germain-en-Laye
Maurice Denis Papers, MSS 9301, 9302, 4322, 4531, 4532, 4534, 12071, 12080, 12081, 12142, 12146, 12159
Private collection, Saint-Germain-en-Laye
Maurice Denis, *Cahier philosophique*, 1887–88

Published Sources

Ahlstedt, Eva, and Pierre Morizet. "Strindberg et Zola." *Les cahiers naturalistes* 35, no. 63 (1989): 27–38.

Albert, Henri. "Auguste Strindberg." *La revue blanche* 7 (December 1894): 481–98.

———. "Les lettres allemandes." *La revue blanche* 9 (15 September 1895): 380–82.

Albury, W. R. "Experiment and Explanation in the Physiology of Bichat and Magendie." *Studies in the History of Biology* 1 (1977): 47–131.

Alexandre, Françoise. "Les carnets de Vuillard: Un peintre au miroir de son écriture intime." *48/14: La revue du Musée d'Orsay*, no. 23 (Fall 2006): 20–33.

———. "Édouard Vuillard, carnets intimes (1888–1905 et 1er janvier 1914–11 novembre 1918): Édition critique avec présentation, notes, annexes et bibliographie." PhD diss., Université Paris VIII, 1998.

André-Nuytz, Louis. "Le spiritualisme et l'école expérimentale." *La philosophie positive* 4 (January 1869): 135–53.

Andrews, Jonathan. "The (Un)dress of the Mad Poor in England, c. 1650–1850, Part 1." *History of Psychiatry* 18 (March 2007): 5–24.

Antoine, Jules. "Critique d'art: Pavillon de la ville de Paris—Exposition des artistes indépendants." *La plume*, no. 49 (1 May 1891): 156–57.

———. "Impressionnistes et synthétistes." *Art et critique*, no. 24 (9 November 1889): 369–71.

Argüelles, José. *Charles Henry and the Formation of a Psychophysical Aesthetic*. Chicago: University of Chicago Press, 1972.

Armstrong, Carol M. *Cézanne in the Studio: Still Life in Watercolors*. Los Angeles: J. Paul Getty Museum, 2004.

———. *Odd Man Out: Readings of the Work and Reputation of Edgar Degas*. 2nd ed. Los Angeles: Getty Research Institute, 2003.

Arnheim, Rudolf. "The Rationale of Deformation." In "The Issue of Caricature," edited by Judith Wechsler, special issue, *Art Journal* 43 (Winter 1983): 319–24.

Aubert, Andreas. "Høstudstillingen: Aarsarbejdet IV; Edvard Munch." *Dagbladet*, 5 November 1890.

Aurier, G.-Albert. "Beaux-arts: Les symbolistes." *Revue encyclopédique* 2 (1 April 1892): 474–86.

———. "Ratiocinations familières, et d'ailleurs vaines, à propos des trois Salons de 1891." *Mercure de France* 3 (July 1891): 30–39.

———. "Le Salon du Champ-de-Mars." *Mercure de France* 1 (July 1890): 248.

———. "Le symbolisme en peinture: Paul Gauguin." *Mercure de France* 2 (March 1891): 155–65.

———. *Textes critiques, 1889–1892: De l'impressionnisme au symbolisme.* Paris: École Nationale Supérieure des Beaux-Arts, 1995.

Balzac, Honoré de. "Le chef d'oeuvre inconnu." In Balzac, *Études philosophiques*, vol. 17, 145–237. Paris: Delloye et Lecou, 1837.

Balzamo, Elena, ed. *August Strindberg*. Paris: L'Herne, 2000.

Bann, Stephen. *Experimental Painting: Construction, Abstraction, Destruction, Reduction.* London: Studio Vista, 1970.

Barrows, Susanna. *Distorting Mirrors: Visions of the Crowd in Late Nineteenth-Century France.* New Haven: Yale University Press, 1981.

Beaune, Jean-Claude. "La notion de 'pathologique' chez Claude Bernard." In *La nécessité de Claude Bernard*, edited by Jacques Michel, 287–98. Paris: Méridiens Klincksieck, 1991.

Becker, Bernard H. "Games and Gamesters." *Fortnightly Review* 36 (July 1884): 75–87.

Becker, Colette, Gina Gourdin-Servenière, and Véronique Lavielle. *Dictionnaire d'Émile Zola: Sa vie, son oeuvre, son époque, suivi du dictionnaire des "Rougon-Macquart" et des catalogues des ventes après décès des biens de Zola.* Paris: Robert Laffont, 1993.

Bélugou, Léon. "Chronique philosophique." *La revue blanche* 7 (November 1894): 453–58.

———. "Chronique philosophique: Cours du Collège de France: M. Th. Ribot." *La revue blanche* 8 (1 February 1895): 182–83.

———. "Chronique philosophique: La philosophie à l'Université." *La revue blanche* 8 (1 January 1895): 69–77.

———. "Chronique universitaire." *Mercure de France* 15 (February 1898): 572–76.

———. "Chronique universitaire." *Mercure de France* 19 (June 1899): 789–94.

———. "Chronique universitaire." *Mercure de France* 27 (February 1902): 502–7.

———. "La philosophie." *La revue blanche* 14 (15 November 1897): 304–7.

——— [Bouvard et Pécuchet, pseud.]. "Le pouvoir de l'imagination chez les enfants: Recherches expérimentales." *Mercure de France* 52 (October 1904): 136–49.

———, ed. "Quelques opinions sur l'oeuvre de Taine." *La revue blanche* 13 (15 August 1897): 263–95.

Benjamin, Roger. "The Decorative Landscape, Fauvism, and the Arabesque of Observation." *Art Bulletin* 75 (June 1993): 295–316.

———. "Expression, Disfiguration: Matisse, the Female Nude, and the Academic Eye." In *In Visible Touch: Modernism and Masculinity*, edited by Terry Smith, 75–106. Chicago: University of Chicago Press, 1997.

———. *Matisse's "Notes of a Painter": Criticism, Theory, and Context, 1891–1908.* Ann Arbor: UMI Research Press, 1987.

Benjamin, Walter. *The Arcades Project.* Edited by Rolf Tiedemann. Translated by Howard Eiland and Kevin McLaughlin. Cambridge: Belknap Press of Harvard University Press, 1999.

Bennett, Tony, Lawrence Grossberg, and Meaghan Morris, eds. *New Keywords: A Revised Vocabulary of Culture and Society.* London: Wiley-Blackwell, 2005.

Berendsohn, Walter A., ed. *Briefe an Strindberg.* Mainz: Florian Kupferberg, 1967.

Bergman-Carton, Janis. "The Medium Is the Medium: Jules Bois, Spiritualism, and the Esoteric Interests of the Nabis." *Arts Magazine* 61 (December 1986): 24–29.

Bergson, Henri. *Essais sur les données immédiates de la conscience.* Paris: Alcan, 1889.

———. "La philosophie de Claude Bernard." 1913. In Bergson, *La pensée et le mouvant: Essais et conférences*, 257–66. Paris: Alcan, 1934.

Berman, Greta. "Strindberg, Painter, Critic, Modernist." *Gazette des beaux-arts* 86 (October 1975): 113–21.

Berman, Patricia G. "Edvard Munch's 'Modern Life of the Soul.'" In *Edvard Munch: The Modern Life of the Soul*, exh. cat., edited by Kynaston

McShine, 35–51. New York: Museum of Modern Art, 2006.

———. "Edvard Munch's *Self-Portrait with Cigarette:* Smoking and the Bohemian Persona." *Art Bulletin* 75 (December 1993): 627–46.

———. "Edvard Munch: Women, 'Woman,' and the Genesis of an Artist's Myth." In *Munch and Women: Image and Myth*, exh. cat., edited by Patricia G. Berman and Jane van Nimmen, 11–41. Alexandria, Va.: Art Services International, 1997.

———. "From Munch's Laboratory." In *Edvard Munch's Laboratory: The Path to the Aula*, exh. cat., edited by Ingebjørg Ydstie, 41–71. Oslo: Munch Museum, 2011.

Berman, Patricia G., and Jane van Nimmen, eds. *Munch and Women: Image and Myth*. Alexandria, Va.: Art Services International, 1997. Exhibition catalogue.

Bernard, Claude. "Étude sur la physiologie du coeur." *Revue des deux mondes* 56 (1 March 1865): 236–52. Reprinted in Bernard, *La science expérimentale*, 316–66. Paris: Baillière, 1878.

———. *L'introduction à l'étude de la médecine expérimentale*. Paris: Baillière, 1865.

———. *Leçons sur la chaleur animale, sur les effets de la chaleur et sur la fièvre*. Paris: Baillière, 1876.

———. *La science expérimentale*. Paris: Baillière, 1878.

Bernard, Émile. *Émile Bernard: Les lettres d'un artiste (1884–1941)*. Edited by Neil McWilliam. Dijon: Presses du Réel, 2012.

———. *Propos sur l'art*. Edited by Anne Rivière. 2 vols. Paris: Séguier, 1994.

Bernier, Georges. *La revue blanche: Ses amis, ses artistes*. Paris: Hazan, 1991.

Berry, Francesca. "Bedrooms: Corporeality and Subjectivity." In *Domestic Interiors: Representing Homes from the Victorians to the Moderns*, edited by Georgina Downey, 129–46. London: Bloomsbury, 2013.

Berson, Ruth, ed. *Reviews*. Vol. 1 of *The New Painting: Impressionism, 1874–1886; Documentation*. San Francisco: Fine Arts Museums of San Francisco, 1996.

Berthelot, Marcellin. *Science et philosophie*. Paris: Calmann-Lévy, 1886.

Bhogal, Gurminder Kaur. *Details of Consequence: Ornament, Music, and Art in Paris*. New York: Oxford University Press, 2013.

Bitker, Brigitte Ranson, and Gilles Genty. *Paul Ranson, 1861–1909: Catalogue raisonné; japonisme, symbolisme, art nouveau*. Paris: Somogy, 1999.

Bjerke, Øivind Storm, and Achille Bonito Oliva, eds. *Munch, 1863–1944*. Rome: Complesso del Vittoriano, 2005. Exhibition catalogue.

Bodelsen, Merete. "Gauguin, the Collector." *Burlington Magazine* 112 (September 1970): 588–615.

Bonillo, Jean-Lucien. *Les Riviera de Charles Garnier et Gustave Eiffel: Le rêve de la raison*. Marseille: Imbernon, 2004. Exhibition catalogue.

Bouillon, Jean-Paul. "Arabesques." In *Lost Paradise: Symbolist Europe*, exh. cat., edited by Pierre Théberge and Jean Clair, 376–84. Montreal: Museum of Fine Arts, 1995.

———. "Denis: Du bon usage des théories." In *Nabis, 1888–1900*, exh. cat., edited by Claire Frèches-Thory and Ursula Perucchi-Petri, 61–67. Paris: Réunion des Musées Nationaux, 1993.

———. "Denis, Taine, Spencer: Les origines positivistes du mouvement Nabi." *Bulletin de la Société de l'histoire de l'art français* (2000): 291–309.

———. "Histoire de l'histoire de l'art: Faut-il brûler Taine?" *Quarante-huit/Quatorze*, no. 1 (1989): 52–62.

———. *Maurice Denis*. Geneva: Skira, 1993.

———, ed. *Maurice Denis (1870–1943)*. Paris: Musée d'Orsay, 2006. Exhibition catalogue.

———. "Le moment symboliste." In "Le moment symboliste," edited by Jean-Paul Bouillon, special issue, *Revue de l'art*, no. 96 (1992): 5–11.

———. "Vuillard et Denis: Le tournant classique de 1898." *48/14: La revue du Musée d'Orsay*, no. 17 (Fall 2003): 80–91.

Bouillon, Jean-Paul, and Catherine Méneux, eds. *Le promenade du critique influent: Anthologie de la critique d'art en France, 1850–1900*. 2nd rev. ed. Paris: Hazan, 2010.

Bourdieu, Pierre. *The Rules of Art: Genesis and Structure of the Literary Field*. Translated by Susan Emanuel. Stanford: Stanford University Press, 1996.

Bourrelier, Paul-Henri. *La revue blanche: Une génération dans l'engagement, 1890–1905*. Paris: Fayard, 2007.

Boyer, Patricia Eckert, ed. *Artists and the Avant-Garde Theater in Paris, 1887–1900: The Martin and Liane W. Atlas Collection*. Washington, D.C.: National Gallery of Art, 1998. Exhibition catalogue.

Boyle-Turner, Caroline. *Les Nabis*. Translated by Yvonne Rosso. Lausanne: Edita, 1993.

———. *Paul Sérusier*. Ann Arbor: UMI Research Press, 1980.

Brain, Robert Michael. "How Edvard Munch and August Strindberg Contracted Protoplasmania: Memory, Synthesthesia, and the Vibratory Organism in Fin-de-Siècle Europe." *Interdisciplinary Science Reviews* 35 (March 2010): 7–38.

———. *The Pulse of Modernism: Physiological Aesthetics in Fin-de-Siècle Europe*. Seattle: University of Washington Press, 2015.

Brettell, Richard R. *Impression: Painting Quickly in France, 1860–1890*. New Haven: Yale University Press, 2000. Exhibition catalogue.

Brettell, Richard R., Françoise Cachin, Claire Frèches-Thory, and Charles F. Stuckey. *The Art of Paul Gauguin*. Washington, D.C.: National Gallery of Art, 1988. Exhibition catalogue.

Brooks, John I., III. *The Eclectic Legacy: Academic Philosophy and the Human Sciences in Nineteenth-Century France*. Newark: University of Delaware Press, 1998.

———. "Philosophy and Psychology at the Sorbonne, 1885–1913." *Journal of the History of the Behavioral Sciences* 29 (April 1993): 123–45.

Broude, Norma. *Impressionism: A Feminist Reading; The Gendering of Art, Science, and Nature in the Nineteenth Century*. 2nd ed. Boulder: Westview Press, 1997.

Brown, Stephen, and Richard R. Brettell. *Édouard Vuillard: A Painter and His Muses, 1890–1940*. New York: Jewish Museum, 2012. Exhibition catalogue.

Brunetière, Ferdinand. "Revue littéraire—Le roman expérimental." *Revue des deux mondes* 37 (15 February 1880): 935–48.

———. *Le roman naturaliste*. Paris: Calmann-Lévy, 1883.

Bruteig, Magne, Maite van Dijk, and Leo Jansen, eds. *Munch: Van Gogh*. Amsterdam: Van Gogh Museum, 2015. Exhibition catalogue.

Buchloh, Benjamin. "Theorizing the Avant-Garde." *Art in America* 72 (November 1984): 19–21.

Bürger, Peter. *Theory of the Avant-Garde*. Translated by Michael Shaw. Minneapolis: University of Minnesota Press, 1984.

Burhan, Filiz Eda. "Vision and Visionaries: Nineteenth-Century Psychological Theory, the Occult Sciences, and the Formation of the Symbolist Aesthetic in France." PhD diss., Princeton University, 1979.

Burns, C. A. "Henry Céard and His Relations with Flaubert and Zola." *French Studies* 6 (October 1952): 308–24.

Butor, Michel. "Émile Zola, romancier expérimental et la flamme bleue." *Critique* 23 (April 1967): 407–37. Reprinted as the introduction to "Le roman expérimental," in Zola, *Oeuvres complètes*, vol. 10, edited by Henri Mitterand, 1145–71. Paris: Cercle du Livre Précieux, 1968.

Cabanès, Jean-Louis. "Zola et le modèle bernardien." *Romantisme* 23, no. 82 (1993): 83–89.

Caillié, René. "Études cosmologiques: Dieu et univers (suite)." *L'étoile*, no. 5 (July 1890): 156–58.

———. "Spiritualisme expérimental: Le témoignage des faits; Histoire du mouvement spirite (suite)." *L'étoile*, no. 1 (March 1890): 19–26.

Cándida Smith, Richard. *Mallarmé's Children: Symbolism and the Renewal of Experience*. Berkeley: University of California Press, 1999.

Canguilhem, Georges. *L'idée de la médecine expérimentale selon Claude Bernard*. Paris: Palais de la Découverte, 1965.

———. *The Normal and the Pathological*. Translated by Carolyn R. Fawcett in collaboration with Robert S. Cohen. Introduction by Michel Foucault. New York: Zone Books, 1991.

Caro, Elme. *Le matérialisme et la science*. 2nd ed. Paris: Hachette, 1868.

———. *Le pessimisme au XIXe siècle: Leopardi, Schopenhauer, Hartmann*. Paris: Hachette, 1878.

Carroy, Jacqueline. "Hypnose, psychothérapie et expérimentation: L'École de Nancy et l'École de

la Salpêtrière." *Bulletin de psychologie* 38, no. 370 (1985): 537–41.
———. "'Mon cerveau est comme dans un crâne de verre': Émile Zola sujet d'Édouard Toulouse." *Revue d'histoire du XIXe siècle*, nos. 20–21 (2000): 181–202.
———. *Nuits savantes: Une histoire des rêves (1800–1945)*. Paris: Éditions de l'École des Hautes Études en Sciences Sociales, 2012.
Carroy, Jacqueline, and Régine Plas. "How Pierre Janet Used Pathological Psychology to Save the Philosophical Self." *Journal of the History of the Behavioral Sciences* 36 (Summer 2000): 231–40.
———. "La méthode pathologique et les origines de la psychologie française aux XIXe siècle." *Revue internationale de psychopathologie* 12 (1993): 603–12.
———. "The Origins of French Experimental Psychology: Experiment and Experimentalism." *History of the Human Sciences* 9 (February 1996): 73–84.
Cézanne, Paul. *The Letters of Paul Cézanne*. Edited and translated by Alex Danchev. Los Angeles: J. Paul Getty Museum, 2013.
———. *Paul Cézanne correspondance*. Edited by John Rewald. Rev. and augmented ed. Paris: Bernard Grasset, 1978.
———. *Paul Cézanne Letters*. Edited by John Rewald. Translated by Seymour Hacker. Rev. and augmented ed. New York: Hacker Art Books, 1984.
Chamberlin, J. Edward, and Sander L. Gilman, eds. *Degeneration: The Dark Side of Progress*. New York: Columbia University Press, 1985.
Chapoy, Henri. "Choses d'art en 1896: Les Indépendants." *Revue idéaliste*, 15 April 1896, 159.
Chastel, André. *Vuillard, 1868–1940*. Paris: Floury, 1946.
Cheetham, Mark. *The Rhetoric of Purity: Essentialist Theory and the Advent of Abstract Painting*. Cambridge: Cambridge University Press, 1991.
Christin, Anne-Marie. "Un livre double: *Le Voyage d'Urien* par André Gide et Maurice Denis (1893)." *Romantisme* 14, no. 43 (1984): 73–90.
Chu, Petra ten-Doesschate. "Dormeurs et rêveurs dans l'art du XIXe siècle." In *Le sommeil, ou Quand la raison s'absente*, exh. cat., edited by Jörg Zutter, Catherine Lepdor, and Caroline Nicod, 5–17. Lausanne: Musée Cantonal des Beaux-Arts, 1999.
———. "Lecoq de Boisbaudran and Memory Drawing: A Teaching Course Between Idealism and Naturalism." In *The European Realist Tradition*, edited by Gabriel P. Weisberg, 242–89. Bloomington: Indiana University Press, 1982.
Clark, T. J. *Farewell to an Idea: Episodes from a History of Modernism*. New Haven: Yale University Press, 1999.
———. "Madame Matisse's Hat." *London Review of Books* 30 (14 August 2008): 29–32.
———. *Picasso and Truth: From Cubism to Guernica*. Princeton: Princeton University Press, 2013.
Clarke, Jay A. "Art Equals Life? Munch and the Problem of Biography." In *Edvard Munch, 1863–1944*, exh. cat., edited by Mai Britt Guleng, Birgitte Sauge, and Jon-Ove Steihaug, 50–61. Oslo: Munch Museum and National Museum for Art, Architecture, and Design, 2013.
———. *Becoming Edvard Munch: Influence, Anxiety, and Myth*. Chicago: Art Institute of Chicago, 2009. Exhibition catalogue.
Cogeval, Guy. *Édouard Vuillard*. Washington, D.C.: National Gallery of Art, 2003. Exhibition catalogue.
Cogeval, Guy, and Isabelle Cahn, eds. *Bonnard: Peindre l'Acadie*. Paris: Musée d'Orsay, 2015. Exhibition catalogue .
Cogeval, Guy, Claire Denis, and Thérèse Barruel. *Maurice Denis, 1870–1943*. Lyon: Musée des Beaux-Arts, 1994. Exhibition catalogue.
Cogeval, Guy, Claire Frèches-Thory, and Gilles Genty. *The Time of the Nabis*. Montreal: Montreal Museum of Fine Arts, 1998. Exhibition catalogue.
Cogeval, Guy, Sylvie Patry, and Stéphane Guégan, eds. *Van Gogh, Gauguin, Cézanne, and Beyond: Post-Impressionist Masterpieces from the Musée d'Orsay*. San Francisco: Fine Arts Museums of San Francisco, 2010. Exhibition catalogue.
Colrat, Jean. "Eugène Véron: Contribution à une histoire de l'esthétique au temps de Spencer et Monet (1860–1890)." *Revue d'histoire des sciences humaines* 1, no. 18 (2008): 203–28.

Comes, Geneviève. "Le 'Mercure de France' dans l'évolution des arts plastiques, 1890–1895." *Revue d'histoire littéraire de la France* 92 (January–February 1992): 40–55.

———. "*La revue blanche* et le mouvement des idées." PhD diss., Université Paris XII, 1987.

Compagnon, Antoine, Paolo Tortonese, Yves Michaud, Michael F. Zimmermann, and Hippolyte Taine. *Relire Taine*. Paris: École Nationale Supérieure des Beaux-Arts, 2001.

Comte, Auguste. *Auguste Comte and Positivism: The Essential Writings*. Edited by Gertrud Lenzer. New Brunswick: Transaction, 1998.

———. *Système de politique positive, ou Traité de sociologie instituant la religion de l'humanité*. Vol. 1. Paris: Mathias, 1851.

Connelly, Frances S. *The Sleep of Reason: Primitivism in Modern European Art and Aesthetics, 1725–1907*. University Park: Pennsylvania State University Press, 1995.

Constable, John. "Pensées d'un paysagiste." *Magasin pittoresque* 23 (August 1855): 266–71; (October 1855): 342–43.

Coon, Robert L. "The Building of the Rictus Legend." *Papers on Language and Literature* 3 (Winter 1967): 20–33.

Cordulack, Shelly Wood. *Edvard Munch and the Physiology of Symbolism*. Madison: Fairleigh Dickinson University Press, 2002.

Court, Marie. "Vuillard: Les années de jeunesse vues à travers les carnets de l'Institut." Master's thesis, Université Blaise Pascal, Clermont-Ferrand II, 1992.

Crary, Jonathan. "Seurat's Modernity." In *Seurat at Gravelines: The Last Landscapes*, exh. cat., edited by Ellen Wardwell Lee, 61–65. Indianapolis: Indianapolis Museum of Art, 1990.

———. *Suspensions of Perception: Attention, Spectacle, and Modern Culture*. Cambridge: MIT Press, 1999.

Cugini, Carla. *"Er sieht einen Fleck, er malt einen Fleck": Physiologische Optik, Impressionismus und Kunstkritik*. Basel: Schwabe, 2006.

Czarnocka, Anna. "Nietzsche, Przybyszewski, and the Berlin *Bohême* from the Circle of the *Kneipe 'Zum schwarzen Ferkel*.'" In *Totenmesse: Modernism in the Culture of Northern and Central Europe*, edited by Piotr Paszkiewicz and Roman Taborski, 41–50. Warsaw: Institute of Art, Polish Academy of Sciences, 1996.

Dahlkvist, Tobias. "*By the Open Sea*—A Decadent Novel? Reconsidering Relationships Between Nietzsche, Strindberg, and *Fin-de-Siècle* Culture." In *The International Strindberg: New Critical Essays*, edited by Anna Westerståhl Stenport, 195–214. Evanston: Northwestern University Press, 2012.

Daston, Lorraine, and Peter Galison. *Objectivity*. New York: Zone Books, 2007.

Daston, Lorraine, and Elizabeth Lunbeck, eds. *Histories of Scientific Observation*. Chicago: University of Chicago Press, 2011.

Daston, Lorraine, and Katharine Park. *Wonders and the Order of Nature, 1150–1750*. New York: Zone Books, 1998.

Datta, Venita. *Birth of a National Icon: The Literary Avant-Garde and the Origins of the Intellectual in France*. Albany: State University of New York Press, 1999.

———. "The Dreyfus Affair and Anti-Semitism: Jewish Identity at *La revue blanche*." *Historical Reflections/Réflexions historiques* 21 (Winter 1995): 113–29.

———. "*La revue blanche*: Un milieu parisien fin-de-siècle." *French Cultural Studies* 9 (June 1998): 147–65.

Dauberville, Jean, and Henry Dauberville. *Bonnard: Catalogue raisonné de l'oeuvre peint*. 4 vols. Paris: Bernheim-Jeune, 1966–74.

de Brahm, Alcanter. "Un côté inconnu de l'art scandinave—Edward Munch." *La revue de l'Est* (15 February 1895): 828–29.

———. "Salon des Indépendants." *La critique* 2 (20 April 1896): 64.

Décaudin, Michel. "Le *Mercure de France*: Filiations et orientations." *Revue d'histoire littéraire de la France* 92 (January–February 1992): 7–15.

Delay, Jean. "Jacksonism and the Works of Ribot." *Archives of Neurology and Psychiatry* 78 (November 1957): 505–15.

Deleuze, Gilles. *Francis Bacon: The Logic of Sensation*. Translated by Daniel W. Smith. London: Continuum, 2003.

Delille, Damien. "Le troisième genre: Androgynie et trouble de la masculinité dans les arts visuels en France au passage du XXe siècle." PhD diss., Université Paris I Panthéon-Sorbonne, 2015.

Delius, Frederick. *Delius: A Life in Letters.* Edited by Lionel Carley. 2 vols. Cambridge: Harvard University Press, 1983–88.

Denis, Maurice [Pierre Louis, pseud.]. "À M. Alphonse Germain." *Art et critique*, no. 73 (18 October 1890): 667–68.

———. "À propos de l'exposition d'A. Séguin, chez Le Barc de Boutteville." *La plume*, no. 141 (1 March 1895): 118–19.

———. "L'aventure posthume de Cézanne." *Prométhée*, no. 6 (July 1939): 193–96.

———. "Cézanne." *L'Occident*, no. 70 (September 1907): 118–33.

———. "Cézanne—I." Translated by Roger Fry. *Burlington Magazine* 16 (January 1910): 207–19.

———. "Cézanne—II." Translated by Roger Fry. *Burlington Magazine* 16 (February 1910): 275–80.

———. *Le ciel et l'Arcadie.* Edited by Jean-Paul Bouillon. Paris: Hermann, 1993.

———. "Définition du néo-traditionnisme." *Art et critique*, no. 65 (23 August 1890): 540–42; 66 (30 August 1890): 556–58. Reprinted in *Le ciel et l'Arcadie*, 5–21.

———. *Journal.* 3 vols. Paris: La Colombe, 1957–59.

———. *Nouvelles théories sur l'art moderne, sur l'art sacré, 1914–1921.* Paris: Rouart et Watelin, 1922.

———. "Pierre Bonnard." *Le point* 4, no. 24 (1943): 4–5.

———. Preface. In *L'École de Pont-Aven et les Nabis, 1888–1908.* Paris: Galerie Parvillée, 1943. Exhibition catalogue.

———. *Théories, 1890–1910: Du symbolisme et de Gauguin vers un nouvel ordre classique.* Paris: Bibliothèque de l'Occident, 1912.

de Pontmartin, Armand. "M. Émile Zola: Le roman expérimental." *Gazette de France*, 7 November 1880. Reprinted in de Pontmartin, *Souvenirs d'un vieux critique: Première série*, 149–64. Paris: Calmann-Lévy, 1881.

Deryng, Xavier. "Les Variations Popoffsky (Strindberg—Réja—Biegas)." *Artibus et historae* 19, no. 37 (1998): 103–23, 202.

Dessy, Clément. *Les écrivains et les Nabis: La literature au défi de la peinture.* Rennes: Presses Universitaires de Rennes, 2015.

Dévoluy, Pierre. "Lettre ouverte de Pierre Dévoluy à Jules Bois." *Le chimère*, no. 9 (April 1892): 155–56.

Diderot, Denis. *De l'interprétation de la nature.* 1754. In Diderot, *Oeuvres philosophiques*, edited by Paul Vernière, 165–244. Paris: Garnier, 1956.

Diderot, Denis, and Jean Le Rond d'Alembert. *Encyclopédie ou Dictionnaire raisonné des sciences, des arts et des métiers, par une Société de gens de lettres.* 17 vols. Paris: Briasson, 1751–65. Now available at the ARTFL Encyclopédie Project of the University of Chicago, edited by Robert Morrissey. http://encyclopedie.uchicago.edu.

Dierig, Sven. "Con Sordino for Piano and Brain: Bohemian Neuroscience in a 1900 Cultural Metropolis." *Configurations* 9, no. 3 (2001): 413–40.

Dierig, Sven, and Dennis Meyhoff Brink. "Beyond the Temples of Science: Bohemian Neuroscience in Fin-de-siècle Berlin." In *The Virtual Laboratory* (2006). Accessed 19 May 2015. http://vlp.mpiwg-berlin.de/essays/data/art5/art44.

Dombrowski, André. *Cézanne, Murder, and Modern Life.* Berkeley: University of California Press, 2013.

Doran, P. M., ed. *Conversations avec Cézanne.* Paris: Macula, 1978.

Dorra, Henri. "Munch, Gauguin, and the Norwegian Painters." *Gazette des beaux-arts* 88 (November 1976): 175–79.

Dostoevsky, Fyodor. *The Gambler.* Edited by Gary Saul Morson. Translated by Constance Garnett. New York: Modern Library, 2003.

Druick, Douglas W., and Peter Zegers. "Scientific Realism: 1873–1881." In *Degas*, exh. cat., edited by Jean Sutherland Boggs, 197–211. New York: Metropolitan Museum of Art, 1988.

Druick, Douglas W., Peter Zegers, Britt Salvesen, Kristen Hoermann Lister, and Mary C. Weaver. *Van Gogh and Gauguin: The Studio of the South.* Chicago: Art Institute of Chicago, 2001. Exhibition catalogue.

D'Souza, Aruna. "Paul Cézanne, Claude Lantier, and Artistic Impotence." *Nineteenth-Century Art Worldwide* 3 (Autumn 2004). Accessed 18

April 2016. http://www.19thc-artworldwide.org/autumn04/67-autumn04/autumn04article/295-paul-cezanne-claude-lantier-and-artistic-impotence.
Duncan, Carol. "The MoMA's Hot Mamas." In "Images of Rule: Issues of Interpretation," edited by Thomas DaCosta Kaufmann, special issue, *Art Journal* 48 (Summer 1989): 171–78.
Duranty, Edmond [L. Présurier, pseud.]. "Idées de Constable, peintre anglais." *Réalisme*, no. 2 (15 December 1856): 29–30.
———. "Sur l'outillage dans l'art." *L'artiste*, 1 July 1870, 11.
Easton, Elizabeth W. "Édouard Vuillard's Interiors of the 1890s." PhD diss., Yale University, 1989.
———. *The Intimate Interiors of Édouard Vuillard*. Houston: Museum of Fine Arts, 1989. Exhibition catalogue.
———, ed. *Snapshot: Painters and Photography, Bonnard to Vuillard*. Amsterdam: Van Gogh Museum, 2011. Exhibition catalogue.
Eggum, Arne, and Rodolphe Rapetti, eds. *Munch et la France*. Paris: Musée d'Orsay, 1991. Exhibition catalogue.
Ellenberger, Henri F. *The Discovery of the Unconscious: The History and Evolution of Dynamic Psychiatry*. New York: Basic Books, 1970.
Endresen, Signe. "Serial Experiments: Close-Readings of Edvard Munch's *Det grønne værelset* (1907)." PhD diss., University of Oslo, 2015.
Evers, Franz. "Psychischer Naturalismus." *Sphinx* 18 (April 1894): 319–21.
"Experimente übersinnlicher Gedanken-Übertragung in London." *Sphinx* 1 (January 1886): 34–41.
Faber, D. P. "Théodule Ribot and the Reception of Evolutionary Ideas in France." *History of Psychiatry* 8 (December 1997): 445–58.
Facos, Michelle. *Symbolist Art in Context*. Berkeley: University of California Press, 2009.
Facos, Michelle, and Thor J. Mednick, eds. *The Symbolist Roots of Modern Art*. Burlington, Vt.: Ashgate, 2015.
Fahlgren, Margaretha. "Strindberg and the Woman Question." In *The Cambridge Companion to August Strindberg*, edited by Michael Robinson, 20–33. Cambridge: Cambridge University Press, 2009.
Falcone, Francesca. "The Arabesque: A Compositional Design." *Dance Chronicle* 19, no. 3 (1996): 231–53.
———. "The Evolution of the Arabesque in Dance." *Dance Chronicle* 22, no. 1 (1999): 71–117.
Fauvel, Aude. "Aliénistes contre psychiatres: La médecine mentale en crise (1890–1914)." *Psychologie clinique*, no. 17 (Summer 2004): 61–76.
———. "Les mots des sciences de l'homme: Psychiatrie." *Bulletin de la Société française pour l'histoire des sciences de l'homme*, no. 29 (Spring–Summer 2006): 43–51.
Fay, Caroline Marie. "Stories of the Sleeping Body: Literary, Scientific, and Philosophical Narratives of Sleep in Nineteenth-Century France." PhD diss., University of Virginia, 2002.
Ferdas, René. *La physiologie expérimentale et le "Roman expérimental": Claude Bernard et Monsieur Zola*. Paris: Hurtau, 1881.
Feuk, Douglas. *August Strindberg: Paradisbilder, Infernomåleri*. Hellerup: Bløndel, 1991.
Flaubert, Gustave. *Correspondance, 1877–1880*. Vol. 16 of Flaubert, *Oeuvres complètes*. Edited by the Société des études littéraires françaises. Paris: Club de l'Honnête Homme, 1976.
Flournoy, Théodore. *Des Indes à la planète Mars: Étude sur un cas de somnambulisme avec glossolalie*. 3rd ed. Paris: Alcan, 1900.
Flugsrud, Sverre. "En dåres försvarstal: Et Strindberg-manuskript på vandring." *Nordisk Tidskrift* 50, no. 3 (1974): 125–37.
Foa, Michelle. *Georges Seurat: The Art of Vision*. New Haven: Yale University Press, 2015.
———. "On the Spaces of Painting and Perception: Seurat and Helmholtz." In *Georges Seurat: Figure in Space*, exh. cat., edited by Christoph Becker, 112–23. Ostfildern: Hatje Cantz, 2009.
Foucault, Michel. *The Archaeology of Knowledge*. Translated by A. M. Sheridan Smith. Reprint, London: Routledge, 1989.
Frangne, Pierre-Henry. *La négation à l'oeuvre: La philosophie symboliste de l'art (1860–1905)*. Rennes: Presses Universitaires de Rennes, 2005.
Frank, Hilmar. "Arabesque, Cipher, Hieroglyph: Between Unending Interpretation and Loss of Meaning." In *The Romantic Spirit in German Art*,

1790–1990, exh. cat., edited by Keith Hartley, 147–54. London: Thames and Hudson, 1994.
Frèches-Thory, Claire, and Ursula Perucchi-Petri. *Nabis: 1888–1900*. Paris: Réunion des Musées Nationaux, 1993. Exhibition catalogue.
Frèches-Thory, Claire, and Antoine Terrasse. *The Nabis: Bonnard, Vuillard, and Their Circle*. New York: Harry N. Abrams, 1991.
Fried, Michael. *Manet's Modernism, or The Face of Painting in the 1860s*. Chicago: University of Chicago Press, 1996.
Fry, Roger. *Cézanne: A Study of His Development*. London: L. & V. Woolf, 1927.
———. *A Roger Fry Reader*. Edited by Christopher Reed. Chicago: University of Chicago Press, 1996.
Frydenberg-Flaatten, Hans-Martin. "Edvard Munch's 'Skrik': En studie av maleriets kunstteoretiske og litteraere bakgrunn i perioden 1891–92." Master's thesis, University of Oslo, 2004.
Galison, Peter, and Alexi Assmus. "Artificial Clouds, Real Particles." In *The Uses of Experiment: Studies in the Natural Sciences*, edited by David Gooding, Trevor Pinch, and Simon Schaffer, 225–74. Cambridge: Cambridge University Press, 1989.
Gamboni, Dario. "Confondre avec méthode: Remarques sur l'incertitude chez Vuillard." In *Édouard Vuillard: La porte entrebâilliée*, exh. cat., edited by Jean-Paul Monery and Jörg Zutter, 53–61. Milan: Skira, 2000.
———. "'Dieses Schillern der Eindrücke freute mich . . .': August Strindberg und unabsichtliche Bilder im Paris der 1890er Jahre." In *Die Unvermeidlichkeit der Bilder*, edited by Gerhart von Graevenitz, Stefan Rieger, and Felix Thürlemann, 173–86. Tübingen: Gunter Narr, 2001.
———. "'Fabrication of Accidents': Factura and Chance in Nineteenth-Century Art." In "Factura," edited by Joseph Leo Koerner, special issue, *Res: Anthropology and Aesthetics* 36 (Autumn 1999): 205–25.
———. *Potential Images: Ambiguity and Indeterminacy in Modern Art*. London: Reaktion Books, 2002.
———. "Vuillard and Ambiguity." In Cogeval, *Édouard Vuillard*, exh. cat., 405–22. Washington, D.C.: National Gallery of Art, 2003.
Garb, Tamar. "Cézanne's Late Bathers: Modernism and Sexual Difference." In Garb, *Bodies of Modernity: Figure and Flesh in Fin-de-Siècle France*, 196–219. London: Thames and Hudson, 1998.
———. *Sisters of the Brush: Women's Artistic Culture in Late Nineteenth-Century Paris*. New Haven: Yale University Press, 1994.
Gauguin, Paul. *Lettres de Gauguin à sa femme et à ses amis*. Edited by Maurice Malingue. Paris: Bernard Grasset, 1946.
———. "Notes sur l'art à l'Exposition universelle." *Le moderniste*, no. 11 (4 July 1889): 84–86; no. 12 (13 July 1889): 90–91.
Gavel Adams, Ann-Charlotte. "Strindberg et l'occultisme en France." In *August Strindberg*, edited by Elena Balzamo, 173–85. Paris: L'Herne, 2000.
———. "Strindberg som Ockultismens Zola." *Strindbergiana* 7 (1992): 123–38.
Geelhaar, Christian. "The Painters Who Had the Right Eyes: On the Reception of Cézanne's Bathers." In *Paul Cézanne: The Bathers*, exh. cat., edited by Mary Louise Krumrine, 275–303. London: Thames and Hudson, 1990.
Geffroy, Gustave. *La vie artistique*. 8 vols. Paris: Dentu, 1892–1903.
Gendron, Pierre. *Claude Bernard: Rationalité d'une méthode*. Paris: Vrin, 1992.
Genova, Pamela. *Symbolist Journals: A Culture of Correspondence*. Aldershot: Ashgate, 2001.
Germain, Alphonse. "À l'exposition des Indépendants: Neo-luminaristes et néo-traditionnistes." *Le moniteur des arts*, 27 March 1891, 543–44.
———. "Critique d'art: Sur un tableau refusé; Théorie du symbolisme des teintes." *La plume*, no. 50 (15 May 1891): 171–72.
———. *Du beau moral et du beau formel*. Paris: Girard, 1895.
———. "Du symbolisme dans la peinture." *Art et critique*, no. 58 (5 July 1890): 417–20.
———. "Du temperament peintre." *Art et critique*, no. 70 (27 September 1890): 618–20.

———. "Le modernisme et le beau." *La plume*, no. 46 (15 March 1891): 115–16.

———. *Pour le beau: Essai de kallistique*. Paris: Girard, 1893.

———. "Théorie des déformateurs." *La plume*, no. 57 (1 September 1891): 289–90.

Gessmann, Gustav. "Somnambules Zeichnen." *Sphinx* 6 (August 1888): 100–105.

Gibbs, Walter. "Museum Reopens Without Its 'Scream.'" *New York Times*, 18 June 2005.

Gilman, Sander L. *Difference and Pathology: Stereotypes of Sexuality, Race, and Madness*. Ithaca: Cornell University Press, 1985.

Gisbourne, Mark. "Le spiritisme chez Victor Hugo, Justinus Kerner et quelques autres." In *L'âme au corps: Arts et sciences, 1793–1993*, exh. cat., edited by Jean Clair, 488–99. Paris: Réunion des Musées Nationaux, 1993.

Goldstein, Jan. "The Advent of Psychological Modernism in France: An Alternative Narrative." In *Modernist Impulses in the Human Sciences, 1870–1930*, edited by Dorothy Ross, 190–209. Baltimore: Johns Hopkins University Press, 1994.

———. *Console and Classify: The French Psychiatric Profession in the Nineteenth Century*. 2nd ed. Chicago: University of Chicago Press, 2001.

———. "Neutralizing Freud: The Lycée Philosophy Class and the Problem of the Reception of Psychoanalysis in France." *Critical Inquiry* 40 (Autumn 2013): 40–82.

———. *The Post-Revolutionary Self: Politics and Psyche in France, 1750–1850*. Cambridge, Mass.: Harvard University Press, 2005.

———. "The Uses of Male Hysteria: Medical and Literary Discourse in Nineteenth-Century France." *Representations* 34 (Spring 1991): 134–66.

Goldwater, Robert. *Symbolism*. New York: Harper and Row, 1979.

Gombrich, Ernst. "Experiment and Experience in the Arts." In *Art, Science, and Human Progress: The Richard Bradford Trust Lectures Given Between 1975 and 1978 Under the Auspices of the Royal Institution*, edited by R. B. McConnell, 145–73. London: John Murray, 1983.

Gooding, David, Trevor Pinch, and Simon Schaffer, eds. *The Uses of Experiment: Studies in the Natural Sciences*. Cambridge: Cambridge University Press, 1989.

Gordon, Rae Beth. "From Charcot to Charlot: Unconscious Imitation and Spectatorship in French Cabaret and Early Cinema." In *The Mind of Modernism: Medicine, Psychology, and the Cultural Arts in Europe and America, 1880–1940*, edited by Mark S. Micale, 93–124. Stanford: Stanford University Press, 2004.

Goulet, Andrea. *Optiques: The Science of the Eye and the Birth of Modern French Fiction*. Philadelphia: University of Pennsylvania Press, 2006.

Granath, Ollé, David Campany, Helen Sainsbury, and Göran Söderström. *August Strindberg: Painter, Photographer, Writer*. London: Tate Gallery, 2005. Exhibition catalogue.

Grande Covián, Francisco, and Maurice B. Visscher, eds. *Claude Bernard and Experimental Medicine*. Cambridge, Mass.: Schenkman, 1967.

Greenberg, Clement. *The Collected Essays and Criticism*. Edited by John O'Brian. 4 vols. Chicago: University of Chicago Press, 1986–93.

Grimal, Sophie. "L'intuition du tout et l'objectivité expérimentale: La pensée scientifique d'August Strindberg." *Europe: Revue littéraire mensuelle* 78 (October 2000): 164–77.

Grmek, Mirko D. *Raisonnement expérimental et recherches toxicologiques chez Claude Bernard*. Geneva: Droz, 1973.

Grogin, R. C. *The Bergsonian Controversy in France, 1900–1914*. Calgary: University of Calgary Press, 1988.

Groom, Gloria. *Beyond the Easel: Decorative Painting by Bonnard, Vuillard, Denis, and Roussel, 1890–1930*. Chicago: Art Institute of Chicago, 2001. Exhibition catalogue.

———. *Édouard Vuillard: Painter-Decorator; Patrons and Projects, 1892–1912*. New Haven: Yale University Press, 1993.

Guedj, Aimé. Preface to *Le roman expérimental*, by Émile Zola, 17–44. Paris: Garnier-Flammarion, 1971.

Guieu, Jean-Max. "Les caricatures antidreyfusards de H. Lebourgeois." In *Mimesis et semiosis: Littérature et représentation; Miscellanées offertes à Henri Mitterand*, edited by Philippe Hamon and Jean-Pierre Leduc-Adine, 435–46. Paris: Nathan, 1992.

Guleng, Mai Britt, ed. *eMunch.no—Text and Image.* Oslo: Munch Museum, 2011.

Guleng, Mai Britt, Birgitte Sauge, and Jon-Ove Steihaug, eds. *Edvard Munch: 1863–1944.* Oslo: Munch Museum and National Museum for Art, Architecture, and Design, 2013. Exhibition catalogue.

Hannoosh, Michèle. "The Poet as Art Critic: Laforgue's Aesthetic Theory." *Modern Language Review* 79 (July 1984): 553–69.

Hansen, Uffe. "Strindberg und die dynamische Psychiatrie der 80er Jahre: Überlegungen zur frühen deutschen Strindbergrezeption." In *Fin de Siècle: Zu Naturwissenschaft und Literatur der Jahrhundertwende im deutsch-skandinavischen Kontext*, edited by Klaus Bohnen, Uffe Hansen, and Friedrich Schmöe, 39–61. Copenhagen: Wilhelm Fink, 1984.

Harrington, Anne. "Hysteria, Hypnosis, and the Lure of the Invisible: The Rise of Neo-Mesmerism in *Fin-de-siècle* French Psychiatry." In *The Anatomy of Madness: Essays in the History of Psychiatry*, vol. 3, edited by W. F. Bynum, Roy Porter, and Michael Shepherd, 226–46. London: Routledge, 1988.

Harris, Ruth. "Murder Under Hypnosis in the Case of Gabrielle Bompard: Psychiatry in the Courtroom in Belle Époque Paris." In *The Anatomy of Madness: Essays in the History of Psychiatry*, vol. 2, edited by W. F. Bynum, Roy Porter, and Michael Shepherd, 197–241. London: Tavistock, 1985.

Harrison, Charles. *Painting the Difference: Sex and Spectator in Modern Art.* Chicago: University of Chicago Press, 2005.

Hartmann, Eduard von. *Philosophie des Unbewussten: Versuch einer Weltanschauung.* 3 vols. Berlin: Carl Duncker, 1869.

Hatfield, Gary. "Remaking the Science of the Mind: Psychology as Natural Science." In *Inventing Human Science: Eighteenth-Century Domains*, edited by Christopher Fox, Roy Porter, and Robert Wokler, 184–231. Berkeley: University of California Press, 1995.

Hedström, Per, ed. *August Strindberg: Painter and Photographer.* New Haven: Yale University Press, 2001. Exhibition catalogue.

Heller, Reinhold. "The Art Work as Symbol." In *Fernand Khnopff and the Belgian Avant-Garde*, exh. cat., edited by Cheryl A. Kempler, 20–15. New York: Barry Friedman, 1983.

———. "Concerning Symbolism and the Structure of Surface." In "Symbolist Art and Literature," edited by Sharon L. Hirsh, special issue, *Art Journal* 45 (Summer 1985): 146–53.

———. "'Could Only Have Been Painted by a Madman'—Or Could It?" In *Edvard Munch: The Modern Life of the Soul*, exh. cat., edited by Kynaston McShine, 17–33. New York: Museum of Modern Art, 2006.

———. "Edvard Munch's 'Life Frieze': Its Beginnings and Origins." PhD diss., Indiana University, 1969.

———. "Edvard Munch's Night, the Aesthetics of Decadence, and the Content of Biography." *Arts Magazine* 53 (October 1978): 80–105.

———. "Expressionism's Ancients." In *Parallel Visions: Modern Artists and Outsider Art*, exh. cat., edited by Maurice Tuchman and Carol S. Eliel, 78–93. Los Angeles: Los Angeles County Museum of Art, 1992.

———. "'In Part Myself': Edvard Munch and Vincent van Gogh Compared." In *Munch: Van Gogh*, exh. cat., edited by Magne Bruteig, Maite van Dijk, and Leo Jansen, 62–80. Amsterdam: Van Gogh Museum, 2015.

———. "Making a Painting Scream." In *Edvard Munch, The Scream: Impressionist and Modern Art Evening Sale*, 28–30. New York: Sotheby's, 2012.

———. *Munch: His Life and Work.* Chicago: University of Chicago Press, 1984.

———. "'Das schwarze Ferkel' and the Institution of an Avant-Garde in Berlin, 1892–1895." In *Künstlerischer Austausch/Artistic Exchange*, edited by Thomas W. Gaehtgens, 3:509–19. Berlin: Akademie, 1993.

———. *The Scream.* London: Allen Lane, 1973.

Hennequin, Émile. "Beaux-Arts: Les expositions des arts libéraux et des artistes indépendants." *La revue littéraire et artistique*, 11 March 1882, 155.

———. "L'esthétique de Wagner et la doctrine Spencérienne." *La revue Wagnérienne* 1 (8 November 1882): 282–86.

Higonnet, Anne. "Imaging Gender." In *Art Criticism and Its Institutions in Nineteenth-Century France*, edited by Michael Orwicz, 146–61. Manchester: Manchester University Press, 1994.

Hilton, Alison. "The Exhibition of Experiments in St. Petersburg and the Independent Sketch." *Art Bulletin* 70 (December 1988): 677–98.

———. "*Le messager de l'Europe*: Zola's Art Criticism Beyond Paris." In *Émile Zola and the Arts*, edited by Jean-Max Guieu and Alison Hilton, 61–72. Washington, D.C.: Georgetown University Press, 1988.

Hirsh, Sharon L. *Symbolism and Modern Urban Society*. Cambridge: Cambridge University Press, 2004.

Hirst, Paul Q. *Durkheim, Bernard, and Epistemology*. London: Routledge and Kegan Paul, 1975.

Hofmann, Karl-Ludwig, and Christmut Praeger. "Bilder aus Klecksen: Zu den Klecksographien von Justinus Kerner." In Justinus Kerner, *Justinus Kerner: Nur wenn man von Geistern spricht; Briefe und Klecksographien*, edited by Andrea Berger-Fix, 124–52. Stuttgart: Erdmann in K. Thienemanns Verlag, 1986.

Høifødt, Frank. "Kvinnen, Kunsten, Korset: Edvard Munch Anno 1900." PhD diss., University of Oslo, 1995.

Holzapfel, Amy Strahler. "Strindberg as Vivisector: Physiology, Pathology, and Anti-Mimesis in The Father and Miss Julie." *Modern Drama* 51 (Fall 2008): 329–52.

Homer, William Innes. *Seurat and the Science of Painting*. Cambridge, Mass.: MIT Press, 1964.

Houe, Poul, Sven Hakon Rossel, and Göran Stockenström, eds. *August Strindberg and the Other: New Critical Approaches*. Amsterdam: Rodopi, 2002.

Hübbe-Schleiden, Wilhelm. "Esoterische und exoterische Naturen." *Sphinx* 5 (June 1888): 393–96.

———. "Sphinx, Aufruf und Vorwort." *Sphinx* 1 (January 1886): i–iv.

Hunter, Matthew C. *Wicked Intelligence: Visual Art and the Science of Experiment in Restoration London*. Chicago: University of Chicago Press, 2013.

Huret, Jules. *Enquête sur l'évolution littéraire*. Paris: Charpentier, 1891.

Huysmans, Joris-Karl. *L'art moderne*. Paris: Charpentier, 1883.

———. "Trois peintres: Cézanne, Tissot, Wagner." *La cravache parisienne*, 4 August 1888. Reprinted in Huysmans, *Certains*, 38–43. Paris: Tresse et Stock, 1889.

Isaacson, Joel. "Constable, Duranty, Mallarmé, Impressionism, Plein Air, and Forgetting." *Art Bulletin* 76 (September 1994): 427–50.

———. "Observation and Experiment in the Early Work of Monet." In *Aspects of Monet: A Symposium on the Artist's Life and Times*, edited by John Rewald and Frances Weitzenhoffer, 14–35. New York: Harry N. Abrams, 1984.

Jackson, A. B. *La revue blanche, 1889–1903: Origine, influence, bibliographie*. Paris: Lettres Modernes, 1960.

James, Tony. *Dream, Creativity, and Madness in Nineteenth-Century France*. Oxford: Clarendon Press, 1995.

Janet, Paul. "Une chaire de psychologie expérimentale et comparée au Collège de France." *Revue des deux mondes* 86 (April 1888): 518–49.

Janet, Pierre. *L'automatisme psychologique: Essai de psychologie expérimentale sur les formes inférieures de l'activité humaine*. Paris: Alcan, 1889.

Jaspers, Karl. *Strindberg und van Gogh: Versuch einer pathographischen Analyse unter vergleichender Heranziehung von Swedenborg und Hölderlin*. Munich: R. Piper, 1949.

Jay, Martin. *Songs of Experience: Modern American and European Variations on a Universal Theme*. Berkeley: University of California Press, 2005.

Jirat-Wasiutyński, Vojtěch. "Paul Gauguin's 'Self-Portrait with Halo and Snake': The Artist as Initiate and Magus." In "Mysticism and Occultism in Modern Art," edited by Linda Dalrymple Henderson, special issue, *Art Journal* 46 (Spring 1987): 22–28.

Jourdain, Frantz. "Petites expositions." *La patrie*, 30 May 1896.

Jumeau-Lafond, Jean-David. "Le néo-impressionnisme idéaliste d'Alexandre Séon." In *Du romantisme à l'art déco: Lectures croisées; Mélanges offertes à Jean-Paul Bouillon*, edited by Rosella

Froissart Pezone, Laurent Houssais, and Jean-François Luneau, 63–76. Rennes: Presses Universitaires de Rennes, 2011.

———. "Symbolismes." In *Le mystère et l'éclat: Pastels du Musée d'Orsay*, exh. cat., edited by Serge Lemoine and Guy Cogeval, 128–81. Paris: Musée d'Orsay, 2008.

Junod, Philippe. *Transparence et opacité: Essai sur les fondements théoriques de l'art moderne; Pour une nouvelle lecture de Konrad Fiedler*. 1976. Nîmes: Éditions Jacqueline Chambon, 2004.

Kahn, Gustave. "Chronique de la littérature et de l'art." *La revue indépendante* 8 (August 1888): 302–16.

———. "De l'esthétique du verre polychrome." *La vogue* 1 (18 April 1886): 54–65.

———. "Difficulté de vivre." *Le symboliste*, 15 October 1886, 1.

———. "Réponse des symbolistes." *L'événement*, 28 September 1886.

Kahn, Sholom J. *Science and Aesthetic Judgment: A Study in Taine's Critical Method*. Westport: Greenwood, 1953.

Kalba, Laura Anne. *Color in the Age of Impressionism: Commerce, Technology, and Art*. University Park: Penn State University Press, 2017.

Kardec, Allan. *Spiritisme expérimental: Le livre des médiums, ou Guide des médiums et des évocateurs*. Paris: Didier, Ledoyen, 1861.

Kearns, James. *Symbolist Landscapes: The Place of Painting in the Poetry and Criticism of Mallarmé and His Circle*. London: Modern Humanities Research Association, 1989.

Kennedy, Ian. "Monet's Boulevard des Capucines: Kansas City or Moscow?" *Apollo Magazine* 165 (March 2007): 69–71.

Kerner, Justinus. *Kleksographien*. Stuttgart: Deutsche Verlags-Anstalt, 1890.

Keshavjee, Serena. "'*L'art inconscient*' and '*L'esthétique des esprits*': Science, Spiritualism, and the Imaging of the Unconscious in French Symbolist Art." PhD diss., University of Toronto, 2002.

Krahmer, Catherine, ed. *Eine deutsch-französische Brieffreundschaft: Richard Dehmel–Henri Albert, Briefwechsel, 1893–1898*. Herzberg: Traugott Botz, 1998.

Krämer, Felix. *Das unheimliche Heim: Zur Interieurmalerei um 1900*. Cologne: Böhlau, 2007.

Krisch, Monika. *Die Munch-Affäre—Rehabilitierung der Zeitungskritik: Eine Analyse ästhetischer und kulturpolitischer Beurteilungskriterien in der Kunstberichterstattung der Berliner Tagespresse zu Munchs Ausstellung, 1892*. Berlin: Tenea, 1997.

Kroker, Kenton. *The Sleep of Others and the Transformations of Sleep Research*. Toronto: University of Toronto Press, 2007.

Kromm, Jane. *The Art of Frenzy: Public Madness in the Visual Culture of Europe, 1500–1850*. London: Continuum, 2002.

Kuenzli, Katherine M. "Aesthetics and Cultural Politics in the Age of Dreyfus: Maurice Denis's *Homage to Cézanne*." *Art History* 30 (November 2007): 683–711.

———. *The Nabis and Intimate Modernism: Painting and the Decorative at the Fin de Siècle*. Burlington, Vt.: Ashgate, 2010.

Lachapelle, Sophie. *Investigating the Supernatural: From Spiritism and Occultism to Psychical Research and Metapsychics in France, 1853–1931*. Baltimore: Johns Hopkins University Press, 2011.

Laforgue, Jules. *Oeuvres complètes*. 3 vols. Edited by David Arkell, Maryke de Courten, Jean-Louis Debauve, Mireille Dottin-Orsini, Daniel Grojnowski, Michèle Hannoosh, Pascal Pia, and Pierre-Olivier Walzer. Lausanne: Éditions l'Âge d'Homme, 1986–2000.

Larson, Barbara. "Curing Degeneration: Health and the Neoclassical Body in Early Twentieth-Century France." In *In Sickness and in Health: Disease as Metaphor in Art and Popular Wisdom*, edited by Laurinda S. Dixon and Gabriel P. Weisberg, 166–86. Newark: University of Delaware Press, 2004.

———. *The Dark Side of Nature: Science, Society, and the Fantastic in the Work of Odilon Redon*. University Park: Penn State University Press, 2005.

———. "Microbes and Maladies: Bacteriology and Health at the Fin de siècle." In *Lost Paradise: Symbolist Europe*, exh. cat., edited by Pierre Théberge and Jean Clair, 385–95. Montreal: Montreal Museum of Fine Arts, 1995.

Lathe, Carla Anne. "The Group Zum schwarzen Ferkel: A Study in Early Modernism." PhD diss., University of East Anglia, 1972.
Leclercq, Julien. "Aux Indépendants." *Mercure de France* 2 (May 1891): 298–300.
Lecomte, Georges. "L'estopsychologie [*sic*]." *La cravache parisienne*, no. 396 (15 September 1888): 1–2.
Leeman, Fred. *Émile Bernard (1868–1941)*. Paris: Wildenstein Institute, 2013.
LeGouis, Catherine. *Positivism and Imagination: Scientism and Its Limits in Émile Hennequin, Wilhelm Scherer, and Dmitrii Pisarev*. Lewisburg: Bucknell University Press, 1997.
Leighten, Patricia. *The Liberation of Painting: Modernism and Anarchism in Avant-Guerre Paris*. Chicago: University of Chicago Press, 2013.
Leja, Michael. *Reframing Abstract Expressionism: Subjectivity and Painting in the 1940s*. New Haven: Yale University Press, 1993.
Lemaître, Jules. *Les contemporains: Études et portraits littéraires; Première série*. Paris: H. Lecène et H. Oudin, 1886.
Lemoine, Serge, ed. *Le Théâtre de l'Oeuvre, 1893–1900: Naissance du théâtre moderne*. Paris: Musée d'Orsay, 2005. Exhibition catalogue.
Léouzon Le Duc, Louis Antoine. "Les Nuits du Somnambule (*Sömngångarnätter*) par August Strindberg." *Le monde poétique* 1 (July 1884): 85–109.
Lethbridge, Robert. "L'homme et *L'oeuvre*: Zola et la déformation romanesque." In *Lire, dé-lire Zola*, edited by Jean-Pierre Leduc-Adine and Henri Mitterand, 135–52. Paris: Nouveau Monde, 2004.
Lewison, Jeremy. "Jackson Pollock and the Americanization of Europe." In *Jackson Pollock: New Approaches*, edited by Kirk Varnedoe and Pepe Karmel, 201–31. New York: Museum of Modern Art, 1999.
Lindström, Hans. *Strindberg och böckerna*. 2 vols. Uppsala: Svenska Litteratursällskapet, 1977–90.
Lindtke, Gustav, ed. *Edvard Munch—Dr. Max Linde, Briefwechsel, 1902–28*. Lübeck: Senat der Hansestadt Lübeck, Amt für Kultur, 1974.
Loran, Erle. *Cézanne's Composition: Analysis of His Form with Diagrams and Photographs of His Motifs*. 2nd ed. Berkeley: University of California Press, 1950.
Lucbert, Françoise. *Entre le voir et le dire: La critique d'art des écrivains dans la presse symboliste en France de 1882 à 1906*. Rennes: Presses Universitaires de Rennes, 2005.
Lucbert, Françoise, and Richard Shyrock, eds. *Gustave Kahn: Un écrivain engagé*. Rennes: Presses Universitaires de Rennes, 2013.
Lugné-Poë, Aurélien. *Le sot du tremplin: Souvenirs et impressions de théâtre*. Paris: Gallimard, 1931.
Lukes, Steven. *Émile Durkheim: His Life and Work; A Historical and Critical Study*. Rev. ed. London: Penguin, 1992.
Luthi, Jean-Jacques. *Émile Bernard: Catalogue raisonné de l'oeuvre peint*. Paris: Éditions SIDE, 1982.
Lüttichau, Mario-Andreas von. "'Crazy at Any Price': The Pathologizing of Modernism in the Run-Up to the 'Entartete Kunst' Exhibition in Munich in 1937." In *Degenerate Art: The Attack on Modern Art in Nazi Germany, 1937*, exh. cat., edited by Olaf Peters, 26–51. New York: Neue Galerie, 2014.
MacGregor, John M. *The Discovery of the Art of the Insane*. Princeton: Princeton University Press, 1989.
Madsen, Børge Gedsø. "The Impact of the French Naturalists and Psychologists on August Strindberg's Plays of the 1880s and 1890s." PhD diss., University of Minnesota, 1958.
Marcussen, Marianne. "Duranty et les impressionnistes II." *Hafnia: Copenhagen Papers in the History of Art*, no. 6 (1979): 27–49.
Marcussen, Marianne, and Hilde Olrik. "Le réel chez Zola et les peintres impressionnistes: Perception and représentation." *Revue d'histoire littéraire de la France* 80 (November–December 1980): 965–77.
Marie, Auguste. "Le Musée de la folie." *Je sais tout* 1 (15 October 1905): 353–60.
Marlais, Michael A. *Conservative Echoes in Fin-de-Siècle Parisian Art Criticism*. University Park: Pennsylvania State University Press, 1992.
Martelli, Diego. *Scritti d'arte*. Edited by Antonio Boschetto. Florence: Sansoni, 1952.
Marx, Roger. "Les petits salons—Les Indépendants: À la galerie des artistes modernes." *Le Voltaire*, 23 April 1891, 2.

Mathews, Patricia. *Passionate Discontent: Creativity, Gender, and French Symbolist Art*. Chicago: University of Chicago Press, 1999.
Mathieu, Pierre-Louis. *La génération symboliste, 1870–1910*. Geneva: Skira, 1990.
Mauner, George L. *The Nabis: Their History and Their Art, 1888–1896*. New York: Garland, 1978.
McDermott, Margaret Milam, and Richard R. Brettell. *Reflections: The Collection of Margaret and Eugene McDermott*. Houston: Bright Sky Press, 2012.
McShine, Kynaston, ed. *Edvard Munch: The Modern Life of the Soul*. New York: Museum of Modern Art, 2006. Exhibition catalogue.
Méheust, Bertrand. *Somnambulisme et médiumnité*. Vol. 2, *Le choc des sciences psychiques*. Le Plessis-Robinson: Institut Synthélabo, 1999.
Melot, Michel. *The Impressionist Print*. Translated by Caroline Beamish. New Haven: Yale University Press, 1997.
Méneux, Catherine. "Les enquêtes artistiques: Vers une nouvelle forme de critique (1894–1905)." *Liame*, no. 22 (2012): 15–34.
Merleau-Ponty, Maurice. "Cézanne's Doubt." In Merleau-Ponty, *Sense and Non-Sense*, trans. Hubert L. Dreyfus and Patricia Allen Dreyfus, 9–25. Evanston: Northwestern University Press, 1964.
Meyer, Michael. *Strindberg: A Biography*. New York: Random House, 1985.
———. "Young Strindberg." *New York Review of Books*, 7 July 1966.
Michaud, Stéphane, and Michèle Le Pavec, eds. *Taine au carrefour des cultures du XIXe siècle*. Paris: Bibliothèque Nationale de France, 1996.
Michel, Jacques, ed. *La nécessité de Claude Bernard*. Paris: Méridiens Klincksieck, 1991.
Millroth, Thomas. *Molards Salong*. Stockholm: Forum, 1993.
Milochevitch, Nikolas. *Nietzsche et Strindberg: Psychologie de la connaissance*. Translated by Zorica Hadji-Vidoïkovitch. Lausanne: Éditions de L'Âge d'Homme, 1997.
Mitterand, Henri. "Textes en intersection: *Le roman expérimental* et *Les Rougon-Macquart*." *Revue de l'Université d'Ottawa* 48, no. 4 (1978): 415–28.
———. "Zola et le réalisme artiste." *Quarante-huit/Quatorze*, no. 1 (1989): 69–77.
Moffett, Charles S., ed. *The New Painting: Impressionism, 1874–1886*. San Francisco: Fine Arts Museums of San Francisco, 1986. Exhibition catalogue.
Monroe, John Warne. *Laboratories of Faith: Mesmerism, Spiritism, and Occultism in Modern France*. Ithaca: Cornell University Press, 2008.
Moréas, Jean. "Le symbolisme." *Le Figaro*, 18 September 1886. Reprinted in *Les premières armes du symbolisme*, edited by Léon Vanier, 31–39. Paris: Léon Vanier, 1889.
Morehead, Allison. "'Are There Bacteria in the Rooms of Monte Carlo?': The Roulette Paintings, 1891–93." In *Munch Becoming "Munch": Artistic Strategies, 1880–1892*, exh. cat., edited by Ingebjørg Ydstie and Mai Britt Guleng, 120–36. Oslo: Munch Museum, 2008.
———. "The *Musée de la folie*: Collecting and Exhibiting *chez les fous*." *Journal of the History of Collections* 23, no. 1 (2011): 101–26.
———. "Symbolism, Mediumship, and the 'Study of the Soul That Has Constituted Itself as a Positivist Science.'" In "Science, Symbolism, and Fin-de-Siècle Visual Culture," edited by Serena Keshavjee, special issue, *RACAR—Revue d'art canadienne/Canadian Art Review* 34 (Spring 2009): 77–85.
Morehead, Allison, and Elizabeth Otto. "Representation in the Age of Mediumistic Reproduction, from Symbolism to the Bauhaus." In *The Symbolist Roots of Modernist Art*, edited by Michelle Facos and Thor J. Mednick, 155–68. Burlington, Vt.: Ashgate, 2015.
Morel, Pierre. *Dictionnaire biographique de la psychiatrie*. Paris: Synthélabo, 1996.
Morice, Charles. *Paul Gauguin*. Paris: Floury, 1919.
Mornand, Pierre. *Vingt-deux artistes du livre*. Paris: Le Courrier Graphique, 1948.
Morowitz, Laura. "Anonymity, Artistic Brotherhoods, and the Art Market in the *Fin de siècle*." In *Artistic Brotherhoods in the Nineteenth Century*, edited by Laura Morowitz and William Vaughan, 185–96. Burlington, Vt.: Ashgate, 2000.
Mørstad, Erik, ed. *Edvard Munch: An Anthology*. Oslo: Oslo Academic Press, 2006.

———. "Edvard Munch under Medusas hode." *En Face: Kunsthistorisk Tiddskrift*, no. 1 (2003): 5–17.

———. "Il linguaggio formale di Edvard Munch: Formule e caricature." In *Munch, 1863–1944*, exh. cat., edited by Øivind Storm Bjerke and Achille Bonito Oliva, 41–56. Rome: Complesso del Vittoriano, 2005.

———. "Responding to *Self-Portrait with Cigarette*: A Case History." In *Edvard Munch: An Anthology*, edited by Erik Mørstad, 107–16. Oslo: Oslo Academic Press, 2006.

Morton, Mary. "Naturalism and Nostalgia: Hippolyte Taine's Lectures on Art History at the École des Beaux-Arts, 1865–1869." PhD diss., Brown University, 1998.

Mucchielli, Laurent. "Origins of Psychology as an Academic Discipline at French Universities (1870–1900): Intellectual Issues, the Political Context, Networks, and Strategies of Alliance Centering Around Théodule Ribot's *Revue philosophique*." *Annals of Science* 55, no. 3 (1998): 263–89.

Müller-Westermann, Iris. *Munch by Himself*. London: Royal Academy of Arts, 2005. Exhibition catalogue.

Munch, Edvard. *Edvard Munch's Writings, Digital Archive*. Oslo: Munch Museum, n.d. Accessed 14 January 2014. http://www.emunch.no/.

Nanteuil, Georges. "Trois expositions à l'Art nouveau: J. L. Forain, L'Exposition canine." *Le soir*, 20 May 1896.

Natanson, Thadée. "Correspondance de Kristiania: M. Edvard Munch." *La revue blanche* 9 (15 November 1895): 477–78.

Nelson, Elizabeth. "Timeknots: Science and Reform at a French Asylum, Circa 1900." PhD diss., Indiana University, 2015.

Newman, Sasha M. "Bonnard, 1900–1920." *Cahiers du Musée national d'art moderne* 13 (July 1984): 96–105.

Nicolas, Serge. "L'introduction de l'enseignement de la psychologie scientifique en France: Théodule Ribot (1839–1916) à la Sorbonne (1885)." *Année psychologique* 100, no. 2 (2000): 285–331.

Nicolas, Serge, and Agnès Charvillat. "Introducing Psychology as an Academic Discipline in France: Théodule Ribot and the Collège de France (1888–1901)." *Journal of the History of the Behavioral Sciences* 37 (Spring 2001): 143–64.

Niess, Robert. *Zola, Cézanne, and Manet: A Study of "L'Oeuvre."* Ann Arbor: University of Michigan Press, 1968.

Nietzsche, Friedrich. *The Will to Power*. Translated by Walter Kaufmann and R. J. Hollingdale. New York: Vintage Books, 1968.

Nordmann, Jean-Thomas. *Taine et la critique scientifique*. Paris: Presses Universitaires de France, 1992.

Nye, Robert. *Crime, Madness, and Politics in Modern France: The Medical Concept of National Decline*. Princeton: Princeton University Press, 1984.

Ockman, Carol. *Ingres's Eroticized Bodies: Retracing the Serpentine Line*. New Haven: Yale University Press, 1995.

Offenstadt, Patrick. *Jean Béraud, 1849–1935: The Belle Époque; A Dream of Times Gone By; Catalogue Raisonné*. Cologne: Taschen, 1999.

Olmsted, J. M. D., and E. Harris Olmsted. *Claude Bernard and the Experimental Method in Medicine*. New York: Henry Schuman, 1952.

Olsson, Ulf. "Going Crazy: Strindberg and the Construction of Literary Madness." In *August Strindberg and the Other: New Critical Approaches*, edited by Poul Houe, Sven Hakon Rossel, and Göran Stockenström, 115–31. Amsterdam: Rodopi, 2002.

———. *Jag blir galen: Strindberg, vansinnet och vetenskapen*. Stockholm: Brutus Östlings Bokförlag Symposion, 2002.

———. "Simulacres: L'histoire de la folie chez Strindberg." In *August Strindberg*, edited by Elena Balzamo, 237–50. Paris: L'Herne, 2000.

Pagès, Alain. "En partant de la théorie du roman expérimental." *Les cahiers naturalistes* 47 (1974): 70–87.

Park, Katharine, and Lorraine J. Daston. "Unnatural Conceptions: The Study of Monsters in Sixteenth- and Seventeenth-Century France and England." *Past & Present* 92 (August 1981): 20–54.

Pastor, Willy. "Edvard Munch." *Frankfurter Zeitung*, 24 January 1894.
Paul, Adolf. *Min Strindbergsbok: Strindbergsminnen och Brev*. Stockholm: Norstedt, 1930.
Paul, Harry W. "The Debate over the Bankruptcy of Science in 1895." *French Historical Studies* 5 (Spring 1968): 299–327.
Péladan, Joséphin. "Un idéalisme expérimental: La philosophie de Léonard de Vinci d'après ses manuscrits." *Mercure de France* 71 (January 1908): 193–214.
Perucchi-Petri, Ursula. "Au lit, im Bett: Zur Entwicklung von Vuillards Frühwerk." In *Édouard Vuillard*, exh. cat., edited by Holger Jacob-Friesen and Astrid Reuter, 62–69. Heidelberg: Kehrer Verlag, 2008.
———. "Les Nabis et le Japon." In *Japonisme in Art: An International Symposium*, edited by Yamada Chisaburo and Omori Tatsuji, 261–77. Tokyo: Committee for the Year 2001, 1980.
Pettersen, Petra. "Berlin 1892—An Attempt at a Reconstruction." In *Munch Becoming "Munch": Artistic Strategies, 1880–1892*, exh. cat., edited by Ingebjørg Ydstie and Mai Britt Guleng, 173–97. Oslo: Munch Museum, 2008.
Pick, Daniel. *Faces of Degeneration: A European Disorder, c. 1848–c. 1918*. Cambridge: Cambridge University Press, 1989.
Pissarro, Camille. *Correspondance de Camille Pissarro*. Edited by Janine Bailly-Herzberg. 5 vols. Paris: Presses Universitaires de France, 1980–91.
Plas, Régine. "De la vivisection morale et intellectuelle: L'hypnotisme comme moyen d'investigation psychologique en France au XIXe siècle." *Phoenix: Revue de nouvelle hypnose et d'hypnothérapie Ericksonienne* 11–12 (1991): 41–44.
Poe, Edgar Allan. *Nouvelles histoires extraordinaires*. Translated by Charles Baudelaire. Paris: M. Lévy Frères, 1857.
Poggioli, Renato. *The Theory of the Avant-Garde*. Cambridge, Mass.: Belknap Press of Harvard University Press, 1968.
Pollock, Griselda. *Avant-Garde Gambits, 1888–1893: Gender and the Colour of Art History*. London: Thames and Hudson, 1992.
Praed, Rosa. "The Casino at Monte Carlo." *Graphic*, 8 November 1890, 525.
Przybyszewski, Stanisław. "Psychischer Naturalismus." *Neue Deutsche Rundschau*, February 1894, 150–56.
———, ed. *Das Werk des Edvard Munch*. Berlin: S. Fischer, 1894.
Puech, Jean-Louis. "Zola et l'Académie." *L'éclair*, 30–31 May 1894.
Rambosson, Yvanhoé. "Le Salon des Indépendants." *La plume*, no. 194 (15 May 1897): 311–13.
———. "3ème exposition des peintres impressionnistes et symbolistes." *La plume*, no. 88 (15 December 1892): 531–32.
Rapetti, Rodolphe. "Munch face à la critique française." In *Munch et la France*, exh. cat., edited by Arne Eggum and Rodolphe Rapetti, 16–31. Paris: Musée d'Orsay, 1991.
———. "Pour une définition du symbolisme pictural: Critique et théories à la fin du XIXe siècle." *Quarante-huit/Quatorze*, no. 5 (1993): 69–82.
———. *Le symbolisme*. Paris: Flammarion, 2005.
Rapetti, Rodolphe, Richard Thomson, Frances Fowle, Ana-Maria von Bonsdorff, and Nienke Bakker. *Dreams of Nature: Symbolism from Van Gogh to Kandinsky*. Amsterdam: Van Gogh Museum, 2012. Exhibition catalogue.
Reith, Gerda. *The Age of Chance: Gambling in Western Culture*. London: Routledge, 1999.
Réja, Marcel. *L'art chez les fous*. Paris: Mercure de France, 1907.
———. "Symbolisme pictural: H. Héran—E. Munch—O. Redon." *La critique* 6 (20 January 1900): 9–11.
Renan, Ernest. *L'avenir de la science: Pensées de 1848*. Paris: Calmann-Lévy, 1890.
Retté, Adolphe. "Septième exposition des artistes indépendants: Notes cursives." *L'Ermitage*, May 1891, 293–301.
———. *Le symbolisme: Anecdotes et souvenirs*. Paris: Vanier, 1903.
Reutersvärd, Oscar. "The 'Violettomania' of the Impressionists." *Journal of Aesthetics and Art Criticism* 9 (December 1950): 106–10.
Rewald, John. *Cézanne, sa vie, son oeuvre, son amitié pour Zola*. Paris: Albin Michel, 1939.

———. *The Paintings of Paul Cézanne: A Catalogue Raisonné*. 2 vols. New York: Harry N. Abrams, 1996.
Reynolds, Dee. *Symbolist Aesthetics and Early Abstract Art: Sites of Imaginary Space*. Cambridge: Cambridge University Press, 1995.
Ribot, Théodule. *Essai sur l'imagination créatrice*. Paris: Alcan, 1900.
———. *Les maladies de la mémoire*. Paris: Baillière, 1881.
———. *Les maladies de la personnalité*. Paris: Baillière, 1885.
———. *Les maladies de la volonté*. 1883. Paris: L'Harmattan, 2002.
———. *La philosophie de Schopenhauer*. Paris: Baillière, 1874.
———. *La psychologie allemande contemporaine*. Paris: Baillière, 1879.
———. *La psychologie anglaise contemporaine*. 1870. 3rd ed. Paris: Alcan, 1896.
Ritter, William. *Edvard Munch*. Edited by Lasse Jacobsen. Translated by Svein Erik Søland. Afterword by Øivind Storm Bjerke. Oslo: Pax-Artes, 2014.
———. "Un peintre norvégien: M. Edvard Munch." In Ritter, *Études d'art étranger*, 6–122. Paris: Société du Mercure de France, 1906.
Robinson, Michael, ed. *The Cambridge Companion to August Strindberg*. Cambridge: Cambridge University Press, 2009.
———. "Finding a New Language: Strindberg and Symbolism." *Scandinavica* 33 (November 1994): 201–15.
———. *"New Arts, New Worlds!" Strindberg and Painting*. Birmingham, UK: University of Birmingham, Department of Drama and Theatre Arts, 1994.
Rosenblum, Robert. *The Paintings of August Strindberg: The Structure of Chaos*. Hellerup: Bløndel, 1995.
Roslak, Robyn. *Neo-Impressionism and Anarchism in Fin-de-Siècle France: Painting, Politics, and Landscape*. Burlington, Vt.: Ashgate, 2007.
Rubin, William, ed. *Cézanne: The Late Work; Essays*. New York: Museum of Modern Art, 1977. Exhibition catalogue.
Rugg, Linda Haverty. "August Strindberg: The Art and Science of Self-Dramatization." In *The Cambridge Companion to August Strindberg*, edited by Michael Robinson, 3–19. Cambridge: Cambridge University Press, 2009.
Russell, John. *Édouard Vuillard, 1868–1940*. Toronto: Art Gallery of Ontario, 1971. Exhibition catalogue.
Salmon, André. *Souvenirs sans fin, 1903–1940*. Rev. ed. Paris: Gallimard, 2004.
Salomon, Antoine, and Guy Cogeval. *Vuillard: Le regard innombrable; Catalogue critique des peintures et pastels*. 3 vols. Milan: Skira Editore and Wildenstein Institute Publications, 2003.
Salomon, Jacques. *Vuillard: Témoignage de Jacques Salomon*. Paris: Michel, 1945.
Saunier, Charles. "L'exposition des Indépendants." *Le Rhône*, no. 117 (28 April 1891): 3.
Schleich, Carl L. *Erinnerungen an Strindberg: Nebst Nachrufen für Ehrlich und von Bergmann*. Munich: Müller, 1917.
Schmidt, Torsten Måtte, ed. *Strindbergs måleri: En monografi*. Malmö: Allhems Förlag, 1972.
Schuré, Édouard. *Les grands initiés: Esquisse de l'histoire secrète des religions*. Paris: Perrin, 1889.
Sérusier, Paul. *ABC de la peinture, suivi de fragments de lettres et propos sur l'histoire, la théorie et la technique artistique*. Paris: Rumeur des Âges, 1995.
———. *ABC de la peinture, suivi d'une étude sur la vie et l'oeuvre de Paul Sérusier par Maurice Denis*. Paris: Floury, 1942.
Servaes, Franz [W., pseud.]. "Das schwarze Ferkel." *Neues Wiener Tagblatt*, 15 June 1901.
Seys, Pascale. *Hippolyte Taine et l'avènement du naturalisme: Un intellectuel sous le Second Empire*. Paris: L'Harmattan, 1999.
Shapin, Steven, and Simon Schaffer. *Leviathan and the Air-Pump: Hobbes, Boyle, and the Experimental Life*. Princeton: Princeton University Press, 1985.
Sharp, Lynn L. *Secular Spirituality: Reincarnation and Spiritism in Nineteenth-Century France*. Lanham, Md.: Lexington Books, 2006.
Sheon, Aaron. "Caricature and the Physiognomy of the Insane." *Gazette des beaux-arts* 88 (October 1976): 156–60.
Shiff, Richard. *Cézanne and the End of Impressionism: A Study of the Theory, Technique, and Critical Evaluation of Modern Art*. Chicago: University of Chicago Press, 1984.

———. *Doubt*. London: Routledge, 2007.

———. "Dream of Abstraction." In *Paths to Abstraction: 1867–1917*, exh. cat., edited by Terence Maloon, 53–69. Munich: Prestel, 2011.

———. "Morality, Materiality, Apples." In *The World Is an Apple: The Still Lifes of Paul Cézanne*, exh. cat., edited by Benedict Leca, 145–95. Hamilton: Art Gallery of Hamilton, 2014.

Showalter, Elaine. "Critical Cross-Dressing: Male Feminists and the Woman of the Year." *Raritan* 3 (Fall 1983): 130–49.

Sickert, Walter. *The Complete Writings on Art*. Edited by Anna Gruetzner Robins. Oxford: Oxford University Press, 2000.

Sidlauskas, Susan. *Body, Place, and Self in Nineteenth-Century Painting*. Cambridge: Cambridge University Press, 2000.

———. "Contesting Femininity: Vuillard's Family Pictures." *Art Bulletin* 79 (March 1997): 85–111.

Silverman, Debora L. *Art Nouveau in Fin-de-Siècle France: Politics, Psychology, and Style*. Berkeley: University of California Press, 1989.

Simpson, Juliet. *Aurier, Symbolism, and the Visual Arts*. Bern: Peter Lang, 1999.

———. "Bourget, Laforgue, and Impressionism's Inside Story." *French Studies* 55 (October 2001): 467–83.

———. "Science, Belief, and the Art of Subjectivity: From Fromentin's to Huysmans's Modern *Primitifs*." In *Between Light and Darkness: New Perspectives in Symbolism Research*, edited by Marja Lahelma, 16–25. Helsinki: The Birch and the Star, 2014.

Singer, Hans W. *Das graphische Werk des Maler-Radierers Paul Herrmann (Henri Héran)*. Berlin: Oskar Rauthe, 1914.

Skredsvig, Christian. "Edvard Munch i Monte Carlo." In Skredsvig, *Dage og Nætter blandt kunstnere*, 111–19. Kristiania: Gyldendalske Boghandel, 1908.

Smith, Paul. *Seurat and the Avant-Garde*. New Haven: Yale University Press, 1997.

Société de psychologie physiologique. *Congrès international de psychologie physiologique, première session*. Paris: Bureau des Revues, 1890.

Söderström, Göran. *Strindberg och bildkonsten*. Stockholm: Forum, 1972.

———. *Strindbergs Maaleri*. Malmö: Allhems, 1972.

Spencer, Herbert. *First Principles of a New System of Philosophy*. London: Williams and Norgate, 1862.

Sprigge, Elizabeth. *The Strange Life of August Strindberg*. London: H. Hamilton, 1949.

Sprinchorn, Evert. "A Vulgar Style." *New York Review of Books*, 22 September 1966.

———. "'The Zola of the Occult': Strindberg's Experimental Method." *Modern Drama* 17 (September 1974): 250–66.

Stang, Ragna. *Edvard Munch: The Man and His Art*. London: Gordon Fraser, 1979.

Stead, Evanghélia. *Le monstre, le singe et le foetus: Tératogonie et décadence dans l'Europe fin-de-siècle*. Geneva: Droz, 2004.

Stenport, Anna Westerståhl. *Locating August Strindberg's Prose: Modernism, Transnationalism, and Setting*. Toronto: University of Toronto Press, 2010.

Stenport, Anna Westerståhl, and Eszter Szalczer. "Strindberg and Radicalism—Strindberg and the Avant-Garde: A Hundred-Year Legacy." In "Strindberg: A Hundred-Year Legacy," edited by Anna Westerståhl Stenport and Eszter Szalczer, special issue, *Scandinavian Studies* 84 (Fall 2012): 235–47.

Stevens, MaryAnne, ed. *Émile Bernard, 1868–1941: A Pioneer of Modern Art*. Amsterdam: Van Gogh Museum, 1990. Exhibition catalogue.

Strindberg, August. *Brev*. 22 vols. Edited by Torsten Eklund and Martin Lamm. Stockholm: Bonniers, 1948–2001.

———. "Des arts nouveaux! ou Le hasard dans la production artistique." *La revue des revues*, 15 November 1894, 265–70.

———. "L'Exposition d'Edward [*sic*] Munch." *La revue blanche* 10 (1 June 1896): 525–26.

———. *Miss Julie and Other Plays*. Translated by Michael Robinson. Oxford: Oxford University Press, 1998.

———. "Om modern drama och modern teater." In *Strindberg om drama och teater*, edited by Göran Lindström, 31–44. Lund: Uniskol, 1968.

———. *Open Letters to the Intimate Theater*. Translated by Walter Johnson. Seattle: University of Washington Press, 1966.

———. "'Salonen' och den franska konsten 1876: En studie af RM." *Post-och Inrike Ridningar*, 15 June 1876 (part 1); *Dagens Nyheter*, 30 November and 9 December 1876 (part 2).

———. *Samlade Verk*. 71 vols. Stockholm: Almqvist & Wiksell, 1981–2013.

———. *Selected Essays*. Edited and translated by Michael Robinson. Cambridge: Cambridge University Press, 1996.

———. "Sensations détraquées." *Le Figaro supplément littéraire*, 17 November 1894, 26 January 1895, 9 February 1895.

———. *The Son of a Servant: The History of the Evolution of a Human Soul, 1849–67*. Translated by Evert Sprinchorn. London: Jonathan Cape, 1967.

———. *Strindberg's Letters*. 2 vols. Edited and translated by Michael Robinson. Chicago: University of Chicago Press, 1992.

———. "Vivisection: De l'infériorité de la femme." *La revue blanche* 8 (1 January 1895): 1–20.

Strindberg, Freda. *Marriage with Genius*. Edited by Frederic Whyte. London: J. Cape, 1940.

Svenaeus, Gösta. *Edvard Munch: Das Universum der Melancholie*. Lund: Gleerup, 1968.

Swedish Academy. *Ordbok öfver Svenska Språket*. Lund: Gleerup, 1918.

Szalczer, Eszter. "Strindberg and the Visual Arts." *PAJ: A Journal of Performance and Art* 25 (September 2003): 42–50.

Taine, Hippolyte. *De l'intelligence*. 1870. 6th ed. 2 vols. Paris: Hachette, 1892.

———. *Philosophie de l'art*. 1865. 13th ed. 2 vols. Paris: Hachette, 1909.

Taussig, Michael T. *Mimesis and Alterity: A Particular History of the Senses*. New York: Routledge, 1993.

Théberge, Pierre, and Jean Clair, eds. *Lost Paradise: Symbolist Europe*. Montreal: Museum of Fine Arts, 1995. Exhibition catalogue.

Thévoz, Michel. "Marcel Réja, découvreur de 'l'art des fous.'" *Gazette des beaux-arts* 107 (May–June 1986): 200–208.

Thomsen, Ingrid Reed. "Skredsvig og Munch på Rivieraen, 1891–1892." *Kunst og Kultur* 82, no. 3 (1999): 163–79.

Thomson, Belinda. *Vuillard*. Oxford: Phaidon Press, 1988.

Thomson, Richard. *Art of the Actual: Naturalism and Style in Early Third Republic France, 1880–1900*. New Haven: Yale University Press, 2012.

Thomson, Richard, and Belinda Thomson. "Maurice Denis's 'Définition du Néo-traditionnisme' and Anti-Naturalism (1890)." *Burlington Magazine* 154 (April 2012): 260–67.

Thue, Sivert. "From Types on the Promenade to Dissecting the Soul." In *Munch on Paper*, exh. cat., edited by Magne Bruteig and Ute Kuhlemann Falck, 112–27. Oslo: Munch Museum, 2013.

Thurmann-Moe, Jan. "Roßkur und Firnis bei Edvard Munch." In *Das 19. Jahrhundert und die Restaurierung: Beiträge zur Malerei, Maltechnik und Konservierung*, edited by Heinz Althöfer, 112–21. Munich: Callway, 1987.

Topalova-Casadiego, Biljana. "Technical Aspects of Munch's Paintings." In Gerd Woll, *Edvard Munch: Complete Paintings*, 2:425–57. London: Thames and Hudson, 2009.

Torjusen, Bente. "The Mirror." In Arne Eggum, Reinhold Heller, Trygve Nergaard, Robert Rosenblum, Ragna Stang, Bente Torjusen, and Gerd Woll, *Edvard Munch: Symbols and Images*, exh. cat., 185–212. Washington, D.C.: National Gallery of Art, 1978.

Toulouse, Édouard. *Enquête médico-psychologique sur les rapports de la supériorité intellectuelle avec la névropathie: I. Introduction générale, Émile Zola*. Paris: Société d'Éditions Scientifiques, 1896.

Treitel, Corinna. *A Science for the Soul: Occultism and the Genesis of the German Modern*. Baltimore: Johns Hopkins University Press, 2004.

"Übersinnlichen Gedanken-Übertragung nach den Untersuchungen der Society for Psychical Research in London." *Sphinx* 1 (February 1886): 105–29.

Uggla, Andrzej Nils. "Przybyszewski and Strindberg." In *Totenmesse: Modernism in the Culture of Northern and Central Europe*, edited by Piotr Paszkiewicz and Roman Taborski, 201–10. Warsaw: Institute of Art, Polish Academy of Sciences, 1996.

Usselmann, Henri. "Strindberg et l'impressionnisme." *Gazette des beaux-arts* 93 (April 1982): 153–62.

Vaisse, Pierre. *La Troisième République et les peintres.* Paris: Flammarion, 1995.

van Dijk, Maite. "International Artists at the Salon des Indépendants in Paris: The Case of Edvard Munch (1896 and 1897)." In *Foreign Artists and Communities in Paris, 1870–1914: Strangers in Paradise*, edited by Karen L. Carter and Susan Waller, 43–52. Burlington, Vt.: Ashgate, 2015.

van Gogh, Vincent. *Vincent van Gogh: The Letters.* Edited by Leo Jansen, Hans Luijten, and Nienke Bakker. Amsterdam and The Hague: Van Gogh Museum and Huygens ING, 2009. Accessed 6 January 2013. http://vangoghletters.org/.

Varvara, Mario. "La justice de l'étranger." *La plume*, no. 129 (1 September 1894): 356–58.

Vaughan, Gerard. "André Gide and Maurice Denis: *Le Voyage d'Urien.*" *Print Quarterly* 1 (September 1984): 173–86.

Verhaeren, Émile. "Le Salon des Indépendants." *L'art moderne*, no. 14 (5 April 1891): 111–12.

Verne, Maurice. "Une lettre de M. Henry Céard: Zola et le prêt du livre de Claude Bernard; Pourquoi Henry Céard avait incité Zola à lire Claude Bernard." *L'information*, 22 July 1918.

Vernon, Pierrette. *Les carnets de croquis de Pierre Bonnard.* Vol. 1. Neuchâtel: Ides et Calendes, 2007.

Véron, Eugène. *Aesthetics.* Translated by W. H. Armstrong. London: Chapman and Hall, 1879.

———. *L'esthétique.* 1878. Preface by Jacqueline Lichtenstein. Paris: Vrin, 2007.

Vincent, Cécile. "Maurice Denis: L'enfance et la formation; Le mouvement Nabi." PhD diss., Université Paris I, 1972.

Virtanen, Reino. "Claude Bernard and the History of Ideas." In *Claude Bernard and Experimental Medicine*, edited by Francisco Grande Covián and Maurice B. Visscher, 9–23. Cambridge, Mass.: Schenkman, 1967.

A Visitor to the Riviera, ed. *Monte Carlo and Public Opinion.* London: Rivingtons, 1884.

Voorhoeve, Paul E. "Contre Zola à la défense de Claude Bernard." *Neophilologus* 79 (April 1995): 217–95.

Waller, Bret, and Grace Seiberling. *Artists of La revue blanche: Bonnard, Toulouse-Lautrec, Vallotton, Vuillard.* Rochester, N.Y.: Memorial Art Gallery, 1984. Exhibition catalogue.

Ward, Martha. "Camille Pissarro in the 1880s." PhD diss., Johns Hopkins University, 1983.

———. *Pissarro, Neo-Impressionism, and the Spaces of the Avant-Garde.* Chicago: University of Chicago Press, 1996.

Wellek, René, ed. *Dostoevsky: A Collection of Critical Essays.* Englewood Cliffs, N.J.: Prentice Hall, 1962.

Wharton, Edith. *Edith Wharton: Lettres à l'ami français.* Edited by Claudine Lesage. Paris: M. Houdiard, 2001.

Whitfield, Sarah, and John Elderfield. *Bonnard.* New York: Harry N. Abrams, 1998. Exhibition catalogue.

Williams, Raymond. *Keywords: A Vocabulary of Culture and Society.* Rev. ed. London: Fontana Paperbacks, 1983.

Wolffram, Heather. *The Stepchildren of Science: Psychical Research and Parapsychology in Germany, c. 1870–1939.* Amsterdam: Rodopi, 2009.

Woll, Gerd. *Edvard Munch: Complete Paintings.* 4 vols. London: Thames and Hudson, 2009.

———. *Edvard Munch: The Complete Graphic Works.* 2nd ed. London: Philip Wilson, 2012.

———. "Le graveur." In *Munch et la France*, exh. cat., edited by Arne Eggum and Rodolphe Rapetti, 240–75. Paris: Musée d'Orsay, 1991.

———. "The Tree of Knowledge of Good and Evil." In Arne Eggum, Reinhold Heller, Trygve Nergaard, Robert Rosenblum, Ragna Stang, Bente Torjusen, and Gerd Woll, *Edvard Munch: Symbols and Images*, exh. cat., 229–47. Washington, D.C.: National Gallery of Art, 1978.

Woźniakowski, Jacek. "De l'arabesque romantique." *Polish Art Studies* 3 (1982): 57–69.

Wright, Alastair. *Matisse and the Subject of Modernism.* Princeton: Princeton University Press, 2004.

Yarborough, Bessie Rainsford. "Exhibition Strategies and Wartime Politics in the Art and Career of Edvard Munch, 1914–1921." PhD diss., University of Chicago, 1995.

Ydstie, Ingebjørg, ed. *Edvard Munch's Laboratory: The Path to the Aula.* Oslo: Munch Museum, 2011. Exhibition catalogue.

———. "Edvard Munch's Painting *Inheritance—Madonna Inside Out*." In *eMunch.no—Text and Image*, edited by Mai Britt Guleng, 201–17. Oslo: Munch Museum, 2011.

Ydstie, Ingebjørg, and Mai Britt Guleng, eds. *Munch Becoming "Munch": Artistic Strategies, 1880–1892*. Oslo: Munch Museum, 2008. Exhibition catalogue.

Young, Marnin. "The Death of Georges Seurat: Neo-Impressionism and the Fate of the Avant-Garde in 1891." In "New Directions in Neo-Impressionism," edited by Tania Woloshyn and Anne Dymond, special issue, *RIHA Journal* (July 2012). Accessed 29 August 2012. http://www.riha-journal.org/articles/2012/2012-jul-sep/special-issue-neo-impressionism/young-death-of-seurat.

———. *Realism in the Age of Impressionism: Painting and the Politics of Time*. New Haven: Yale University Press, 2015.

Zarobell, John. "A Year in Paris: Edvard Munch's *Mermaid*." In "Edvard Munch's *Mermaid*," special issue, *Philadelphia Museum of Art Bulletin* 93 (Summer 2005): 6–23.

Zemel, Carol. *The Formation of a Legend: Van Gogh Criticism, 1890–1920*. Ann Arbor: UMI Research Press, 1980.

Zimmermann, Michael F. *Seurat and the Art Theory of His Time*. Antwerp: Fonds Mercator, 1991.

Zola, Émile. *Correspondance*. 11 vols. Edited by Bard H. Bakker, Henri Mitterand, and Owen Morgan. Montreal: Les Presses de l'Université de Montréal, 1978–2010.

———. *Écrits sur l'art*. Edited by Jean-Pierre Leduc-Adine. Paris: Gallimard, 1991.

———. *Mes haines: Causeries littéraires et artistiques*. Paris: Charpentier, 1879.

———. *Le naturalisme au théâtre: Les théories et les exemples*. 3rd ed. Paris: Charpentier et Fasquelle, 1895.

———. *Oeuvres complètes*. 21 vols. Edited by Henri Mitterand, Armand Lanoux, and Robert Abirached. Paris: Nouveau Monde, 2002–10.

———. *Le roman expérimental*. Paris: Charpentier, 1880.

Index

abstract expressionism, 15, 180 n. 50
abstraction
antinaturalist, 52, 96, 125, 127, 135, 154
arguments over, 48, 187 n. 12
distortion and, 48, 192 n. 100
expressive, 157
modernist, 32, 48, 114, 170, 183 n. 32, 191 n. 86
symbolist, 12, 19, 173, 180 n. 45, 191 n. 86
Académie Julian, 70, 85, 193 n. 19, 195 n. 41
Academy, French. *See* École des Beaux-Arts
Academy, French, dictionary of, 4–5, 34, 177–78 n. 14, 196 n. 70
Adam, Paul, 34, 183–84 n. 50, 186 n. 84
Albert, Henri, 180 n. 53, 180 n. 54
anarchism, 36, 41, 185 n. 65
Angelico, Fra, 71, 108
Anquetin, Louis, 67
anticolonialism, 169
antinaturalism
rhetoric of, 22, 65, 89
visual practice of, 9, 12, 22, 52–53, 62, 160
antipositivism, 44, 176
antipsychiatry, 169
Antoine, André, 116, 199 n. 29
Antoine, Jules, 62–64
arabesque
Denis's use of, 53, 97, 99
history of, 97, 196 n. 70
musical, 97, 99, 196 n. 73
Nabi use of, 81, 99
Romanticism and, 97
symbolist, 99
Taine on, 90
Vuillard's theorization of, 89, 100–101
Vuillard's use of, 9, 22, 82, 85, 96–97, 99, 101–4
Asch, Max, 154, 200 n. 59, 204 n. 49
At the Black Piglet. *See* Black Piglet, At the
Aubé, Benjamin, 181 n. 8, 182 n. 11
Aurier, G.-Albert
antipositivism of, 1–2, 44
on Béraud, 143
on Gauguin, 16, 51, 62–63, 190 n. 53
on naturalism, 1–2, 143
on symbolism, 1–2, 44, 62–63, 72
on van Gogh, 16, 140, 142, 192 n. 95
automatism
avant-garde uses of, 99, 114, 202 n. 87
gender and, 109
psychological studies of, 43, 83, 93
Ribot on, 93, 106
Strindberg's use of, 113–14, 127, 132–33, 135, 202 n. 87
in Vuillard's approach to nature, 85, 87, 92–95, 108
in Vuillard's arabesques, 95–96, 99–102, 104, 108, 193 n. 20
in Vuillard's habitual work, 106, 108
avant-garde
alterity and, 8–9, 163, 174
antiaestheticism of, 9
automatism and, 99, 114, 202 n. 87
Black Piglet, At the, as an, 22, 113, 121, 130, 152
Denis as, 59–60, 62, 77
as experimental, 1–2, 5, 14, 25, 30, 32–33, 174, 176, 177 n. 1, 183 n. 32
history of, 177 n. 1
modernism and, 112
Munch as, 160
Nabis as, 109, 130, 195 n. 41
Neoimpressionism as, 1, 186 n. 84
primitivism and, 9
psychiatric, 140, 160–61, 163, 169
science and, 25, 121, 161, 186 n. 84
Scream as, 9, 157, 160
Strindberg as, 133–35
symbolist, 1–2, 20–21, 44, 62, 111, 163
theater, 99
unconscious and, 82, 180 n. 50

baccalaureate, philosophy, 18, 70, 83–84, 93, 191 n. 83
Bacon, Francis, 69, 72, 111, 178 n. 19
Balzac, Honoré, 111
Barc de Boutteville, gallery of, 194 n. 22, 197 n. 83
Barrès, Maurice, 18, 180, n. 59, 180 n. 63
Baudelaire, Charles, 51, 73, 189 n. 23
Beaunis, Henri, 119
Bélugou, Léon
circle of, 180–81 n. 61
pathological method, promoter of, 19–21
psychology, promoter of, 15, 19–21
Ribot's courses, attendance and notes on, 19–22, 93
Taine, survey on, 18–19
Benjamin, Roger, 82, 192 n. 107
Benjamin, Walter, 143–44, 203 n. 22
Béraud, Jean, 140–43, 153, 203 n. 21
Bergh, Richard, 128, 205 n. 55
Bergson, Henri, 3, 18, 194 n. 32
Berliner Gesellschaft für Experimental-Psychologie, 154
Berman, Patricia G., 139, 203–4 n. 36
Bernard, Claude
on doubt, 107
experimental method, promoter of, 3–4, 7, 26, 177 n. 11
induction and, 3, 48, 69–71
heart, on the, 26, 29
idealism and, 41–44
Introduction to the Study of Experimental Medicine: a priori idea, role in, 25, 28, 36; influence of, 3–4, 25, 39, 42–43, 112, 183 n. 33; Zola's use of, 21, 26–29, 36, 39
Laforgue, admiration for, 35–36, 38, 185 n. 74
Lessons on Animal Heat, 5
painting of, 3–4
pathological method of, 6–7, 27
physiology and, 3, 26–27, 29
positivism and, 28, 48
unconscious and, 38
Bernard, Émile, 33, 47, 60–62, 189 n. 50
Berthelot, Marcellin, 39
Bethlem Hospital, 164
Bhogal, Gurminder Kaur, 99, 196 n. 75
billiards, 140–42, 202, n. 12, 202 n. 13
Binet, Alfred, 20–21
Bjørnson, Bjørnstjerne, 115
Black Piglet, At the
avant-garde, as, 22, 113, 121, 130, 161
Munch, support for, 121, 139, 151–54
intellectual milieu, as a, 113, 127, 130, 156, 204 n. 36, 205 n. 55
scientific milieu, as a, 121
Sphinx, contributions to, 154
Blavet, Émile, 34–35
Boirac, Émile, 69, 191 n. 84
Bois, Jules, 42, 186 n. 93
Bonnard, Pierre
criticism of, 67
Décor, ownership of, 57, 187 n. 6
Denis on, 75–76
Indépendants, at, 85, 91
Mantelpiece, 57–58, 75, 189 n. 39–42
Nabi, associations with, 52, 85, 103, 189 n. 29, 192 n. 104, 193 n. 8
Revue blanche, association with, 16, 193 n. 8
Seated Nude, 192 n. 104
Bonnat, Léon, 49
Bouillon, Jean-Paul, 69, 179 n. 42, 191 n. 82, 192 n. 93
Bourget, Paul, 180 n. 63, 181 n. 2
Boutroux, Émile, 18, 180 n. 61
Brahm, Alcanter de, 160
Brain, Robert M., 180 n. 47, 204 n. 43
Brandes, Georg, 116, 120, 152, 198 n. 5
Broude, Norma, 182 n. 28, 183 n. 34, 184 n. 52
Broussais, Victor, 178 n. 20
Brunetière, Ferdinand, 27–28, 35, 115, 184 n. 58
Bürger, Peter, 177 n. 1
Burhan, Filiz, 69, 191 n. 82
Burty, Philippe, 34

Caillié, René, 41–44
Canguilhem, Georges, 6, 178 n. 23
caricature
of Cézanne, 60
deformation and, 189 n. 37
of Flaubert, 119
by Mirliton (of work by Vuillard), 79–81, 101, 109
in Munch, 149, 164, 203 n. 32
in Vuillard, 99, 101, 197 n. 78, 197 n. 88, 197 n. 92
of Zola, 1–2, 21, 30–31, 35
Caro, Elme Marie, 25, 39, 181 n. 1, 185 n. 74
Carroy, Jacqueline, 7

Catholicism, 18, 180 n. 60, 188 n. 33
Céard, Henry, 27–28, 182 n. 12
Cézanne, Paul
- *Achille Emperaire*, 59–60
- art, statements on, 33, 87, 183 n. 45, 183 n. 46
- *Bathers*, 60, 189 n. 51
- caricature of, 60
- criticism on, 32, 47, 67, 73–77
- deformation and, 47–48, 59–60, 67, 74, 192 n. 100
- Denis on, 60, 73–74
- doubt and, 104
- influence of, 60
- psychology and, 189 n. 46, 206 n. 2
- symbolism and, 206 n. 2
- *Wife of the Garbage Collector*, 59–60, 62
- Zola and, 30, 35, 185 n. 60

Champ-de-Mars, Salon du, 143, 203 n. 21
Charcot, Jean-Martin, 34, 119, 134, 164, 179 n. 33
Chimère, Le, 42
Clair, Jean, 12
Clark, T. J., 15
classicism, 37–38, 68, 73, 189 n. 39, 190 n. 76, 192 n. 5
cloisonnism, 11, 81
Cogeval, Guy, 192 n. 3, 193 n. 20, 194 n. 32, 195 n. 51
Collège de France, 7, 18–20, 106
colonialism, 9, 169
Comte, Auguste, 3, 6–8, 38, 41, 178 n. 20–21
Condorcet, Lycée, 18, 83, 193 n. 19
Condorcet, Nicolas de, 178 n. 19
Constable, John, 31, 182–83 n. 30
Cordulack, Shelly Wood, 203–4 n. 36
Courbet, Gustave, 79, 125, 201 n. 70
Cousin, Victor, 7–8, 178 n. 25, 179 n. 31
Crary, Jonathan, 180, n. 47

Darwin, Charles, 18, 37, 70
Daston, Lorraine, 5, 51
Daudet, Alphonse, 35, 182 n. 11, 184 n. 58
decadence, 19, 38, 43, 50, 132
decoration, 52, 72, 97, 188 n. 25
"Définition du néo-traditionnisme." *See* Denis
deformation
- caricature and, 189 n. 37
- Cézanne and, 47–48, 59–60, 67, 74, 192 n. 10
- critiques of, 48, 50, 62–63, 65, 67–68, 77
- in Degas, 49–50
- Denis's theory of, 21, 47–48, 69, 71–73, 76
- in Denis's work, 21, 47, 52–54, 57, 62–64, 68
- etymology of, 49, 187–88 n. 19
- in Gauguin's work, 189 n. 52
- impressionism and, 49–50, 62
- in literature, 49–51, 187 n. 15, 188 n. 21
- modernism and, 21, 47–48, 59–60, 62, 73–74, 109, 192 n. 100
- in Munch's work, 149
- naturalist, 47, 50
- objective, 48, 51, 69, 72–73, 77
- pathology and, 48–49
- subjective, 21, 47–51, 59, 63–64, 69, 72–73, 77
- symbolism and, 48–51, 57
- Zola and, 49–50, 62, 74

Degas, Edgar, 33–34, 49–50, 57
Degenerate Art Exhibition, 171
degeneration, 27, 48, 67, 181 n. 6, 187 n. 10, 190 n. 76
Dehmel, Richard, 121, 151, 154, 156
Deleuze, Gilles, 48
Delius, Fritz, 163
Denis, Maurice
- anti-Semitism, 193 n. 8
- arabesque and, 53, 97, 99
- *Catholic Mystery*, 63–64
- on Bonnard, 75–76
- Catholicism and, 180, n. 60, 188 n. 33
- on Cézanne, 60, 73–74
- *Décor*: as avant-garde statement, 59–60, 62, 77; Bonnard's ownership of, 57, 187 n. 6; criticism of, 48, 64, 67–68; as decoration, 52; "Définition," as corollary to, 47; *Mantelpiece*, representation in, 57; studies in charcoal and pastel for, 52, 54–56, 188–89, n. 35 ; studies in oil for, 9–10, 57; as "study in subjective deformation," 49, 51, 63–64; Vuillard, connection to, 75
- decoration and, 52, 72
- "Définition du néo-traditionnisme": as anti-academic statement, 52; as antinaturalist statement, 52, 65; arabesque in, 97; formalism and, 47; modernism and, 48; pathological language in, 47, 76; reception of, 65, 67; Sérusier's influence on, 70–71, 107; Vuillard's understanding of, 85, 89–90, 95

deformation, theory of, 21, 47–48, 69, 71–73, 76
deformation, in work of, 21, 47, 52–54, 57, 62–64, 68
education, artistic, of, 52, 188 n. 26
education, philosophical, 69–70, 83–85, 191 n. 82
Germain, correspondence with, 65, 190 n. 67
ideal love and, 190–91 n. 77
Indépendants, 47, 51–52, 62, 64–65, 85, 91, 194 n. 22
modernism and, 47–49
naturalism and, 47, 52–53, 93, 188 n. 27, 188 n. 28
pathological method and, 69–70
Portrait of a Young Girl in an Evening Décor, 68
positivism and, 48, 74, 77, 93
psychology and, 69–71
Revue blanche, La, association with, 15–16
Sagesse drawings, 51, 64
Sérusier, influence of, 70–71, 107–8
symbolism and, 47–48, 51, 57, 69, 70–71, 73, 82
Theme of Trinitarian Evening by Ad. Retté: Ornamental Painting, 64–66, 68
Voyage of Urien, illustrations for, 54
Vuillard, correspondence with, 67, 74, 82–85, 188 n. 29
Descartes, René, 3
Dévoluy, Pierre, 42
Diderot, Denis, 1
divisionism, 186 n. 84. *See also* neoimpressionism *and* pointillism
Dombrowski, André, 206 n. 2
Dostoevsky, Fyodor, 145, 147, 203 n. 23, 203 n. 31
Dreyfus Affair, 30, 193 n. 8
Druet, Eugène, 187 n. 6
Duranty, Edmond, 33–35, 183 n. 30, 184 n. 52
Durkheim, Émile, 18, 180 n. 61
dynamogeny, 186 n. 84, 195 n. 48

Easton, Elizabeth, 192 n. 3
eclecticism, 7, 83, 179 n. 31. *See also* Cousin, Victor
École des Beaux-Arts, 52, 60, 68, 91, 188 n. 26
empiricism, 3, 39, 177 n. 11, 177 n. 13
Encyclopédie, 97
Enlightenment, 173
Ericson, Ida, 161
esotericism, 3, 25, 41, 43–44, 130
Essen, Siri von, 118–19, 175, 200 n. 44
Étoile, L', 41
Evers, Franz, 154, 204 n. 51
expérience
definitions of, 4–5, 33–34, 71, 112, 135, 177–78 n. 14
etymology of, 4, 177 n. 13
impressionism and, 32–35
symbolism and, 177 n. 5
experimental art
as discursive formation, 1–2, 5, 14, 25–26, 30–32, 44, 183 n. 32
etymology of, 178 n. 16
experimentalism
aesthetics and, 36, 38–39, 44
Bernard, Claude, promoter of, 3–4, 7, 26, 177 n. 11
doubt and, 104
as epistemological ideal, 21, 24–25, 40, 44, 71, 174
impressionism and, 31, 33
Munch and, 140, 160, 169
naturalism and, 26, 29, 93–94
pathological method and, 7, 14, 22, 38, 170
psychology and, 7, 41
scientific, 32, 177 n. 10, 177 n. 11
Strindberg and, 112–13, 120, 134
symbolism and, 10, 15, 19, 21, 29, 41, 44, 174
truth and, 14, 26, 41–42, 44–45
experimental method. *See* experimentalism
experimentation
in art, 1, 31–32, 140
avant-garde and, 1–2, 5, 14, 25, 30, 32–33, 174, 176, 177 n. 1, 183 n. 32
observation and, 18
in psychology, 7, 112, 132, 153
in science, 5–6, 30, 32–34, 41, 114, 177 n. 10
self, on the, 8, 129, 134, 170
Strindberg and, 112–14, 121, 129, 132, 134–35
truth and, 174

Facos, Michelle, 11–12, 179 n. 41
fauvism, 77
Fechner's Law, 36, 185 n. 63
feminism, 12, 174
Fénéon, Félix, 36, 39–40 passim, 186 n. 84
Figaro, Le, 127, 184 n. 59
Flaubert, Gustave, 20, 28, 119, 181 n. 71

Flournoy, Théodore, 84
formalism, 32, 47, 49
Fort, Paul, 99
Foucault, Michel, 23, 178 n. 23
Fouquier, Henry, 34
Freud, Sigmund, 83, 179 n. 37, 203 n. 31
Fry, Roger, 32, 47, 73–74, 187 n. 1, 192 n. 100

Galison, Peter, 5, 51
gambling, 139, 142–51, 153, 203 n. 19
Gamboni, Dario, 99, 104, 194 n. 32, 197 n. 92
Garnier, Charles, 137
Gauguin, Paul
  Aurier on, 16, 51, 62–63, 190 n. 53
  Café at Arles, 72–73
  collection of, 59
  criticism on, 16, 51, 62, 67, 72, 190 n. 53
  deformation and, 189 n. 52
  Denis, correspondence with, 191 n. 78
  Denis on, 48, 72–73
  influence of, 70, 85, 87–88, 142, 202 n. 2
  Loss of Virginity, 59, 62, 191 n. 52
  memory, on working from, 87–88
  Munch, possible influence on, 202 n. 2
  Nude Breton Boy, 189 n. 52
  primitivism of, 108, 133
  Strindberg, correspondence with, 133, 201 n. 78
  syphilis of, 163
Geffroy, Gustave, 33, 64
Géricault, Théodore, 152–53
Germain, Alphonse, 48, 65–69
Gessmann, Gustav, 154–57
Gide, André, 54
Goethe, Johann Wolfgang von, 94
Goldstein, Jan, 83, 178 n. 19, 201 n. 80
Goldwater, Robert, 11
Gombrich, Ernst, 31–32, 183 n. 32, 187 n. 97
Graphic, 143, 203, n. 20, 203 n. 21
Greenberg, Clement, 1, 74
Groom, Gloria, 192 n. 3, 194 n. 21, 195 n. 53

Hamsun, Knut, 203 n. 23
Hansson, Ola, 152
Harrison, Charles, 57, 189 n. 42
Hartmann, Eduard von, 36–38, 185 n. 67, 185 n. 74
Harvey, William, 177 n. 10
Hauge, Alfred, 205 n. 70
Havard, Henry, 34
Hegel, Georg Wilhelm Friedrich, 39, 185 n. 67
Heller, Reinhold, 14, 139, 202 n. 1, 202 n. 13, 204 n. 46, 204 n. 47
Helmholtz, Hermann von, 36–37, 185 n. 64, 186 n. 84
Hennequin, Émile, 35, 40, 50, 186 n. 85
Henry, Charles, 36, 38–41, 185–86 n. 75, 195 n. 48
Héran, Henri (Paul Herrmann), 163, 205 n. 67
Herrmann, Paul (pseud. Henri Héran), 163, 205 n. 67
Hirsh, Sharon L., 12
Hogarth, William, 164
Hôpital Saint-Louis, 163–64, 205 n. 70
Hulst, Maurice Le Sage d'Hauteroche d', 18
Huret, Jules, 41
Huysmans, Joris-Karl, 34, 206 n. 1
hypnosis, 119, 156

Ibsen, Henrik, 170, 206 n. 84
idealism
  in aesthetics, 35–41
  in art, 2, 44, 59, 74, 187 n. 3
  Bernard, Claude, and, 41–44
  experimentalism and, 24
  in literature, 26, 44, 49, 118
  in philosophy, 18, 41–43
  positivism and, 38–39, 42
  in science, 38–42, 44
impressionism
  aesthetics of, 34, 36–38
  in art history, 25, 30–33, 35, 182 n. 28, 183 n. 34
  criticism on, 29–30, 33–35, 40, 73, 87
  deformation and, 49–50, 62
  expérience and, 32–35
  experimentalism and, 31, 33
  gender and, 175, 182 n. 28, 183 n. 34
  Munch and, 137
  naturalism and, 35, 87, 174, 182 n. 13, 182 n. 20
  neoimpressionism and, 184 n. 50
  objectivity and, 30–31, 184 n. 52
  science and, 30–31, 34, 36, 183 n. 34, 184 n. 52, 184 n. 54
  Strindberg and, 117, 127, 199 n. 33
  in Vuillard's work, 82, 84, 86–87, 94, 173
Indépendants, Salon des
  Bonnard at, 85, 91
  Denis at, 47, 51–52, 62, 64–65, 85, 91, 194 n. 22
  Munch at, 161, 164, 202 n. 13, 205 n. 56

Ingres, J.-A.-D., 64, 71, 191 n. 91
"interior impressionism," 84, 94, 173
Introduction to the Study of Experimental Medicine. See Bernard, Claude

Jackson, John Hughlings, 8
Jacobson, Daniel, 169
Janet, Pierre, 18–20
Jaspers, Karl, 112–13
Jullien, Jean, 89
Jumeau-Lafond, Jean-David, 54

Kahn, Gustave, 2, 36, 39–41, 50–51, 186 n. 90
Kearns, James, 190 n. 53
Kerner, Justinus, 122–23, 201 n. 64
Koeber, Raphael von, 130
Krohg, Christian, 137, 184 n. 54
Kuenzli, Katherine M., 197 n. 79

Laforgue, Jules, 35–39
landscape painting, 31, 49, 111, 123, 125, 130
Larousse dictionary, 4, 183 n. 48
Larsen, Tulla, 204 n. 49
Larson, Barbara, 179 n. 40
Lebourgeois, H., 2, 30–31, 35
Leclercq, Julien, 64
Leja, Michael, 180 n. 50
Lemaître, Jules, 49, 197 n. 90
Lemot, Achille, 119
Leroy, Louis, 30
Lhermitte, Léon, 3–4
Lie, Jonas, 115
Littré, Émile
- dictionary of, 4–5, 34, 183 n. 48, 196 n. 70
- retrospective medicine, 119

Lombroso, Cesare, 18
Londe, Albert, 164, 167
Lorrain, Claude, 182–83 n. 30
Lugné-Poë, Aurélien-Marie
- drawings representing, 100, 103
- Nabis, association with 52, 65, 89, 99, 188 n. 29
- portrait (painted) of, 9, 11, 82, 85, 101, 103–4, 109, 197 n. 83

Maeterlinck, Maurice, 198 n. 5
Mallarmé, Stéphane, 18, 34, 184 n. 50
Manet, Édouard, 30, 34, 59, 77, 184 n. 59
Marandon de Montyel, Évariste, 166
Marie, Auguste, 163, 165–67, 169, 206 n. 79
Martelli, Diego, 184 n. 54
Marx, Roger, 64
Mathews, Patricia, 12
Mathieu, Pierre-Louis, 179 n. 42
Matisse, Henri, 73, 77, 174, 192 n. 107
Maupassant, Guy de, 28
Maxwell, James Clerk, 185 n. 64
medievalism, 97
mediumism, 120–21, 154, 156–57
Meier-Graefe, Julius, 163
Meissonier, Ernest, 47
memory, 86–88, 128–29, 150–51
Mercure de France, 15–20
Merleau-Ponty, Maurice, 104
Mesmer, Franz Anton, 123
Meunier, Paul Gaston (pseud. Marcel Réja), 161, 163, 165–67
Meurier, Marthe, 68, 190 n. 77
modernism
- abstraction and, 32, 48, 114, 170, 183 n. 32, 191 n. 86
- in art, 12–13, 60, 62, 73, 173–75
- art histories of, 11, 13, 15, 25, 32, 48, 77, 99, 180 n. 45
- avant-garde and, 112
- in criticism, 13–14, 48, 60, 65, 74
- deformation and, 21, 47–48, 59–60, 62, 73–74, 109, 192 n. 100
- Denis on, 47–49
- Munch and, 171
- Strindberg and, 112–13, 131, 135, 173
- symbolism and, 13, 22, 114, 173–75, 180 n. 45
- Vuillard and, 99, 109

Molard, William, 161, 163
Monet, Claude, 30, 32, 35, 87, 183 n. 33
monism, 139, 200 n. 59
Mont, Elie de, 34
Monte Carlo, casino, 139, 142–50, 203 n. 21
Moréas, Jean, 10, 50–51
Moreau, Gustave, 10, 12
Moreau, Jacques-Joseph, 179 n. 31
Moreau de la Sarthe, Louis-Jacques, 178 n. 19
Morosov, Ivan, 192 n. 94
Mozart, Wolfgang Amadeus, 94
Munch, Edvard

as anatomist, 169–70
in art history, 139
At the Roulette Table in Monte Carlo (painting), 147–50, 203 n. 28
as avant-garde, 160
billiards and, 140–42
biography and, 139
Black Piglet, At the, support for, 121, 139, 151–54
caricature and, 149, 164, 203 n. 32
criticism on, 15–16, 133, 152–53, 160–61
Death in the Sickroom, 150–51, 157, 160–61
deformation and, 149
Frieze of Life, 125, 170
Gamblers in Monte Carlo, 144, 146–47
Gauguin, possible influence of, 202 n. 2
Ibsen and, 170
impressionism and, 137
Indépendants, at, 161, 164, 202 n. 13, 205 n. 56
Inheritance, 164, 205–6 n. 70
In the Men's Clinic, 168–69
madness and, 22, 139–40, 147, 150, 153–54, 159–60, 171
Man's Head/Art, 157–58
Marcel Réja, 163
mediumism and, 157, 159
modernism and, 171
in Monte Carlo, 139–40, 142–51
naturalism and, 9, 22, 137, 142–44, 147–48, 153, 157, 168, 170, 173
pathological method and, 140, 161, 170
primitivism and, 133, 153
psychiatry and, 160–61, 163–65, 167, 169
"psychic naturalism" and, 125, 151–54, 170, 173
psychology and, 140, 152–53, 160–61
Revue blanche, La, association with, 16–17
roulette and, 22, 142–50
Roulette I, 143, 145, 203 n. 28
Roulette Table (etching), 149
Scream (lithograph), 16–17
Scream (painting): as avant-garde act, 9, 157, 160; madness and, 153, 157, 159–60, 175, 204 n. 47; as mediumistic, 157, 159–60; criticism on, 160; as symbolist, 13, 139, 150–51, 153
Self-Portrait, 137–39
symbolism and, 124, 137, 139–40, 142, 147, 150–53, 161, 167
synthetism and, 137, 202 n. 2
Tree of Knowledge of Good and Evil, 206 n. 85
Van Gogh, possible influence of, 202 n. 13
Verein Berliner Künstler, exhibition at, 121, 147, 152
Weeping Woman, 169–70
Women in Hospital (painting), 139, 161–65, 167–70
Women in the Hospital (lithograph), 164–65
writings of, 145–48, 151, 153, 170–71, 206 n. 82
Musée du Luxembourg, 59

Nabis
arabesque, 81, 99
art of, 57, 85, 95
avant-garde, as, 109, 130, 195 n. 41
Bonnard as a, 52, 85, 103, 188 n. 29, 192 n. 104, 193 n. 8
criticism on, 67
education of, 83, 180 n. 57
decoration and, 188 n. 25
divisions within, 193 n. 8
formation of, 52, 85
gender and, 189 n. 33
influences on, 42
Lugné-Poë, association with, 52, 65, 89, 99, 188 n. 29
primitivism and, 71, 97, 99
Revue blanche, association with, 16–18, 193 n. 8
theater and, 196–97 n. 78
theories of, 69, 81–82, 84, 86, 91, 94, 100–101, 108. *See also* Denis, Maurice
Vuillard as a, 75, 84–85, 89, 193–94 n. 21
Nahasch. See Denis, *Décor*
Natanson, Thadée, 16
National Gallery, Norway, 159
naturalism
Aurier on, 1–2, 143
Bernard, Émile, and 62
deformation and, 47, 50
Denis and, 47, 52–53, 93, 188 n. 27, 188 n. 28
experimentalism and, 26, 29, 93–94
impressionism and, 35, 87, 174, 182 n. 13, 182 n. 20
Munch and, 9, 22, 137, 142–44, 147–48, 157, 168, 170, 173
novel, in the, 26–29, 35–36, 115, 119
positivism and, 36, 39, 42
psychic, 124–25, 151–54, 170, 173

naturalism (*cont.*)
representation and, 31
science and, 21, 182 n. 13
Strindberg and, 22, 112–13, 115–17, 127, 129, 133–134, 173
symbolism and, 44, 113–14, 129, 134, 206 n. 2
in theatre, 114, 117
Vuillard and, 84–96, 99, 173
Zola and, 2, 26–29, 35, 44, 114, 116, 118
natural philosophy, 31, 178 n. 19, 183 n. 34, 206 n. 5
Nazi Party, 171
neoidealism, 187 n. 3
neoimpressionism
as avant-garde, 1, 186 n. 84
criticism on, 39–40, 50, 184 n. 50
impressionism and, 184 n. 50
psychophysiology and, 39, 186 n. 84
science and, 39–40, 186 n. 81, 186 n. 84
symbolism and, 1, 11, 40, 186 n. 84, 190 n. 59
Neoplatonism, 130, 191 n. 86,
neotraditionism. *See* Denis, "Définition du néo-traditionnisme"
Nietzsche, Friedrich
influence of, 139, 152, 180 n. 53
on pathological method 5, 7, 152
Strindberg and, 120, 200 n. 54
Nordau, Max, 171, 187 n. 10

objectivity
in art, 1, 31, 50, 168–69
in impressionism, 30–31, 184 n. 52
in psychology, 8, 70–71, 134
in scientific discourse, 5, 28–29, 51, 73, 107, 112
in symbolism, 15, 29, 50–51, 72–73, 121, 168–69
observation
"arabesque of," 82, 101, 108
psychological, 7, 18, 92–94, 118–19
scientific, 3–5, 112
Strindberg on, 118–19
Vuillard on, 86, 92–94, 96, 99, 104
occultism, 41–42, 121, 123–24, 130, 134

Paris Salon, 59, 166
Pastor, Willy, 152–53
pathological method
Bélugou as promoter of, 19–21
Bernard, Claude, and, 6–7, 27
Canguilhem's analysis of, 6
Comte and, 6
Denis's understanding of, 69–70
emergence of, 178 n. 19
experimentalism and, 7, 14, 22, 38, 170
Laforgue, Jules, on, 36, 38
Munch's use of, 140, 161, 170
Nietzsche on, 5, 7, 152
in philosophy baccalaureate, 18, 69–70
psychology, as paradigm for, 6–9, 19–20, 42, 70–71, 82, 104
Ribot, Théodule, and, 7–8, 15, 19–21, 94, 106
Strindberg's use of, 113, 128–29, 134
symbolism and, 13–14, 38, 40, 113
Taine, Hippolyte, and, 7–9, 20–21, 38, 133
Péladan, Joséphin, 38, 185 n. 75
physiology
Bernard, Claude, on, 3, 26–27, 29
in aesthetics, 34, 36–37, 119
as science, 3, 7, 26–27, 37, 42–43
Picavet, François, 18
Pinel, Philippe, 164, 166
Plas, Régine, 7
Plato, 39, 42, 130, 191 n. 86
Plume, La, 64–65, 67, 112, 161
Poe, Edgar Allan, 51, 64, 188 n. 23
pointillism, 139. *See also* divisionism, neoimpressionism
Pollock, Jackson, 99, 174
Pont-Aven group, 62, 69, 137, 188 n. 33
positivism
in aesthetics, 40, 69, 176
Bernard and, 29, 48
Comte, according to, 3, 41
critiques of, 25–27, 44
Denis, influence on, 48, 74, 77, 93
as epistemology, 134, 176
idealism and, 38–39, 42
influence of, 7, 35, 41–42, 69
in literature, 27, 41
naturalism and, 36, 39, 42
in philosophy, 3, 25, 48, 69, 83
in psychology, 7, 70, 201 n. 80
Taine, according to, 18, 38, 77, 179 n. 30, 180 n. 60
Zola on, 28, 36
Prel, Carl du, 123

primitivism
avant-garde and, 9
Gauguin and, 108, 133
Munch and, 133, 153
Nabis and, 71, 97, 99
Strindberg and, 129–30, 132–34
Prinzhorn, Hans, 163, 165
Przybyszewski, Stanisław, 151–54, 170, 173
psychiatry, 140, 160–61, 163–66, 169
"psychic naturalism," 125, 151–54, 170, 173
Psychologische Gesellschaft, 154
psychology
automatism, studies of, 43, 83, 93
Cézanne and, 189 n. 46, 206 n. 2
Denis's understanding of, 69–71
as discipline, 41, 44
experimental, 7–9, 19–20, 29–30, 43–44, 48, 70–71, 92, 114–15, 121, 153
experimentalism and, 7, 41
in Germany, 7–8, 36, 154
history of, 179 n. 37, 201 n. 80
Laforgue on, 36, 38–39
Munch and, 140, 152–53, 160–61
objectivity and, 8, 70, 201 n. 80
pathological method of, 6–9, 19–20, 70–71, 82
philosophy and, 7, 83
Ribot and, 8, 19–21, 93–94, 101
scientific, 7–9, 19, 30, 115, 178 n. 28
sleep, of, 101
in *Sphinx*, 121, 154
Strindberg's understanding of, 112–13, 115, 118–20, 129, 134
symbolism and, 6, 13–15, 19, 44
Taine and, 8–9, 18, 116, 120, 133
will, of, 83, 105–6, 108
Zola and, 29–30, 115, 118
psychopathology, 8–9, 121
psychophysiology, 36–37, 39–40, 128, 186 n. 84
Puvis de Chavannes, Pierre, 47, 68

Rambosson, Yvanhoé, 48, 68–69, 160–61
Randon, Gabriel (pseud. Jehan Rictus), 163
Ranson, Paul, 187 n. 94, 193 n. 8, 195 n. 43
Rapetti, Rodolphe, 11
realism, 43, 50, 62–63, 79, 127
Redon, Odilon, 10, 12, 179 n. 40
Réja, Marcel (Paul Gaston Meunier), 161, 163, 165–67
Renan, Ernst, 6–7, 38, 178 n. 24
Retté, Adolphe, 33, 63–65, 68–69
Revue blanche, La
Bonnard and, 16, 193 n. 8
Denis and, 15–16
intellectual ambitions of, 16, 180 n. 58
Laforgue's essays in, 36–38
Munch and, 16–17
Nabi association with, 16–18, 193 n. 8
psychology in, 18–20
Strindberg and, 16–17
symbolist art, support for, 15–18
Vuillard and, 16–17
Revue de psychiatrie, La, 165
Ribot, Théodule
on automatism, 93, 106
Collège de France, courses at the, 7, 18–20, 30, 93, 106
Diseases of the Will, 8, 106
Essay on the Creative Imagination, 20, 94
Evolution of General Ideas, 19
introspection and, 8, 179 n. 31
pathological method, promotion of, 7–8, 15, 19–21, 94, 106
psychology and, 7–8, 19–21, 93–94, 101
Psychology of Emotions, 19
unconscious, on, 20, 84, 93–94, 104, 106
Richet, Charles, 42, 119
Rictus, Jehan (Gabriel Randon), 163
Rimbaud, Arthur, 186 n. 90
Ritter, William, 160–61, 205 n. 59
Robert-Fleury, Tony, 164, 166
Rochefoucauld, Antoine de La, 189 n. 52
romanticism
alterity and, 9
arabesque and, 97
and artist, image of, 51, 72, 121, 134, 139
science and, 183 n. 34
symbolism as renewal of, 99
Rouart, Henri, 34

Sainte-Anne Asylum, 163, 165, 205 n. 66, 205 n. 67
Saint-Louis Hospital, 163–64, 205 n. 70
Salon des Indépendants. *See* Indépendants, Salon des
Salon du Champ-de-Mars, 143, 203 n. 21
Salon, Paris, 59
Salpêtrière, 163–64, 166

Saunier, Charles, 64
Schlegel, Friedrich, 97
Schleich, Carl Ludwig, 121, 151, 200 n. 59, 205 n. 55
Schopenhauer, Arthur, 130, 185 n. 74
Schuré, Édouard, 3, 42–44, 130, 187 n. 94
Scream. *See* Munch
Séguin, Armand, 69, 191 n. 78
Sérieux, Paul, 165
Sérusier, Paul, 74
  Denis, influence on, 70–71, 107–8
  education of, 18, 70, 191 n. 86
  Talisman, 70, 85, 109, 142, 193 n. 21
  theories of, 70–71, 107–8, 191 n. 88
  Vuillard, influence on, 85, 89, 105, 195 n. 41
Servaes, Franz, 133, 152–53
Seurat, Georges, 39–40, 62, 186 n. 84
Shiff, Richard, 48, 74, 182 n. 20, 206 n. 2
Sickert, Walter, 74, 107
Signac, Paul, 180 n. 63, 186 n. 84
Sjåstad, Øystein, 184 n. 54
Société de psychologie physiologique, 154
Society for Psychical Research, 154, 205 n. 54
sociology, 6
Spencer, Herbert, 8, 37, 40, 152, 185 n. 70
Sphinx, 121–23, 154–57, 159
spiritism, 41, 43, 154, 181 n. 1, 186 n. 91
spiritualism, 14, 18, 25, 39–44, 134, 181 n. 1
Spurzheim, J. G., 150
Strindberg, August
  Antibarbarus, 121
  anti-Semitism and, 120
  automatism and, 113–14, 127, 132–33, 135, 202 n. 87
  as avant-garde, 133–35
  "Battle of the Brains," 119–20
  Black Banners, 135
  criticism on, 15–16, 175, 180 n. 54
  "Deranged Sensations," 127–29, 133
  on doubt, 120–21
  experimentalism and, 112–13, 120, 134
  experimentation and, 112–14, 121, 129, 132, 134–35
  Father, 113–14, 116
  Flower on the Shore, 124–25
  Försöksteater, 116–17
  Gauguin, correspondence with, 133, 201 n. 78
  Gothic Rooms, 134
  High Seas, 127, 135
  on impressionism, 117, 127, 201 n. 33
  Inferno, 118, 123, 134–35, 163
  klecksographs, 123–24
  library, 115
  Madman's Defense, 171, 200 n. 44
  Mercure de France, essays in, 16–17
  misogyny, and, 118–20
  Miss Julie, preface to, 116–17
  modernism and, 112–13, 131, 135, 173
  "Mysticism—For Now," 119
  naturalism and, 22, 112–13, 115–17, 127, 129, 133–34, 173
  "naturalistic symbolism," and 111–12, 114, 123, 127, 132, 134, 173
  "New Arts! Or Chance in Artistic Production," 113, 118, 125, 127–33
  Nietzsche and, 120, 200 n. 54
  Night of Jealousy, 9, 12, 111, 124–25, 127, 135
  on observation, 118–19
  Occult Diaries, 123–24
  occultism and, 121, 134
  "On Modern Drama and Modern Theater," 116–17
  paintings of, 201 n. 67
  Palette with Solitary Flower on the Shore, 125–26
  pathological method and, 113, 128–29, 134
  photography and, 123
  primitivism and, 129–30, 132–34
  psychology and, 112–13, 115, 118–20, 129, 134
  Revue blanche, La, essays in, 16–17
  scientific experiments of, 121, 135
  skin condition of, 163
  Somnambulistic Nights in Broad Daylight, 115
  Son of a Servant, 115–16, 118, 199 n. 18
  "Soul Murder (Apropos *Rosmersholm*)," 119–20
  symbolism and, 111–13, 118, 121, 124–25, 127, 133
  telepathy and, 156
  unconscious and, 119–121, 127, 129, 131–32, 134
  Vivisections, 111, 113, 118–19, 127
  as vivisectionist, 200 n. 46
  Weeping Boy, 132
  on women, 200 n. 43
  Wonderland, 132–33
  on Zola, 112, 114–17
  Zola, correspondence with, 116, 199 n. 29

subjectivity, 5, 51, 73, 82, 121, 168–69
surrealism, 15, 99, 135, 174, 202 n. 87
Svenaeus, Gösta, 203 n. 21
symbolism
abstraction and, 12, 19, 173, 180 n. 45, 191 n. 86
anarchism and, 185 n. 65
arabesque and, 99
in art history, 11–13, 23, 179 n. 42, 180 n. 45
Aurier on, 1–2, 44, 62–63, 72
avant-garde and, 1–2, 20–21, 44, 62, 111, 163
Cézanne and, 206 n. 2
criticism on, 1–2, 50–51, 62–63, 87, 161, 163
decoration and, 13, 52, 188 n. 25
deformation and, 48–51, 57
Denis and, 47–48, 51, 57, 69, 70–71, 73, 82
divisionism and, 186 n. 84
elitism of, 130
expérience and, 177 n. 5
Germain on, 65, 68–69
gender and, 177 n. 6, 189 n. 33
impressionism and, 25
journals of, 15–17, 20, 41–42, 112, 161
Kahn on, 2, 10, 25, 40–41, 50–51
in literature, 2, 40, 50–51, 112, 117
Mercure de France, support for, 15–17, 20
modernism and, 13, 22, 114, 173–75, 180 n. 45
Moréas on, 50–51
Munch and, 124, 137, 139–40, 142, 147, 150–53, 161, 167
Nabis and, 94, 101, 107–8
naturalism and, 44, 113–14, 129, 134, 206 n. 2
naturalistic, 111–12, 114, 123, 127, 132, 134, 173
neoimpressionism and, 1, 11, 40, 186 n. 84, 190 n. 59
objectivity and, 15, 29, 50–51, 72–73, 121, 168–69
occultism and, 130, 186 n. 93
in painting, 2, 13, 111, 120, 124, 133, 140
pathological method and, 13–14, 38, 40, 113
psychiatry and, 163
psychology and, 6, 13–15, 19, 44
Revue blanche, La, support for, 15–17, 36
romanticism, as echo of, 99
Scream and, 13, 139, 150–51, 153
Strindberg and, 111–13, 118, 121, 124–25, 127, 133
subjectivity, 51, 73, 82, 168–69
theories of, 10–13, 15, 40, 48, 50, 85–87, 142
truth and, 2, 107
Vuillard and, 82, 85–88, 91, 101, 108–9
Zola and, 112, 117–18
synthetism, 11, 81, 137, 187 n. 3, 202 n. 2
syphilis, 163–64, 170, 205 n. 64, 205–8 n. 70

Taine, Hippolyte
arabesque and, 90
On Intelligence, 7–8, 18, 187–88 n. 19
Laforgue on, 36, 38
pathological method and, 7–9, 20–21, 38, 133
Philosophy of Art, 38, 86, 90–91, 187–88 n. 19
portrait of, 18
positivism and, 18, 38, 77, 179 n. 30, 180 n. 60
psychology and, 7–9, 18, 116, 120, 133
survey on, 18–19
unconscious and, 9, 38
Tanguy, Julien (Père), 59–60
telepathy, 156–57, 159
Théâtre Libre, 99, 116
Théâtre de l'Oeuvre, 99, 196 n. 78
Théberge, Pierre, 12
theosophy, 154, 191 n. 86
Thibout, Anna, 52, 54, 68, 75, 175, 188 n. 29
Thomson, Belinda, 188 n. 27, 193 n. 6
Thomson, Richard, 11, 188 n. 27, 189 n. 37, 193 n. 16, 197 n. 78
Thue, Sivert, 203 n. 23
thought transfer, 156–57, 159
Toulouse, Édouard, 165, 179 n. 30

Uhl, Frida, 111, 124
unconscious
avant-garde and, 82, 180 n. 50
Bernard, Claude, and 38
gender and, 175
Hartmann on, 37–38, 185 n. 67
Munch and, 152–53
psychology of, 9, 83, 108, 179 n. 37
Ribot on, 20, 84, 93–94, 104, 106
Strindberg's explorations of, 119–121, 127, 129, 131–32, 134
Taine on, 9, 38
Vuillard's explorations of, 82–84, 96, 101, 104, 107
Vallotton, Félix, 18, 36, 41, 193 n. 8

Van Gogh, Vincent
Aurier on, 16, 140, 142, 192 n. 95
criticism on, 67
Denis on, 48, 72–73
as mad, 73, 192 n. 95
memory and, 88
Night Café, 72, 140, 142, 202 n. 13
Memory of the Garden at Etten, 194 n. 37
Munch, possible influence on, 202 n. 13
vanguardism. *See* avant-garde
Vaschide, Nicolas, 165
Velázquez, Diego, 160
Verein Berliner Künstler, 121, 147. *See also* Munch
Verhaeren, Émile, 64
Verlaine, Paul, 51, 64, 123, 186 n. 90, 190 n. 59
Véron, Eugène, 40, 50–51, 186 n. 79, 187 n. 19
Villejuif Asylum, 163, 165
Vinci, Leonardo da, 31, 38, 171, 186 n. 75
Vollard, Ambroise, 33
Volpini exhibition, 62, 202 n. 2
vitalism, 3, 178 n. 21
vivisection
anatomical, 6, 20, 170, 181 n. 9
as metaphor, 27, 112, 118, 134, 170, 200 n. 46
psychological, 20–21, 119, 134
Vuillard, Édouard
arabesque in practice, 9, 22, 82, 85, 96–97, 99, 101–4
arabesque in theory, 89, 100–101
automatism and nature, 85, 87, 92–95, 108
automatism and habit, 106, 108
caricature and, 99, 101, 197 n. 78, 197 n. 88, 197 n. 92
carnets. *See* journals
criticism on, 15, 109
decoration and, 13, 197 n. 89
and Denis: correspondence with, 67, 74, 82–85, 189 n. 29; *Décor*, dialogue with, 75; "Définition du néo-traditionnisme," understanding of, 85, 89–90, 95
doubt and, 104–7, 174
Dreyfus, support for, 193 n. 8
education of, 18, 83–84, 108, 193 n. 13, 193 n. 14
gender and, 109
Grisélidis, 100
impressionism and, 82, 84, 86–87, 94, 173
In Bed, 102, 197 n. 88
"interior impressionism," 84, 94, 173
journals of, 82, 84–98, 100, 105–6, 108–9, 192 n. 3
Kettle and Pot, 103, 105
Lugné-Poë, drawings of, 100, 103
Lugné-Poë, painted portrait of, 9, 11, 82, 85, 101, 103–4, 109
memory and, 87–88, 91
modernism and, 99, 109
Nabis, association with, 75, 84–85, 89, 193–94 n. 21
naturalism and, 84–96, 99, 173
on observation, 86, 92–94, 96, 99, 104
Portrait of Marie Vuillard, 88–89
Reclining Nude, 75–76, 79
Revue blanche, La, and, 16–17, 193 n. 8
Self-Portrait with Walking-Stick and Boater, 95
Sérusier, influence of, 85, 89, 105, 195 n. 41
Sleeping Woman, 102–3
symbolism and, 82, 85–88, 91, 101, 108–9
theater, work for, 99–100, 103
unconscious and, 82–84, 96, 101, 104, 107
Woman in Bed, 81, 102
Woman Sleeping, 79–82, 85, 101–3, 109
Vuillard, Marie, 87, 89–91, 175, 197 n. 88

Wagnerianism, 40, 99, 186 n. 83, 197 n. 79
Ward, Martha, 47
Wells, Charles "Monte Carlo," 146–47, 203 n. 25
Werenskiold, Erik, 202 n. 13
Wharton, Edith, 181 n. 63
Willumsen, Jens-Ferdinand, 67
Wundt, Wilhelm, 70
Wyzewa, Théodor de, 36

Young, Thomas, 185 n. 64

Zimmermann, Michael F., 179 n. 43, 186 n. 81
Zola, Émile
Academy, candidacy for, 112, 117
art criticism, 28, 30, 35, 72, 184–85 n. 59
Assommoir, L' 26, 35
Bernard, Claude, influence of, 21, 26–29, 36, 39
caricatures of, 1–2, 21, 30–31, 35
Cézanne and, 30, 35, 185 n. 60
deformation and, 49–50, 62, 74
degeneration and, 27, 181 n. 6

Doctor Pascal, 30–31
Dream, 117
"Experimental Novel": in art history, 30, 32; Bernard, Claude, influence on, 21, 26–29, 36, 39; caricature of, 1–2, 30; criticism of, 26–27, 29, 33, 115, 182 n. 11; Laforgue's admiration for, 36; psychology and, 29–30; Strindberg and, 114–17
idealism and, 44
Masterpiece, 30, 32, 35, 116
Nana, 26, 28
naturalism and, 2, 26–29, 35, 44, 114, 116, 118
Naturalism in the Theater, 114
positivism and, 28, 36
psychology and, 29–30, 115, 118
Rougon-Macquart series, 115
and Strindberg: correspondence with, 116, 199 n. 29; influence on, 112, 114–17
Toulouse, Édouard, subject of, 179 n. 30
symbolism and, 112, 117–18
Thérèse Raquin, 114
temperament and, 21, 27–29, 49–51
Zum schwarzen Ferkel. *See* Black Piglet, At the

REFIGURING MODERNISM

ARTS

LITERATURES

SCIENCES

Refiguring Modernism features cutting-edge interdisciplinary approaches to the study of art, literature, science, and cultural history. With an eye to the different modernisms emerging throughout the world during the twentieth century and beyond, we seek to publish scholarship that engages creatively with canonical and eccentric works alike, bringing fresh concepts and original research to bear on modernist cultural production, whether aesthetic, social, or epistemological. What does it mean to study modernism in a global context characterized at once by decolonization and nation-building; international cooperation and conflict; changing ideas about subjectivity and identity; new understandings of language, religion, poetics, and myth; and new paradigms for science, politics, and religion? What did modernism offer artists, writers, and intellectuals? How do we theorize and historicize modernism? How do we rethink its forms, its past, and its futures?

(A Series Edited By)
Jonathan Eburne

(Other Books in the Series)

David Peters Corbett *The World in Paint: Modern Art and Visuality in England, 1848–1914*

Jordana Mendelson *Documenting Spain: Artists, Exhibition Culture, and the Modern Nation, 1929–1939*

Barbara Larson *The Dark Side of Nature: Science, Society, and the Fantastic in the Work of Odilon Redon*

Alejandro Anreus, Diana L. Linden, and Jonathan Weinberg, eds. *The Social and the Real: Political Art of the 1930s in the Western Hemisphere*

Margaret Iversen *Beyond Pleasure: Freud, Lacan, Barthes*

Stephen Bann, ed. *The Coral Mind: Adrian Stokes's Engagement with Architecture, Art History, Criticism, and Psychoanalysis*

Charles Palermo *Fixed Ecstasy: Joan Miró in the 1920s*

Marius Roux *The Substance and the Shadow*

Aruna D'Souza *Cézanne's Bathers: Biography and the Erotics of Paint*

Abigail Gillman *Viennese Jewish Modernism: Freud, Hofmannsthal, Beer-Hofmann, and Schnitzler*

Stephen Petersen *Space-Age Aesthetics: Lucio Fontana, Yves Klein, and the Postwar European Avant-Garde*

Stefanie Harris *Mediating Modernity: Literature and the "New" Media, 1895–1930*

Michele Greet *Beyond National Identity: Pictorial Indigenism as a Modernist Strategy for Andean Art, 1920–1960*

Paul Smith, ed. *Seurat Re-viewed*

David Prochaska and Jordana Mendelson, eds. *Postcards: Ephemeral Histories of Modernity*

David Getsy *From Diversion to Subversion: Games, Play, and Twentieth-Century Art*

Jessica Burstein *Cold Modernism: Literature, Fashion, Art*

Adam Jolles *The Curatorial Avant-Garde: Surrealism and Exhibition Practice in France, 1925–1941*

Juli Highfill *Modernism and Its Merchandise: The Spanish Avant-Garde and Material Culture, 1920–1930*

Damien Keane *Ireland and the Problem of Information: Irish Writing, Radio, Late Modernist Communication*

Typeset by
REGINA STARACE

Printed and bound by
ASIA PACIFIC OFFSET

Composed in
QUADRAAT AND
DIN NEXT

Printed on
NEO MATTE